AF324299

First LeCosPA Symposium

Towards Ultimate Understanding of the Universe

First LeCosPA Symposium

Towards Ultimate Understanding of the Universe

Taipei, Taiwan, ROC 6 – 9 February 2012

editor

Pisin Chen

Leung Center for Cosmology and Particle Astrophysics
National Taiwan University

World Scientific

NEW JERSEY · LONDON · SINGAPORE · BEIJING · SHANGHAI · HONG KONG · TAIPEI · CHENNAI

Published by

World Scientific Publishing Co. Pte. Ltd.

5 Toh Tuck Link, Singapore 596224

USA office: 27 Warren Street, Suite 401-402, Hackensack, NJ 07601

UK office: 57 Shelton Street, Covent Garden, London WC2H 9HE

British Library Cataloguing-in-Publication Data
A catalogue record for this book is available from the British Library.

TOWARDS ULTIMATE UNDERSTANDING OF THE UNIVERSE
First LeCosPA Symposium

ISBN 978-981-4449-36-6

Printed in Singapore.

PREFACE

To celebrate the 4th anniversary of the establishment of the Leung Center for Cosmology and Particle Astrophysics (LeCosPA) at National Taiwan University (NTU), the First LeCosPA Symposium was held on the NTU campus in Taipei, Taiwan, on February 6-9, 2012. Among the dignitaries who delivered speeches at the Celebration Ceremony were two university presidents (NTU and OIST, Japan), three sister-institution directors or associate directors (KIPAC, Stanford University, Kavli IPMU, University of Tokyo, and ICRANet, University of Rome), and the principal investigators of the experimental projects that LeCosPA has been engaging (ANITA, ARA and UFFO). It was such an honor and so heart-warming to hear the kind words and encouragements from colleagues worldwide.

Equally exciting was the program of the symposium, with the presence of the world's top experts in the fields of cosmology, gravitation, high energy physics, astrophysics and particle astrophysics for the symposium. Among them we had the Fields Medalist Shing-Tung Yau (Harvard University) with the Calabi-Yau manifold fame to deliver the opening talk, while the closing talk was given by one of the founding fathers of the string theory, Gabriele Veneziano (College de France). In between, there were equally inspiring talks one after another. The four-day symposium has covered the entire scope of frontier topics in cosmology, ranging from inflation and CMB in the early universe to the accelerating expansion in the late times, from the dark matter search, the nature of dark energy, to the origin of baryon asymmetry. In particle astrophysics, the presentations covered from the detection of mark matter, ultra-high energy cosmic rays and neutrinos, to gamma ray bursts. In the effort towards an ultimate understanding of the universe, it would be inevitable, or in vain, if without a deeper and more fundamental understanding of gravity and its connection with space and time. We are very pleased to hearing exciting new ideas and insights in these regards.

LeCosPA was formally established on November 13, 2007. In less than 5 years it has grown from an infant to a healthy toddler. It has a vigorous experimental program that ranges from the search of GZK neutrinos through the balloon-borne ANITA experiment in Antarctica and the ARA Neutrino Observatory at the South Pole, to the detection of the GRB prompt signals and initial light-curves with the UFFO-Pathfinder Satellite Telescope, to be launched in Central Asia in summer 2013. Equally active is its theoretical research program spearheaded by 5 Working Groups, ranging from Quantum Cosmology, Dark Energy, Inflation and Early Universe, Large Scale Structure, to Cosmic Neutrinos.

Along with these encouraging developments, there came, a few months after the Symposium, an exciting news that Mr. C. C. Leung, the original donor of the NTU LeCosPA Center, has pledged a second and larger donation of US$19M to NTU on June

21, 2012 to turn LeCosPA into a permanent operation and to erect a new building as the center's permanent home on the NTU campus.

NTU President Si-Chen Lee (front right) shakes hands with Chee-Chen Leung, vice-president and co-founder of Quanta Computer, Inc., after signing the contract on June 21, 2012.

This book documents the exciting proceedings of this meeting. My sincere gratitude goes to all those who came from afar: the university presidents, the sister institution directors, the project PIs, the outstanding colleagues, for their support and encouragement. I am also grateful to the strong camaraderie among my colleagues from different institutions in Taiwan. The generous financial supports of Taiwan's National Science Council (NSC) and NTU College of Science are indispensable and I truly appreciate it. Last but not least, this LeCosPA 4th Anniversary Celebration Ceremony and First LeCosPA Symposium would not have been so successful if without the careful planning and wonderful organization by the excellent team of the LeCosPA staff. I particularly wish to recognize Yen-Ling Anne Lee, Kathy Ho, Tien-Chi Liu, and Greta Chuang for their tireless dedication. In addition to the day-to-day formal events, there were also the unforgettable trip to the Lantern Festival in Pingxi to participate in the launching of Tien-Deng (sky lantern), the Taipei City excursion, the various receptions, and the banquet on the top of Taipei-101, all the results of their flawless execution.

While LeCosPA still has a long way to go, we are determined to strive in achieving academic excellence through major contributions in cosmology and particle astrophysics.

Pisin Chen
September 2012

CONTENTS

LeCosPA's Fourth Anniversary Celebration Addresses

NTU

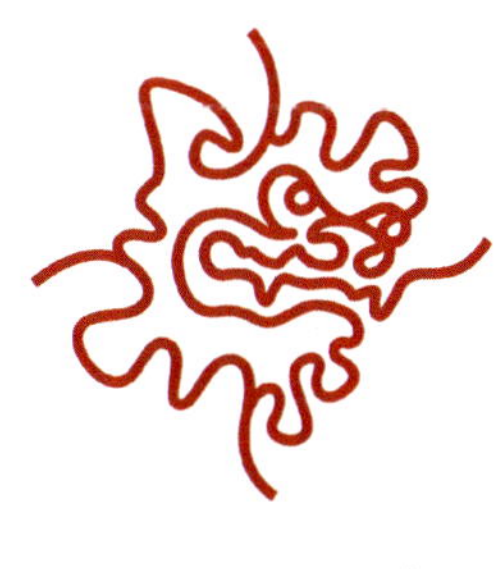

OIST

LeCosPA

KIPAC

IPMU

ICRANet

ANITA

ARA

UFFO

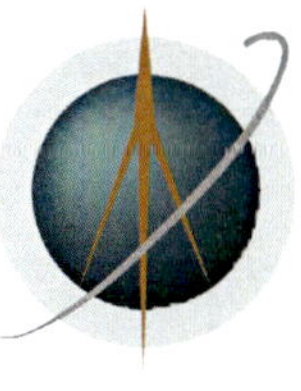

NSPO

LeCosPA's FOURTH ANNIVERSARY CELEBRATION ADDRESS BY PISIN CHEN

Director Pisin Chen, Leung Center for Cosmology and Particle Astrophysics (LeCosPA),
National Taiwan University

Good morning! Welcome all of you to the symposium and this opening ceremony. It is with my tremendous pleasure to share with you the celebration of the fourth anniversary of NTU LeCosPA center. This is a small story about a baby growing into a toddler.

So as I said, this is a small story about a baby growing into a toddler. It starts with the birth. She was born on November 13, 2007. This was a photo taken during the inaugural ceremony on the signing of the donation contract between NTU University President Si-Chen Lee, who's also on the stage here, and our very distinguished alumnus who's also a physics major, in fact, my classmate Mr. C. C. Leung, the co-founder of the Quanta Computers and CEO. So, it was a grand entrance. She was further blessed by her sisters. We have three sisters here, the IPMU of the University of Tokyo. We had a collaboration MOU signed in November 2008, then followed by another signing of a collaboration MOU with the Kavli Institute of Particle Astrophysics and Cosmology at Stanford University, and this was in July, 2009. Further, we signed a collaboration MOU with the ICRANet, the International Center for Relativistic Astrophysics Network in October 2009.

She was further nurtured by lots of distinguished visitors, series of workshops, annual retreats among us, day and night. Now, even babies need to learn to work and we have various working groups on cosmic neutrinos, dark energy, dark matter, early universe, gamma ray bursts, large scale structure, string cosmology. And, these mostly meet every week, very active. Well, baby also needs to learn to relax. Here is what we do. We have what's called Café LeCosPA. This is in a nearby building, a new Astro-math building on the eighth floor. So, there we have a rather interesting lobby, we call it COBE lobby. Because the famous COBE sky map was painted on the ceiling. If you are interested, please go and take a visit. It's on the eighth floor in the next high-rise building nearby us. On the lower left is our master of coffe, Dr. Je-An Gu, who serves us gourmet coffee every week.

Inaugural Ceremony on the Signing of the Donation Contract
between NTU University President Si-Chen Lee and Mr. C. C. Leung

She kept growing, and by now we have all the ten junior fellows. And, she began to walk and talk to express herself and explore the world. This was one of our early findings by one of our junior fellows, Dr. Jiwoo Nam. This was found, from the ANITA data, it was found that it was in addition to neutrinos, this detector can also discover very high energy cosmic neutrinos. This was originally found out of serendipity. We were curious, we further explored our territory and go out. This was a new project launched two years ago, three by now. Three years ago by the name ARA, Askaryan Radio Array, and our PI of this project, Prof. Albrecht Karle would say more later. And, this is in the South Pole. Taiwan is to contribute ten out of 37 antenna stations of this project, playing a rather key role.

This was a photo of myself at the South Pole back in December. So, it's quite an adventure. As if South Pole is not venturous enough for us, we also venture into the space, outer space. This is the project by the name UFFO, Ultra Fast Flash Observatory, which intends to observe the very early prompt signal of gamma ray burst within one second time after the event. This UFFO satellite is to be launched by the Russians this coming June this year. It's a nine-nation collaboration.

ANITA: Discovery of UHECR by Serendipity

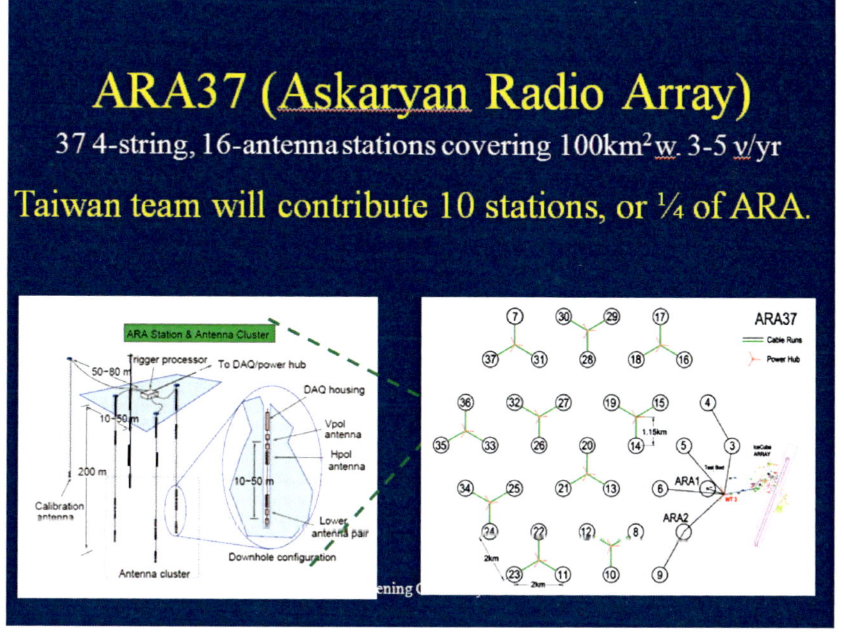

ARA37 (Askaryan Radio Array)

One of the major activities of the preparation of the UFFO was last half a year ago, in July in collaboration with NSPO of Taiwan, National Space Program, whose director general is also here on stage. So, we had successfully made various tests of this UFFO in preparation. Like every baby, there is also growing pain. Here you go.

UFFO Project in Collaboration with National Space Organization (NSPO)

We were very honored and pleased to have the university's strong support by giving us the old dormitory 13 for remodeling into the LeCosPA building and this project was in preparation in the past several years. To our great disappointment, after the construction was finally started last fall, it was found that the structure was unsafe. So, here are a couple of photos. If you are curious, you can take a look, just right next to this building, between the two high-rises, between this Physics building and the Astro-math building. So, this building has to be demolished, to our great disappointment.

But, out of the bad, there was the good, so there's breaking news. Mr. C.C. Leung is now pledging a new donation to construct a brand new LeCosPA building from ground zero. We are very pleased. Since, you know, there're still details to be negotiated, unfortunately we are unable to disclose to you more details, but they will come.

So, here is the brief history of this baby and I thank you all for the tender loving care of her. This baby, born four years ago, has now grown into a healthy toddler. Happy birthday to her! Thank you very much.

LeCosPA's FOURTH ANNIVERSARY CELEBRATION ADDRESS
BY SI-CHEN LEE

President Si-Chen Lee, National Taiwan University

Director Chen, distinguished guests, ladies and gentlemen, good morning. On behalf of National Taiwan University, I would like to welcome all of you to Taiwan and to our campus for the 1st LeCosPA Symposium held in Taipei.

It's great to see so many distinguished scholars gathering here for this exceptional event. During the past decade, there has been remarkable advancement in our understanding of the universe through the efforts of cosmologists and particle astrophysicists worldwide. Yet, the ultimate understanding of the nature of the dark matter and dark energy, the Big Bang, and the beginning of time itself as well as various other fundamental questions still wait for a final answer. It is, therefore, befitting that this 1st LeCosPA Symposium is entitled "Towards Ultimate Understanding of the Universe."

Aspired by the science potential of cosmology and particle astrophysics, National Taiwan University launched the 1st "Leung Center for Cosmology and Particle Astrophysics" with the generous donation by its distinguished alumnus, co-founder and CEO of the Taiwan- based Quanta Computers, Inc. By the way, it's the largest notebook PC manufacturer in the world. We were fortunate to have inaugurated Director of LeCosPA, Pisin Chen. In merely four years, LeCosPA has helped initiate and played key roles in several new international projects, including the ARA Cosmic Neutrinos Observatory, you just saw in the previous Director Chen's introduction, at South Pole and the UFFO satellite GRB telescope. Incidentally, Director Chen's trip to an activity in the South Pole for the ARA project last December has induced a chain reaction of coverage in Taiwan's news media. LeCosPA was also being active in the theoretical understanding of dark matter and dark energy. While celebrating the 4th anniversary of LeCosPA and pleased with the fact that the efforts at NTU have taken the research environment of cosmology in Taiwan to a new level.

As the opening ceremony of the 1st LeCosPA symposium, we feel honored to have the presence of President Jonathan Dorfan from Okinawa Institute of Science and Technology, OIST, who, by the way, was Director Pisin Chen's former boss at Stanford. We are also very honored to have the presence of LeCosPA's sister institution directors and LeCosPA's collaborating project principal investigators. We also feel honored to have so many prominent scientists attending this symposium. I am very sure you will have a very enlightening and productive meeting. I'd like to take this opportunity to briefly introduce our National Taiwan University.

Taiwan university was founded 84 years ago in 1928. It is the oldest, largest and most comprehensive, I would not reluctant to say, the most prominent university in this country. We have 32,000 students, of which 17,000 are undergraduates, more than 10,000 are master degree students and the rest closed to 6,000 PhD students. We have 11 colleges, 54 departments, 102 graduate institutes and about 1,900 faculty members. We were ranked by UK QS company, No. 87 in the world university ranking last year. So, we are among one of the top 100 research universities in the world. According to the Shanghai Jiao Tong University world university ranking, although we were ranked only 123, it is still among the best in the chinese-speaking society, including China, Hong Kong, Taiwan, and Macao. I hope, with, maybe, a new donation of a new CosPA building in the future, we are getting even better in the world ranking. Finally, ladies and gentlemen, thank you for your participation. I wish you enjoy the symposium. As you know, today is the Lantern Festival, I wish you enjoy the trip to Pinxi for the Lantern Festival tonight. Thank you very much.

LeCosPA's FOURTH ANNIVERSARY CELEBRATION ADDRESS BY JONATHAN DORFAN

President Jonathan Dorfan, Okinawa Institute of Science and Technology (OIST) Graduate University

President Lee, Distinguished guests, Director Pisin Chen and participants at this workshop. It is indeed an honor and a pleasure to be here as part of this birthday celebration. Pisin, if you don't mind me using your familiar name, is an old friend and in his introduction he spoke of the birth and growth of this Center. I am going to take you back to the gestation period: back to mid 1999, when I conceived the idea of a particle astrophysics and cosmology institute at Stanford. As I was developing this concept, I sought input from the physicists at SLAC and Stanford. I vividly recall a phone call from Pisin requesting an opportunity to discuss the topic. So it was about twelve years ago, that Pisin and I sat on a bench at SLAC to discuss the founding of what has now become known as the Kavli Institute of Particle Astrophysics and Cosmology (KIPAC). Pisin's inputs in this regard were many. But foremost perhaps was his enthusiasm for helping to raise private funds for the Institute -- part of your training Pisin for what you are doing now. Indeed, Roger Blandford holds the title as Pehong and Adele Chen Director of KIPAC. Without the involvement and enthusiasm of Pisin and the Chen family, KIPAC may well never have come to be. Pisin went on, of course, to be one of the important members of KIPAC. Professor Tune Kamae, who is here today, recently retired after 12 years with SLAC and KIPAC and Greg Madejski is here today to represent KIPAC. But little did I know then, twelve years ago, that this was just the beginning of Pisin's leadership in particle astrophysics and cosmology. NTU saw the scientific opportunity and hired Pisin as the founding director of its LeCosPA Center. In four short years, Pisin and his colleagues here at LeCosPA have brought the institution into the center of the worldwide activity in this exciting field. A vibrant program now exists at NTU with a strong emphasis on scientific excellence and on promoting youth. My warm congratulations go to Pisin, to his colleagues at LeCosPA and to President Lee for what has been achieved until today. It is wonderful to see this additional gift that will lead to a new building: again I think back to how KIPAC was put together. I look forward with excitement as your Okinawan neighbor to watch your further emergence into the teenage years and into the full-grown adult. I also look forward to being a partner with you. Thank you again for inviting me and congratulations.

LeCosPA's FOURTH ANNIVERSARY CELEBRATION ADDRESS BY GREG MADEJSKI

Associate Director Greg Madejski,
Kavli Institute for Particle Astrophysics and Cosmology (KIPAC), Stanford University

Distinguished guests, ladies and gentlemen, I am honored to be able to represent the Kavli Institute for Particle Astrophysics and Cosmology, or KIPAC, here at this glorious occasion. I am the assistant director for scientific program of KIPAC and I am particularly honored to be here because of the many connections that our institutes have. You have already heard from Professor Dorfan that those connections are deeply rooted, and that it really was probably more than just the vision but also a firm realization that particle physics and cosmology - as well as particle physics and astrophysics - have a tremendous potential for joint discovery. That potential has been conveyed by Professor Chen in that very important conversation with Professor Dorfan, aimed at bringing the world-leading particle physicists and astrophysics under one roof. That very conversation resulted in what we now witness as a tremendously successful institution operating as an umbrella for the particle astrophysics and cosmology effort at SLAC and Physics Department at Stanford Campus.

I know Prof. Chen personally: for long time, we both have worked very closely together in trying to bring the scientific aspects of cosmology and astrophysics together to the broader particle physics community. I indeed think this effort has been successful with many particle physicists at SLAC - but also in the greater community. This is witnessed by the excitement about the recent discoveries from new astrophysics instruments coming online. And now, Prof. Chen is repeating the success of bringing the two communities together at the National Taiwan University.

I want to congratulate you for being able to lure Prof. Chen from his full-time appointment at SLAC. But as you know, he is very much still a member of KIPAC. He continues to spend a fair amount of time at Stanford. We are delighted to hear about what exciting news our coming from this institute here, this wonderful growing new institution, and very much look forward to continuing exchange of scientists between our two institutions.

I am also here representing the KIPAC director, Professor Roger Blandford, whom you have seen in one of the slides during the signing ceremony of the memorandum of the understanding. I also represent Professors Persis Drell, the current director of SLAC as well as David MacFarlane, the director of Particle Physics and Cosmology at SLAC National Accelerator Center. All three send their warmest wishes for the continuing success of the LeCOSPA Institute.

We all wish the institute tremendous success, anticipate tremendous and many new discoveries, and hope that the growth as we have seen on the slide presented by Prof. Chen will continue. It really is the very important vision shared by or two institutes from which we all can learn: we must build bridges. I think that that's what I find tremendously wonderful about Prof. Chen that he is so terrific in bringing people from many different disciplines and trying to find what can make two plus two equals seven, or ten, or twenty. That very "synthetic" approach, as well as his enthusiasm I find to be so terrific about him. And I really think that this is my pleasure - and pleasure of all of us at KIPAC - to continue on this cooperation. So, thank you very much again, Prof. Chen, for all your steps towards making our institute a reality. All of us at KIPAC really look forward to many future collaborations!

LeCosPA's FOURTH ANNIVERSARY CELEBRATION ADDRESS BY HIROAKI AIHARA

Deputy Director Hiroaki Aihara, Kavli Institute for the Physics and Mathematics of the Universe (IPMU), the University of Tokyo

President Lee and distinguished guests:

It is my privilege to represent the Institute for Kavli Physics and Mathematics of the Universe at the University of Tokyo (Kavli IPMU) to congratulate you on the fourth anniversary of LeCosPA. Although this photo of snow-covered IPMU looks quite different from LeCosPA in the warm weather in Taiwan, LeCosPA and Kavli IPMU are very close sisters. As Kavli IPMU also celebrated the fourth anniversary in October 2011, LeCosPA and Kavli IPMU are nearly twins.

Institute for the Physics and Mathematics of the Universe (IPMU)

14

I would also like to convey heartiest congratulations from Kavli IPMU Director, Hitoshi Murayama, on LeCosPA, who unfortunately could not make it this celebration. As Director Chen mentioned, LeCosPA and Kavli IPMU already have the memorandum of understanding for collaboration on cosmology and particle astrophysics. We share the common vision of science. We address questions: How did the universe begin? What is its fate? What is it made of? What are its fundamental laws? We would like to understand the universe, in particular, the evolution of the universe. At Kavli IPMU researchers of astrophysics, mathematics and physics work together to address these big questions.

Greetings from all the members at IPMU

Furthermore, LeCosPA and Kavli IPMU share the important mission, that is, to promote the visibility of scientific research in Asia. We are looking forward to working with you.

Once again, congratulations on the fourth anniversary of LeCosPA.

Thank you very much.

LeCosPA's FOURTH ANNIVERSARY CELEBRATION ADDRESS
BY REMO RUFFINI

Director Remo Ruffini, International Center for Relativistic Astrophysics Network (ICRANet),
University of Rome "Sapienza"

It was such a great pleasure for me to celebrate with Prof. Pisin Chen the opening of the LeCosPA center in Taipei, 30 October 2009. I am proud to say the ICRANet is one of the first centers which signed the sister agreement with the LeCosPA, in addition to many other centers all over the world.[1] The LeCosPA is now an important scientific research center that strongly collaborates with important scientific institutions in the Mainland China and all over world.

Signature of the Agreement between ICRANet and LeCosPA (30 Oct, 2009)

The ICRANet, in addition to the headquarter in Pescara[2] and the University of Rome Sapienza[3] in Italy, is also developing its new centers in Nice[4] and Rio de Janeiro.[5]

We expect to strengthen the collaboration between the ICRANet and LeCosPA in the following three topics:

ICRANet Coordinating Center in Pescara (Italy)

(1) To establish the joint exchange programs of graduate students and faculty members in the IRAP PhD program (see Refs. 6 and 7), developed by ICRANet in collaboration with the Freie Universität Berlin, Brazilian Center for Space Research, University of Ferrara, Indian Center for Space Physics, Université de Nice Sophia Antipolis, Sapienza - Università di Roma, Université de Savoie, Shanghai Observatory, Stockholm University, and with the following Research Centers: Institute in Potsdam, ICRA and ICRANet, Observatoire de la Côte d'Azur and Tartu Observatory, recently joined also by INFN.[8]

(2) To promote the organization of the yearly international meetings Galileo-Xu Guangqi[9] and of the Marcel Grossmann meetings which are held once every three years.[10] The Galileo-Xu Guangqi meetings have been created to have a forum for strategic exchanges between eastern and western science at the highest level dealing with relativistic astrophysics and related fundamental theoretical, experimental and observational fields. The aim is to enlarge the audience from the one, strictly Chinese and Italian, to one embracing European and western scientific interests together with the eastern ones. Therefore a broader participation from Asia is encouraged, as well as a participation of scientists from Europe and the Americas. The Marcel Grossmann meetings[10] have been organized in order to provide opportunities for discussing recent advances in general relativity and fundamental field theories, emphasizing mathematical foundations, physical predictions and experimental tests. The objective of these meetings is to elicit exchange among scientists that may deepen our understanding of spacetime structures, as well as to review the

status of ongoing experiments aimed at testing Einstein's theory of gravitation either from the ground or from space.

(3) To develop common research projects directed:

    a. To foster collaboration with the LeCosPA and many Chinese universities on the theoretical and observational properties of White Dwarfs, Neutron Stars and Black Holes. Such topics will involve Pulsars, Supernovae, GRBs, Active Galactic Nuclei, observations in all wavelengths regime, radio optical X and Gamma Rays;

    b. To foster collaboration on the theoretical studies on dark matter, cosmological structures and high energy astroparticle physics between ICRANet, LeCosPA and INFN, with special attention to theoretical interpretations;

    c. Development of shared facilities of data analysis in all wavelengths. We recall that ICRANet is currently developing in its new Seat of Rio de Janeiro a transfer of the Data Center of Frascati. This facility will be in due time assemble all data not only of Space missions in the X and Gamma Rays but also in Optical wave length and astroparticle physics;

    d. ICRANet interest in collaborating with international institutions in France, Russia, China and USA working on ultra-high energy laser pulse leading to the creation of an electron positron plasma in the Earth bound experiments. Such knowledge is acquired in ICRANet by the comprehension of the Astrophysical processes occurring in gravitational collapses leading to the largest impulsive energy sources in the Universe: Gamma Ray Bursts.

I finally express my best wish for a fruitful collaboration between the ICRANet and LeCosPA.

References

1. Memorandum of Understanding between International Center for Relativistic Astrophysics Network (ICRANet) and Leung Center for Cosmology and Particle Astrophysics (LeCosPA) http://www.icranet.org/docs/accordoLeCosPa.pdf
2. ICRANet Pescara Coordinating Center http://www.icranet.org/SeatPescara
3. ICRANet Rome http://www.icranet.org/SeatRome
4. Villa Ratti, Nice http://www.icranet.org/SeatNice
5. Urca, Rio de Janeiro http://www.icranet.org/SeatRio
6. International Relativistic Astrophysics (IRAP) PhD Program http://www.irap-phd.org/
7. IRAP PhD – Announcement http://www.icranet.org/irapphd
8. Istituto Nazionale di Fisica Nucleare (INFN) http://www.infn.it/indexen.php
9. The sun, the stars, the universe and General Relativity: The First Galileo-Xu Guangqi Meeting http://www.icranet.org/docs/GXMeetings.pdf
10. Marcel Grossmann Meetings on General Relativity http://www.icra.it/MG/

LeCosPA's FOURTH ANNIVERSARY CELEBRATION ADDRESS
BY ALBRECHT KARLE

Professor Albrecht Karle, Principal Investigator, Askaryan Radio Array (ARA) Observatory,
University of Wisconsin-Madison

Dear Pisin, Dear Guests,

Thank you very much for the invitation. It's really a privilege and an honor to be here for this event to celebrate the LeCosPA center. It was easy to make this visit a priority especially because of the excellent dumplings that I had last time when I was here. I hope I get another opportunity.

In 1912, Victor Hess discovered the existence of cosmic rays, through experiments conducted on a hot air balloon. He used an electroscope and took it on balloon flights, where he noticed that, after an initial decrease, the radiation increased with higher elevation, and concluded that the cosmic rays were originating from outer space. Today, 100 years later, the origin of highest energy cosmic rays still remains unresolved.

Another event happened at almost exactly the same time, when Roald Amundsen and his crew arrived as the first explorers at the South Pole on December 14, 1911. So, as you might guess, I am not trying to give a history lesson here. But we often wonder about what happened in the past and how it relates to where we stand 100 years later. One event that was very important to me was the successful completion of the IceCube neutrino detector. Another key event was Pisin Chen's visit to the South Pole. I think this visit really symbolizes the expansive nature of the international scientific collaboration under the umbrella of the United States Antarctic Program.

Askaryan Radio Array (ARA) project
- a very large large radio neutrino detector at the South Pole

Scientific Goal

- Discover and determine the flux of highest energy cosmic neutrinos.
- Understanding of highest energy cosmic rays, other phenomena at highest energies.

Method:

Monitor the ice for radio pulses generated by interactions of cosmic neutrinos with nuclei of the 2.8km thick and radio transparent ice sheet at the South Pole

Askaryan Radio Array (ARA) Project

What we are about to undertake now is a new, very large cosmic neutrino detector, the Askaryan Radio Array. It is not all there yet, but we have started and it is our goal to build this detector in the coming years. The scientific goal is to determine the origin of highest energy cosmic rays. We plan to do that by measuring neutrinos of highest energy levels and trying to understand the cosmic rays and other phenomena at extremely high energy ranges. The method employs radio antennas to look into the deep ice. The ice turns out to be not only the world's clearest medium in optical photons but also in radio waves. With radio waves we can see all the way down, kilometers deep.

ARA- Collaboration

- ARA is an international Collaboration.
- LeCosPA at NTU has been a critical collaborator and founding member of this collaboration.
- The commitments to the Proposal have been critical for the progress made today, and they will be for the further development of the project.
- Pisin Chen's participation at the South Pole workshop in Washington 2011 was helpful to advance the relation between NTU, Taiwan as a whole and run by the US National Science Foundation's Antarctic program.

Design and Initial Performance of the Askaryan Radio Array
Prototype EeV Neutrino Detector at the South Pole

P. Allison,[1] J. Auffenberg,[2] R. Bard,[3] J. J. Beatty,[1] D. Z. Besson,[4] S. Böser,[5] C. Chen,[6] P. Chen,[6] A. Connolly,[1] J. Davies,[7] M. DuVernois,[2] B. Fox,[8] P. W. Gorham,[8] E. W. Grashorn,[1] K. Hanson,[9] J. Haugen,[2] K. Helbing,[10] B. Hill,[8] K. D. Hoffman,[3] M. Huang,[6] M. H. A. Huang,[6] A. Ishihara,[11] A. Karle,[12] D. Kennedy,[4] H. Landsman,[2] A. Laundrie,[2] T. C. Liu,[6] L. Macchiarulo,[8] K. Mase,[11] T. Meures,[9] R. Meyhandan,[8] C. Miki,[8] R. Morse,[8] M. Newcomb,[2] R. J. Nichol,[7] K. Ratzlaff,[13] M. Richman,[3] L. Ritter,[8] B. Rotter,[8] P. Sandstrom,[2] D. Seckel,[14] J. Touart,[3] G. S. Varner,[8] M. -Z. Wang,[6] C. Weaver,[12] A. Wendorff,[4] S. Yoshida,[11] and R. Young[13]

(ARA Collaboration)

[1] Dept. of Physics, Ohio State Univ., Columbus, OH 43210, USA.
[2] IceCube Research Center, Univ. of Wisconsin, Madison WI 53703
[3] Dept. of Physics, Univ. of Maryland, College Park, MD, USA
[4] Dept. of Physics and Astronomy, Univ. of Kansas, Lawrence, KS 66045, USA
[5] Dept. of Physics, Univ. of Bonn, Bonn, Germany
[6] Dept. of Physics, Grad. Inst. of Astrophys. & Leung Center for Cosmology and Particle Astrophysics, National Taiwan Univ., Taiwan
[7] Dept. of Physics and Astronomy, Univ. College London, London, United Kingdom
[8] Dept. of Physics and Astronomy, Univ. of Hawaii, Manoa, HI 96822, USA.
[9] Dept. of Physics, Univ. Libre de Bruxelles, Belgium
[10] Dept. of Physics, Univ. of Wuppertal, Wuppertal, Germany
[11] Dept. of Physics, Chiba university, Tokyo, Japan
[12] Dept. of Physics, Univ. of Wisconsin Madison, Madison, WI, USA.
[13] Instrumentation Design Laboratory, Univ. of Kansas, Lawrence, KS 66045, USA
[14] Dept. of Physics, Univ. of Delaware, Newark, DE 19716

Introduction to ARA Collaboration

Today we are working in an international collaboration. The idea became solid about three years ago and the first journal publication documents first results and the details of the plans. The LeCosPA at NTU has been a critical collaborator and one of the founding members of this collaboration. The commitments to our plan and to the project have been critical for the progress made to date. There will be further progress to be expected.

Pisin Chen's participation at the South Pole workshop last spring in Washington was also useful for increasing the collaboration within the science community in the U.S. On behalf of the Wisconsin IceCube Particle Astrophysics Center at UW–Madison as well as the Askaryan Radio Array collaboration I am wishing Pisin and all of the LeCosPA best of luck and success and we look forward to a great and successful collaboration.

Testbed site & hardware

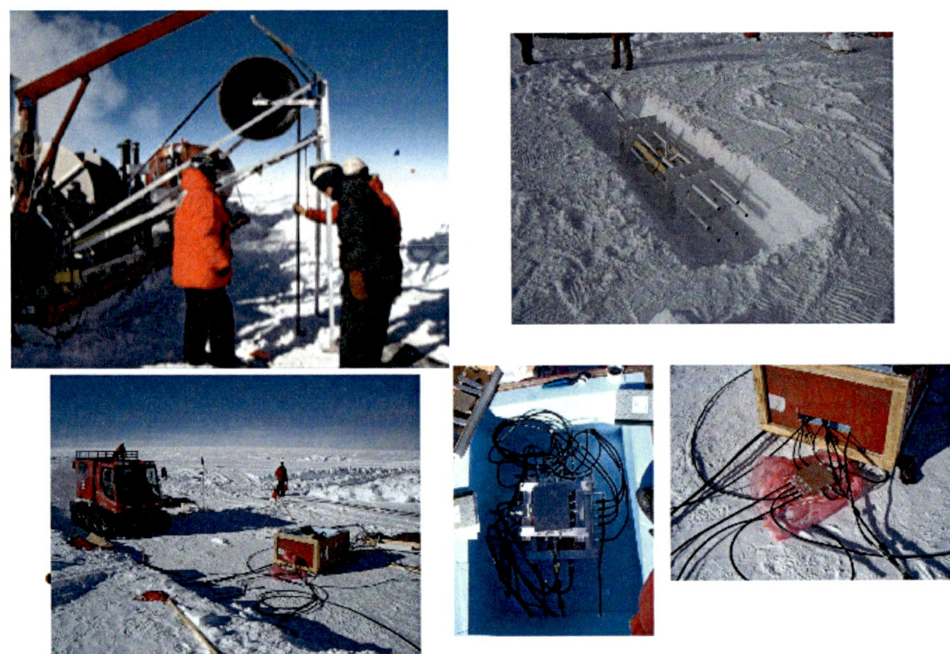

ARA Field Activities on the Testbed Site at the South Pole

LeCosPA's FOURTH ANNIVERSARY CELEBRATION ADDRESS
BY PETER GORHAM

Professor Peter Gorham, Principal Investigator, Antarctic Impulsive Transient Antenna (ANITA),
University of Hawaii

Please allow me first to thank the organizers of this meeting for the honor of allowing me to address you. It has been a privilege for me to share Particle Astrophysics research interests and goals with the Leung center, and in particular let me thank Prof. Pisin Chen for his enthusiasm and support in our collaborative efforts over the last decade. I would like to briefly highlight one such effort here, the Antarctic Impulsive Transient Antenna, or ANITA project, which dates back a bit less than ten years, during which LeCosPA faculty, including Prof. Chen and Prof. Jiwoo Nam, have played important roles for almost the entire duration.

ANITA is a NASA long-duration balloon payload which flies an array of 40 antennas above Antarctica with a goal of detecting radio bursts from ultra-high energy particles interacting in the Antarctic atmosphere and ice sheets. Ultra-high energy cosmic ray particles collide with air molecules and produce an extensive cascade of relativistic particles in the air above the continent. This cascade then forms a radio impulse as it accelerates through the geomagnetic field, and ANITA was the first to detect these radio bursts at the very highest cosmic ray energies. Work done with Prof. Chen at the Stanford Linear Accelerator Center in 2006 was instrumental in determining the precise calibrations necessary to accomplish this.

An equally exciting goal of ANITA is the detection of particle cascades within the ice sheets that are initiated by ultra-high energy cosmogenic neutrinos.

These neutrinos have not yet been detected, but must be present at some level according to the standard model of physics and by simple cosmological arguments. Because we can monitor a million cubic kilometers from our float altitude in Antarctica,

ANITA has proven to be the most sensitive experiment to date in the ongoing search for these expected neutrinos, LeCosPA faculty, postdoctoral fellows, and students have been centrally involved in analyzing data for ANITA's ongoing neutrino search, they have often led the way in developing new ideas and initiatives in this critical analysis.

The third flight of ANITA, to take place late next year, will likely be the most exciting flight yet, with many hundreds of cosmic ray events likely to be detected, the largest ultra-high energy radio-detected sample of any instrument. For neutrinos, ANITA's third flight will improve the sensitivity by up to a factor of three, to a level where the first detection of these cosmogenic neutrinos is quite possible. In all of these efforts we are honored to have the opportunity to work closely with Prof. Chen and our colleagues at LeCosPA, and we look forward to sharing the pleasure of achieving these scientific goals with them.

Thank you again for hosting this wonderful conference, and we wish you all the best of fortune in this auspicious new year of the water dragon -- or perhaps for us, the ice dragon!

LeCosPA's FOURTH ANNIVERSARY CELEBRATION ADDRESS
BY GUEY-SHIN CHANG

Director General Guey-Shin Chang, National Space Organization (NSPO), Taiwan

President Lee, Professor Chen, and all distinguish guests: it is my great honor to be here to celebrate the ceremony for the fourth anniversary of LeCosPA. The National Space Organization (NSPO) is pretty honored to be involved in the UFFO project. So, I would like to show your some charts with the works that have been performed at NSPO. Established in 1991, the National Space Organization has executed the Taiwan's space program for past two decades. We have dual roles and responsibilities: First, we act as the national space agency in Taiwan. Second, we have a mission to build-up NSPO as a space technology R&D organization. For the past twenty years, we have launched three satellite programs. Currently, we are working on two major programs: FORMOSAT-5 and FORMOSAT-7. Our vision is to become a center of innovation and excellence for space technology in Taiwan and also to conduct space programs with Taiwan's strengths and global competitiveness.

This is our operations model. We have four major facilities in support of our space program. We have a very complete satellite I&T facility. We have a very comprehensive ground segment including a Mission Operation & Control Center where the whole missions are operated and controlled from here. We also have a Remote Sensing Image Processing facility. In addition, we have more than ten spacecraft technology R&D laboratories to support our programs.

For the core technologies, they include Satellites Systems Engineering, Spacecraft Development, Remote Sensing Instrument, Satellite Operation & Control, and Mission Data Processing including science data operations.

National Space Organization (NSPO): Role, Responsibilities, and Vision

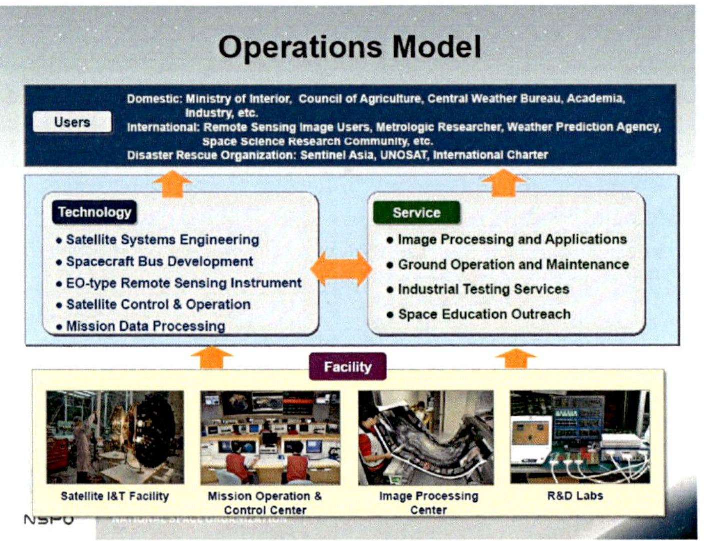

NSPO Operations Model

I'd like to share with you some activities that have been done for the UFFO project. UFFO's and NSPO's teams performed the thermal vacuum test during July 12th to 15th, 2011. This team is formed by four organizations: NSPO, Korean Ewha Woman's University, and National Taiwan University. Sorry I didn't mention National United University.

We have spent lots of effort to perform the most comprehensive thermal vacuum test with 24 hours operations in our facilities. And all the instruments have been successfully conducted and tested with very good performance during the hot and cold space environments.

Follow that, we also performed the vibration test and shock test at NSPO facility. Although we have a lot of works and challenges ahead us, the team worked very well and completed the test on 27th July, 2011. I am glad to know UFFO is going to be launched in June this year.

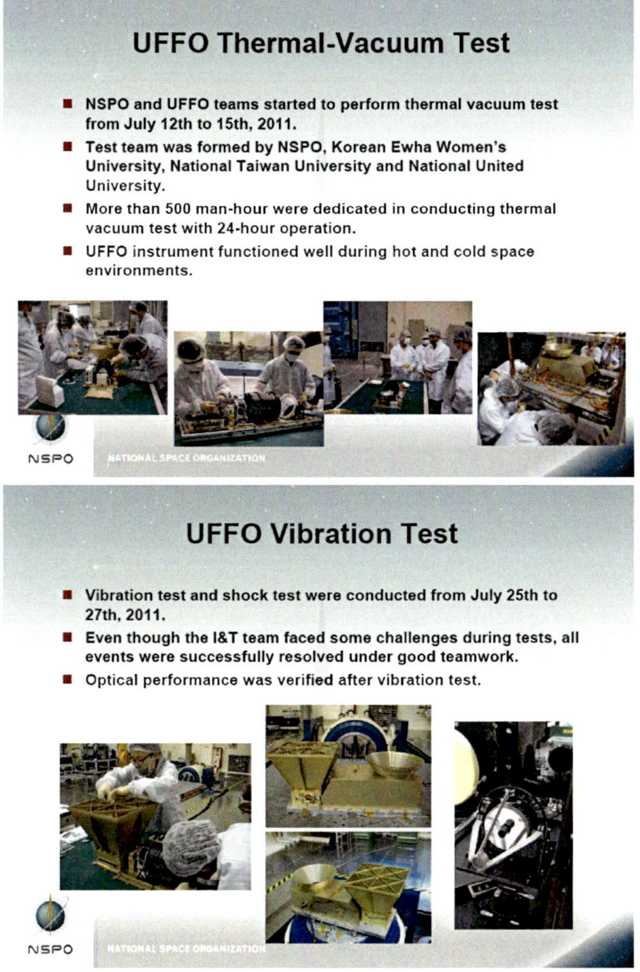

UFFO Tests at NSPO, Hsinchu, Taiwan, July 2012

I think NSPO would be glad to be a partner of the LeCosPA. I believe the UFFO mission is one of the best science researches in Taiwan. NSPO is also committed to cooperate with all the space community in Taiwan and the world as well.

Section I

Overview of Institutional Research Programs

CURRENT AND FUTURE RESEARCH PROGRAMS AT STANFORD'S KAVLI INSTITUTE FOR PARTICLE ASTROPHYSICS AND COSMOLOGY

GREG MADEJSKI

Kavli Institute for Particle Astrophysics and Cosmology, Stanford University
Menlo Park, CA 94025, USA
E-mail: madejski@slac.stanford.edu

The Kavli Institute of Particle Astrophysics and Cosmology, or KIPAC, at Stanford University, and the LeCosPA Institute at the Taiwan National University were sibling institutions even before their respective official births. The existence of both institutes was to a great extent facilitated by the foresight of Prof. Pisin Chen, current director of LeCosPA, and we fully envision the vibrant on-going collaboration between the two institutes for the years to come.

This presentation highlights the current research direction of KIPAC, including the wide range of programs in particle astrophysics and cosmology. Of the on-going projects, the main current effort at KIPAC is the operation of, and the analysis of data from the Large Area Telescope on-board the space-borne Fermi Gamma-ray Space Telescope, which is described in more detail in the article by Prof. Kamae in these proceedings. That article focuses on the instrument, and the results gleaned from observations of our own Galaxy. Here, the second part of this article also includes the highlights for astrophysics of jets emanating from the vicinity of black holes, which are prominent gamma-ray sources detected by Fermi: this is the area of research of the article's author.

Keywords: Astrophysics; Gamma-rays.

1. KIPAC and Its Research Programs

Kavli Institute was envisioned as a research organization bringing together the expertise of particle physicists, and astrophysicists working in the area of particle astrophysics and cosmology. The vision behind the formation of the Institute has been (and still is!) to explore the synergy of the Cosmic Frontier with the Energy and Intensity Frontiers. The research at KIPAC[a] is accurately described by its name: Particle Astrophysics includes studies of very energetic particles in the Cosmos, how they are accelerated, how they interact with the ambient photons and other particles, and how they are directly or indirectly detected. The Cosmology part includes the studies of the origin, the content, the large-scale structure, and the history of the Universe. In both parts, KIPAC is engaged via three somewhat overlapping areas. First, this encompasses the development and construction of new

---

[a]http://kipac.stanford.edu/kipac/

Fig. 1. The Fred Kavli building at SLAC, hosting the Kavli Institute for Particle Astrophysics and Cosmology.

experimental facilities. The second is the observational research taking advantage of the existing facilities operated by national agencies. The third area includes theoretical work - both direct interpretation of observational and experimental data, but also work on entirely new, fundamental theories, aimed to account for a broad range of astrophysical observations.

Just as KIPAC has been envisioned to facilitate an interaction between the researchers engaged in the experimental, observational, and theoretical efforts, it is intended to build bridges between Stanford's Physics Department, and Stanford Linear Accelerator Center (now SLAC National Accelerator Laboratory). This bridging is extensive: KIPAC researchers work in both locations, and participate in twice-weekly all-hands meetings alternating between the Physics Dept. and SLAC. KIPAC's SLAC home is illustrated in Fig. 1. KIPAC is a member of a larger network of Kavli Institutes, and one of the goals is to collaborate closely with other Kavlis, both in the US and abroad. KIPAC members also collaborate closely with researchers at several "sister institutions" - and LeCosPA is one of them.

One of the main future experimental / observational efforts with KIPAC's extensive involvement is the study of dark energy. This elusive entity, manifesting itself as the acceleration of the expansion of the Universe, can be studied via a number of different experimental approaches. Currently, the KIPAC researchers are measuring the properties of Dark Energy using clusters of galaxies, mainly via studies of the distribution of their masses as inferred from their X-ray emission: this effort uses space-borne X-ray observatories such as Chandra X-ray Observatory, XMM-Newton Observatory, and the Suzaku X-ray satellite. KIPAC is a part of the Dark Energy

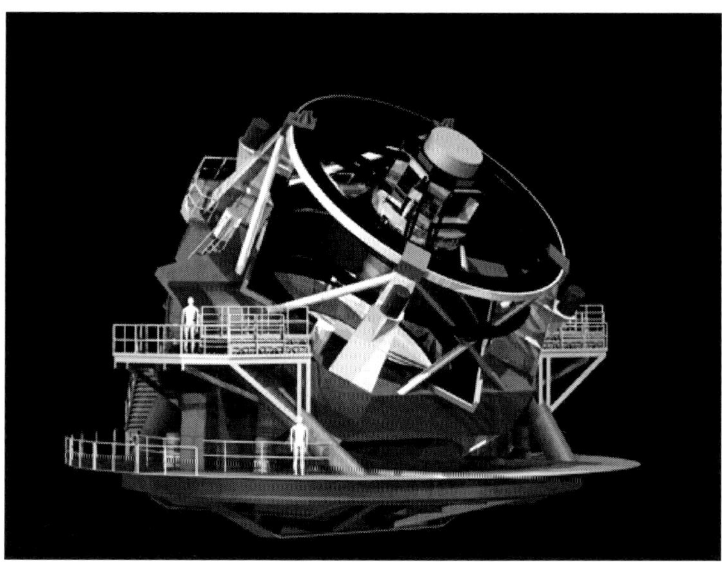

Fig. 2.   Schematic drawing of the LSST telescope. Credit: LSST Corporation

Survey (DES)[b] collaboration, led by the Fermi National Accelerator Laboratory, and formed to deploy and operate and optical telescope facility at a Chilean mountaintop. The DES, now becoming operational, will provide excellent data allowing a multi-prong approach to the study of dark energy. Further in the future, the dark energy studies will involve the Large Synoptic Survey Telescope, with science goals described in Ref. 1. LSST[c] is illustrated in Fig. 2, and is a significantly larger, dedicated telescope facility under construction, also in Chile, slated for data taking in about seven years. This instrument has been in development for a number of years, and currently it is an approved project, funded jointly by the National Science Foundation, Department of Energy, and international partners, as well as private grants. It will allow detailed studies of distribution of the gravitating material in the Universe. This will be accomplished mainly by the use of gravitational lensing by such material. The careful design of the telescope features excellent angular resolution, which, when coupled with a large field of view, will be essential to measure over a large part of the sky the distortion of shapes of the background galaxies resulting from gravitational lensing by intervening matter. Of course this is only one of many areas of astrophysics and cosmology addressed by the LSST mission: such an ambitious, wide-field instrument, designed to scan the sky at a good cadence, will be sensitive to a wide range of transient phenomena. The discovery potential is truly enormous. SLAC is responsible for the development and construction of the heart of the facility - the telescope's main camera.

---

[b]http://www.darkenergysurvey.org/science/
[c]http://www.lsst.org/lsst/

Fig. 3.   Image of one of the CDMS detectors. Credit: CDMS collaboration

KIPAC researchers are engaged in an ambitious, multi-pronged study of the nature of dark matter. This includes direct searches for signatures of the interaction of the putative dark matter particles with ordinary matter: this effort continues via a low-temperature, Cryogenic Dark Matter Search (CDMS)[d] experiment. This experiment, consisting of a number of custom-designed Germanium sensors (illustrated in Fig. 3), is located in an underground laboratory to reduce the effects of interaction of cosmic rays with the detector. Another avenue includes an indirect search for products of annihilation of putative Weakly Interacting Massive Particles, or WIMPs. Such annihilation is expected to ultimately result in a production of two photons. Since the most compelling (on theoretical grounds) mass range for such a WIMP is in the range of between a fraction of GeV and many TeV, the annihilation should produce gamma-rays, which in turn should be detectable by the Fermi Gamma-ray Space Telescope[e]. Searches are conducted via studies of gamma-ray emission from large concentrations of dark matter, but those where one does not expect strong astronomical gamma-ray signal: clusters of galaxies, small galaxies that are satellites to our own Milky Way, or the vicinity of the Galactic Center. Fermi LAT has not yet detected any such signal, but has provided very stringent limits. The search in the GeV range continues with Fermi, but the next frontier is the higher mass range which will be enabled by the next generation ground-based Cerenkov Telescope Array, or CTA[f], sensitive in the energy range upwards of (roughly) 50 GeV. KIPAC researchers are extensively involved in the development of cameras aimed to detect the Cerenkov light in such planned arrays.

---

[d]http://cdms.berkeley.edu/
[e]http://fermi.gsfc.nasa.gov/ssc/
[f]http://www.cta-observatory.org/

There are many additional efforts with strong involvements of KIPAC researchers. One area experiencing strong growth involves detailed measurements of the Cosmic Microwave Background. There, the most ambitious is the study of B-mode polarization of the CMB. Most notable is the involvement of KIPAC researchers in the ongoing experiments QUIET and POLAR. So far, there are no detections, but the future experiments considered at KIPAC will have the reach to probe the signatures of inflation, proposed to describe the very early phase of growth of the Universe.

2. Selected Results from Studies of Astrophysical Jets with the Fermi Large Area Telescope

While the Fermi LAT has not conclusively detected (as yet) any signal resulting from annihilation of the putative dark matter particles, it has been a tremendously productive facility to detect and measure the gamma-ray emission from astrophysical sources, presumably produced by well-understood processes. The telescope, described in more detail in the article by Prof. Kamae (this volume) is a testimony to the extremely successful and fruitful collaboration between particle physicists and astrophysicists. Fermi LAT is essentially a particle tracker, capable of measuring the direction of arrival, and the energy of a celestial gamma-ray, and is described in detail in Ref. 2. It is sensitive to photons arriving from any direction of roughly 2 steradians of the sky in the energy range of ~ 0.02 to ~ 300 GeV, so it is essentially an all-sky gamma-ray monitor. The telescope geometry and basic design can localize the direction of arrival of a gamma-ray to (roughly) 1 degree, depending on its energy. Importantly, while designed as a gamma-ray detector, Fermi is capable of measuring the spectrum of cosmic ray electrons (Ref. 3). Such measurements have *not* confirmed the previous claims of a sharp spectral feature in their spectrum. Interestingly, it is possible for Fermi-LAT to distinguish the cosmic positrons from cosmic electrons using the Earth's magnetic field, and Fermi has been able to measure the spectrum of cosmic ray positrons up to ~ 200 GeV (Ref. 9), confirming the results derived from the PAMELA data (Ref. 10).

Fermi LAT studies of the celestial sources of gamma-rays confirmed the findings of its predecessor, the EGRET instrument on-board of the Compton Gamma-ray Observatory, that our own Galaxy produces copious gamma-rays in a form of relatively uniform, diffuse emission: this emission is significantly enhanced in the Galactic plane. Our Galaxy hosts numerous distinct supernova remnants, and those are also prominent, clearly identified emitters of gamma-rays. There, the kinetic energy of ejecta interacting with the ambient interstellar medium results in shocks, and those shocks are the most compelling sites for acceleration of particles to energies required to produce the gamma-rays. In conjunction with data from radio, optical, X-ray, and TeV-range gamma-ray observatories, Fermi paints a picture where the gamma-ray emission from supernova remnants originates via multiple channels. One is the inverse Compton process by energetic electrons - presumably the same population that is producing radio emission via the synchrotron process, and the other is the proton-proton interaction, resulting in pions and eventually in gamma-rays. Since it requires an acceleration of protons to energies in the multi-GeV range, this last process would indeed implicate the supernova remnants as the sources of Galactic cosmic rays, and indirectly, as the source of energy for the diffuse Galactic emission mentioned above.

Another clearly identified Galactic gamma-ray emitters are rotation-powered pulsars. While such rotation-powered pulsars were known to be strong gamma-ray emitters in the EGRET days, a notable and important discovery by Fermi LAT was the detection of pulsed gamma-ray emission from millisecond pulsars - also rotating neutron stars, but located in stellar binary systems. This discovery will significantly contribute to our understanding of evolution of binary stars in our Galaxy.

2.1. *Fermi and Jets in Active Galaxies*

Perhaps the most prominent, discrete sources of gamma-rays outside of our own Milky Way Galaxy are active galaxies. Nuclei of such active galaxies are bright, point-like emitters of electromagnetic radiation detected in all accessible bands, with measured flux many times stronger than the total output of all stars in the entire host galaxy. We now believe that such nuclear activity is powered by the release of gravitational energy of the galaxian material falling - or accreting - onto a supermassive black hole, with a mass of many million times greater than the Sun. Fermi has detected strong and variable gamma-ray emission from many such active galaxies, but there appears to be a clear correlation of gamma-ray emission with the presence of relativistic jets, although generally only if such jets are pointing close to our line of sight. Such jets, emanating from the vicinity of the accreting black hole, are clearly imaged by radio interferometric observations, and often also seen in high-resolution optical and X-ray images. We now believe that the gamma-ray emission arises from such relativistic jets, but to have a complete picture, including the details of the source of energy and its transport to the region where radiation is produced, it is necessary to study these jets in all accessible bands of electromagnetic spectrum, from radio waves, through the microwave, infrared, optical, and ultraviolet emission, up to the X-ray and gamma-ray bands. Since the emission is variable, on time scales from days to several months, the studies require simultaneous measurements, but the time variability provides an additional handle on the size of the emitting region, as well as the relationship of regions responsible for emission in various bands.

Such broad-band spectra of their electromagnetic radiation are well measured for many extragalactic jets. They generally consist of two very broad "humps," with the low energy one peaking in the far-IR to UV range, and another peaking in the range of $10 - 300$ MeV: this is illustrated in Fig. 4 for a representative, well-studied active galaxy 3C279. Since the distances to those active galaxies, and fluxes of radiation detected from them are well measured, it is possible to estimate the physical conditions in such jets. The considerations of the extremely high particle

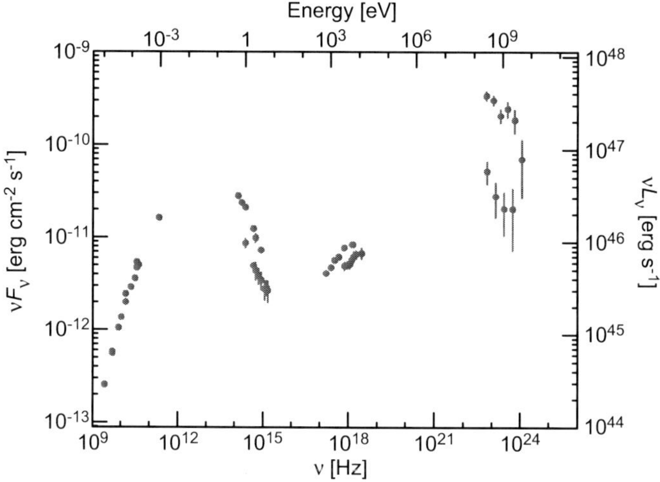

Fig. 4. Broad-band electromagnetic spectral energy distribution for the well-known jet-dominated active galaxy 3C279. The red and blue colors represent spectra at different epochs (from Ref. 4).

and photon energy density in the emitting regions provide a compelling argument that the emission region is moving with the relativistic speeds close to our line of sight, and is subject to strong relativistic effects. Such relativistic motion is in fact directly observed via time-resolved interferometric imaging studies of astrophysical jets in the radio-frequency regime. The consequence of relativistic motion is Doppler-boosting of radiation towards our line of sight (and thus it appears highly collimated, jet-like), and foreshortening of the apparent variability time scales. In reality, the co-moving size of the emitting region is greater, by a factor of $\delta = (\Gamma_{\mathrm{jet}})^{-1}(1 - \beta_{\mathrm{jet}}\cos\theta)^{-1}$ (where $\beta_{\mathrm{jet}} = v/c$, Γ_{jet} is the Lorentz factor of the bulk motion of the emitting material in the jet, typically $\sim 10-20$, and θ is the angle of motion to the line of sight). Likewise, the total (spectrum-integrated) emitted flux as measured in the co-moving frame is lower, by a factor of δ^4 than that inferred from observations without accounting for the relativistic effects.

We now have a reasonably good, compelling picture of the radiation processes responsible for the broad-band emission in extragalactic jets. Since the radiation in the low-energy peak, covering the emission from radio to the UV bands is polarized, the most compelling process that can account for polarization over such a large photon energy range is the synchrotron process. The individual particles are extremely energetic, with Lorentz factors up to $\sim 10^6$. Since the radiation is highly collimated, the magnetic field, already implicated in the radiation process, is very likely to play an important role in the jet's collimation. The high energy peak, covering the gamma-ray band, is due to the inverse Compton process, where lower energy photons are inverse-Compton-scattered by the same electron population. Those "soft" photons can originate in the jet by the synchrotron process or can be external to the jet, associated with the accretion disk or the host galaxy.

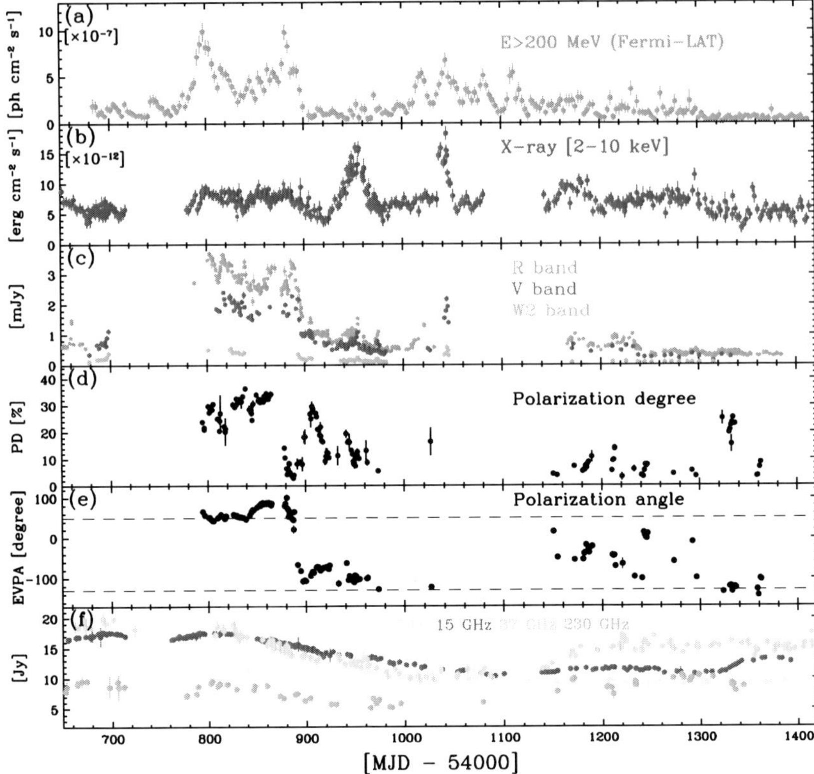

Fig. 5. Light curve (time series) of the electromagnetic emission in various bands, measured from a jet-dominated active galaxy 3C279 (from Ref. 5). Most notable are panels (a) and (e), the former showing the gamma-ray flares measured by Fermi, and the latter showing the rotation of the polarization angle measured in the optical band.

The total power associated with a jet in some active galaxies appears comparable to the total power delivered by the material losing its gravitational energy in the process of accretion, so the conversion of gravitational energy must be very efficient. A very important question regarding the jet structure and formation regards the location of the region where energy is dissipated into radiation. From simple considerations, one would expect that the gravitational energy is released relatively close to the black hole, since the potential energy depends as 1/distance. The light-crossing time for a black hole in a jet-dominated active galaxy such as 3C279 - with the mass of several hundred million Solar masses - is on the order of an hour. Wherever the radiation is released, the energy must be efficiently delivered from the immediate vicinity of the black hole to such "dissipation" region.

Recent observations motivated by the deployment of Fermi indicate that the location of the emission region seems more distant than one would naively expect. Optical observations indicate that the gamma-ray flares are essentially contemporaneous with the optical flares: this is illustrated in Fig. 5 (from Ref. 4). A remarkable

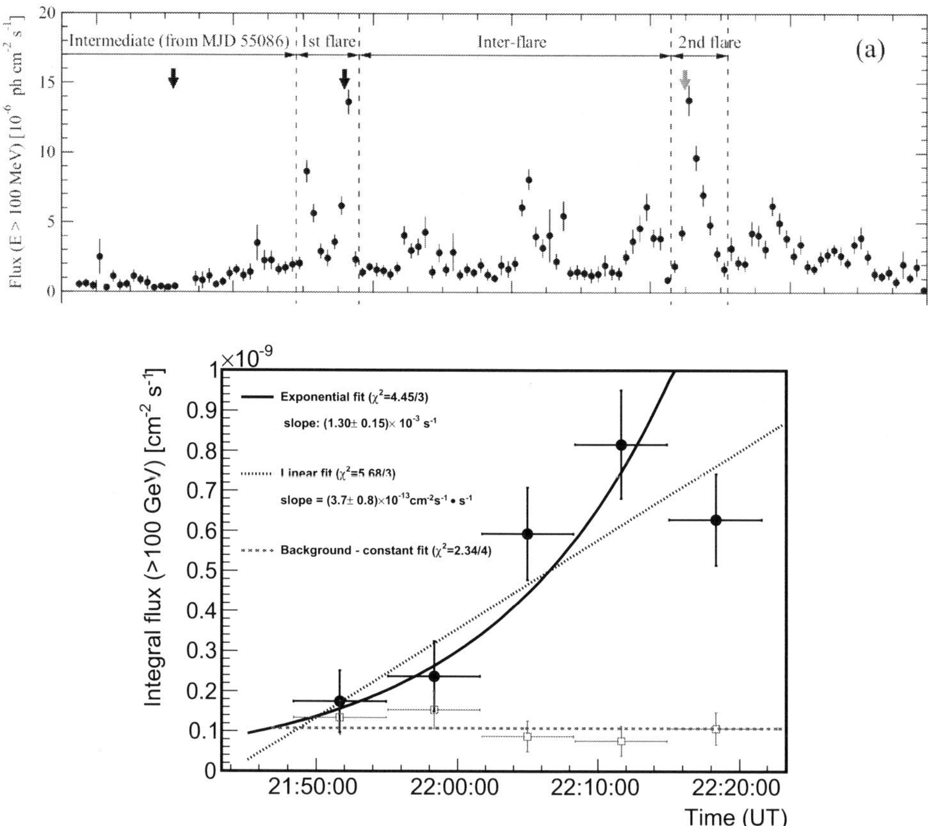

Fig. 6. Top: Light curve (time series) of the gamma-ray - bright active galaxy PKS 1222+21, as measured by the Fermi LAT instrument. The X-axis covers approximately 120 days, and illustrates frequent rapid flares measured in this object (from Ref. 6). Bottom: time series for the TeV-range gamma-ray emission from PKS1222+21, measured by the MAGIC TeV telescope. Here, the X-axis covers about one hour (from Ref. 7).

observation using optical polarimetry indicates that such optical / gamma-ray flares are generally associated with the rotation of polarization angle, which takes place over tens of days. If one correctly considers the relativistic effects of the emitting material moving with relativistic speeds, the dissipation region must extend over at least tens of light-days, and possibly as far as a parsec away from the black hole.

Perhaps the most compelling observation of relativistic jets weighing in on the location of the dissipation region was motivated by Fermi, but conducted by a ground-based, TeV-sensitive gamma-ray observatory MAGIC. It so happens that many of those jets are strong emitters of gamma-rays in the TeV regime. A prominent relativistic jet in an active galaxy known as PKS 1222+21 was discovered by the Fermi satellite to flare rapidly (see Fig. 6, top; Ref. 6). This triggered observations with the MAGIC TeV telescope, which in turn detected strong TeV flux, varying rapidly, on a time scale of hours (Ref. 7 and Fig. 6, bottom). This observation suggested a compact dissipation region, but required that the emission region must be at a considerable distance from the black hole. This is because the TeV gamma-rays are subject to opacity against pair production from interaction with UV and soft X-ray flux. Such flux is clearly measured in this (and many other) active galaxies, and is believed to be produced in the accreting material within a few light-months from black hole. Now if the gamma-rays were to originate close to the black hole, they would have not been able to escape the very dense radiation field associated with the accretion. Therefore, the conversion of the jet's energy into radiation must take place beyond the realm where the soft UV and X-ray radiation is produced. In summary, Fermi LAT is providing important new evidence that active galaxies are very efficient in coupling the accretion power to their relativistic jets, but that before the conversion to radiation, the energy must be efficiently transported over many orders of magnitude in distance. This is a complex and difficult problem, requiring sophisticated magneto-hydrodynamical simulations (see, e.g., Ref. 8).

3. Summary

The Kavli Institute is engaged in a very vibrant research program in particle astrophysics and cosmology. The current operating facility, the Large Area Telescope on-board of the Fermi Gamma-ray Space Telescope, is providing excellent data on the sources of most energetic radiation detected in the Universe. Over the next few years, the Kavli Institute will take advantage of data from several new facilities aimed to answer pressing questions on cosmology, addressing the questions of dark matter and dark energy. We all look forward to the continuing collaboration between KIPAC and LeCosPA!

Acknowledgments

The author is grateful to Prof. Chen for the invitation to the Symposium, and wishes to thank Profs. Roger Blandford and Marek Sikora for comments on the manuscript.

References

1. LSST Science Collaborations and LSST Project, LSST Science Book, Version 2.0, arXiv:0912.0201, http://www.lsst.org/lsst/scibook (2009).
2. W. Atwood, et al., *Astrophysical Journal*, **697**, 1071 (2009).
3. A. A. Abdo, et al. *Phys Rev. Letters*, **103**, 1101A (2009).
4. A. A. Abdo, et al. *Nature*, **463**, 919 (2009).
5. M. Hayashida, et al. *Astrophysical Journal*, in press (2012).
6. Y. T. Tanaka, et al. *Astrophysical Journal*, **733**, 19 (2011).
7. J. Aleksic, et al. *Astrophysical Journal*, **730**, L8 (2011).
8. J. McKinney and R. Blandford, *Mon. Not. Royal Astron. Soc*, **394**, 126 (2009).
9. M. Ackermann, et al. *Phys. Rev. Letters*, **108**, 11103 (2012).
10. O. Adriani, et al. *Astropart Phys.*, **34**, 1 (2010).

COSMOLOGY AND PARTICLE ASTROPHYSICS AT KAVLI IPMU

HIROAKI AIHARA

Kavli Institute for the Physics and Mathematics of the Universe,
The University of Tokyo
5-1-5 Kashiwanoha, Kashiwa, 277-8583, Japan
E-mail: aihara@phys.s.u-tokyo.ac.jp

Kavli Institute for the Physics and Mathematics of the Universe (Kavli IPMU) currently undertakes two large-scale projects in cosmology and particle astrophysics. One is Subaru Measurement of Images and Redshifts, the Sumire project. It observes images and redshifts of the galaxies using Subaru telescope to study cosmology and astronomy. The other is XMASS experiment aiming to detect the cold dark matter using liquid Xenon. We provide a brief introductory description of these projects.

Keywords: Observational Cosmology; Particle Astrophysics; Dark Energy; Dark Matter.

1. Subaru Measurement of Images and Redshifts

Subaru measurement of images and redshifts (SuMIRe) will produce wide-field deep astronomical imaging and spectroscopic data of the northern sky using Hyper Suprime Cam (HSC)[1] and Subaru Prime Focus Spectrograph (PFS)[2] mounted on the 8.2-meter Subaru telescope at the summit of Mauna Kea in Hawaii. HSC is a wide-field (1.5-degree in diameter or 1.77 square degree) camera consisting of 116 pieces of $2K \times 4K$ CCDs, totaling 928 Megapixels. The CCD, developed and manufactured by Hamamatsu Photonics, is fully depleted and its sensitivity to the longer wave length is greatly improved. The cross section of HSC, consisting of the wide field corrector, the focal plane, the CCD dewar, and the filer exchanger, is illustrated in Fig. 1 and a photograph of the focal plane CCD array is shown in Fig. 2.

One of the most powerful techniques for constraining cosmological models is that of weak gravitational lensing, by which light rays from distant galaxies are systematically bent by the gravitational potential of the material between the galaxy and the observer, imprinting a coherent distortion pattern in shapes of different galaxies. This subtle effect (typically a few % effect in ellipticity of an individual galaxy) can only be measured statistically by averaging over a large number of lensed galaxies. Because it is sensitive to the gravitational potential, gravitational lensing measures the clustering of all matter, both luminous and dark. Measuring this signal as a function of redshift explores the evolution of the clustering of dark matter, which is sensitive to cosmological parameters including the equation of state parameter w of dark energy. The combination of the large aperture of the telescope, the large field of view of HSC and the excellent image quality (routinely FWHM of 0.7 arcseconds) of the site and the telescope enables us to obtain multi-color ($grizy$-band or $0.4 - 1.05$ micron) imaging data over 1400 square degree solid angle

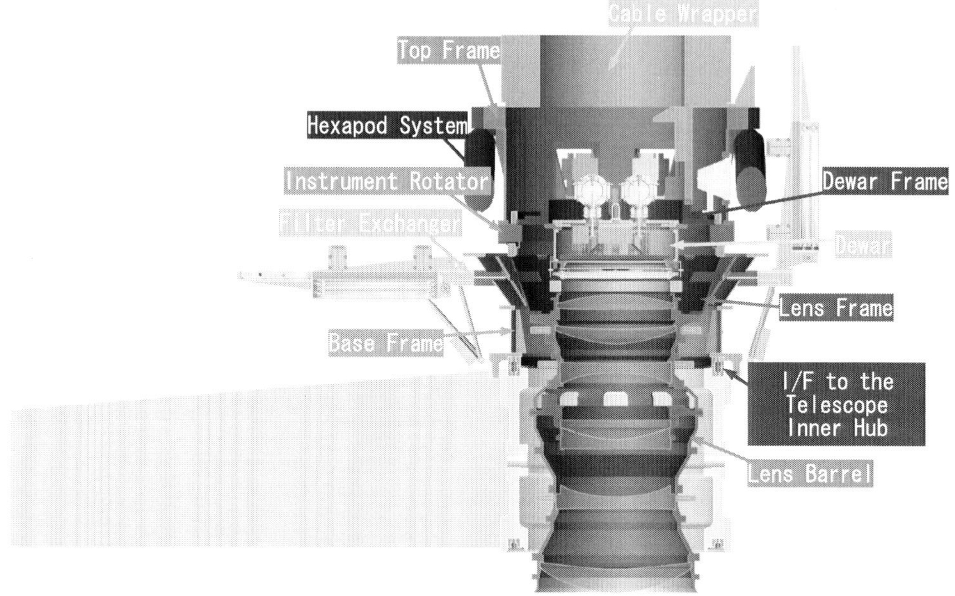

Fig. 1.   Crosssectional view of Hyper Suprime Cam (HSC).

Fig. 2.   The focal plane CCD array of HSC.

with a sufficiently-high statistical precision for the weak lensing observables in a 300-night survey. The expected sensitivity to the equation of state parameter of dark energy, $w(a) = w_0 + w_a(1 - a)$ where a is the scale factor of the Universe, is shown in Fig. 3. HSC is expected to see first light in 2012 and the survey will follow afterwards. Astronomers from Japan, Taiwan and Princeton University of the US are involved in the HSC survey.

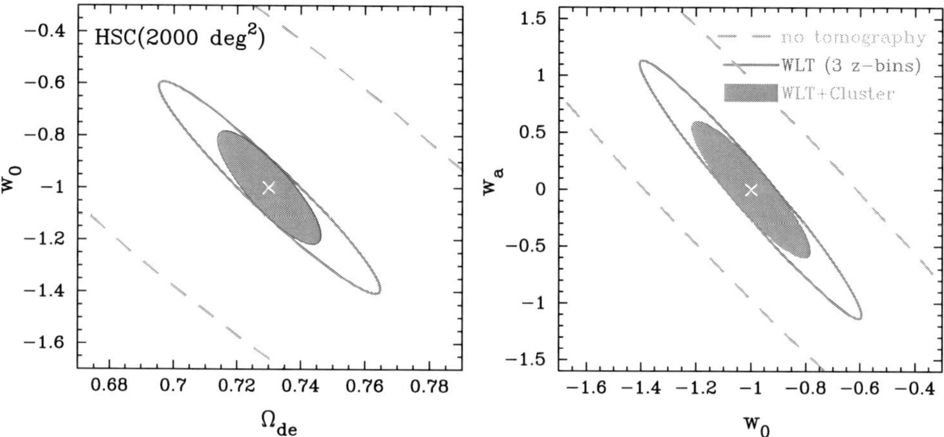

Fig. 3. The expected sensitivity of HSC, with 2000 square degree solid angle survey, to the equation state parameter of dark energy $w(a) = w_0 + w_a(1 - a)$ and the dark energy density parameter Ω_{DE}.

PFS is designed to allow simultaneous low and intermediate-resolution spectroscopy of 2400 astronomical objects over a 1.3 degree hexagonal field. It shares with HSC some instruments including the wide field corrector and has an array of 2400 optical fibers. Each fiber tip position is controlled in plane by a two-stage piezo electric fiber positioner and covers a particular region. These regions, in total, fully sample the 1.3 degree field. A fiber connector relays light to four identical fixed-format 3-arm twin-dichroic all-Schmidt spectrographs providing continuous wavelength coverage from 0.38 micron to 1.3 micron. The blue $(0.38 - 0.67$ micron) and red $(0.65 - 1.0$ micron) arms use two Hamamatsu CCDs (as in HSC). The near-infrared $(0.97 - 1.3$ micron) arm uses a new Teledyne 4RG 4K×4K HgCdTe 1.7-micron cut-off array. PFS enables us to survey over 1400 square degree solid angle and to sample galaxies within a comoving volume of 9 $(\mathrm{Gpc}/h)^3$ over $0.8 \leq z \leq 2.4$ in a 100-night survey.

The scientific goals of the PFS survey are to : measure the Hubble expansion rate and the angular diameter distance to 3% fractional accuracies in each of 6 redshift bins over $0.8 < z < 2.4$ using the baryonic acoustic oscillation (BAO) method, use the distance measurements for determining the dark energy density parameter $\Omega_{\mathrm{de}}(z)$ to $\sim 7\%$ accuracy in each redshift bin, use the geometrical constraints to determine the curvature parameter Ω_K to 0.3% accuracy, and measure the redshift-space distortions in order to reconstruct the growth rate of large-scale structure to 6% accuracy since a redshift $z = 2.4$. Fig. 4 shows the expected fractional errors of determining the angular diameter distance and the Hubble expansion rate in each redshift bin. The PFS prediction can be compared with the precisions of the existing and ongoing SDSS/BOSS surveys.

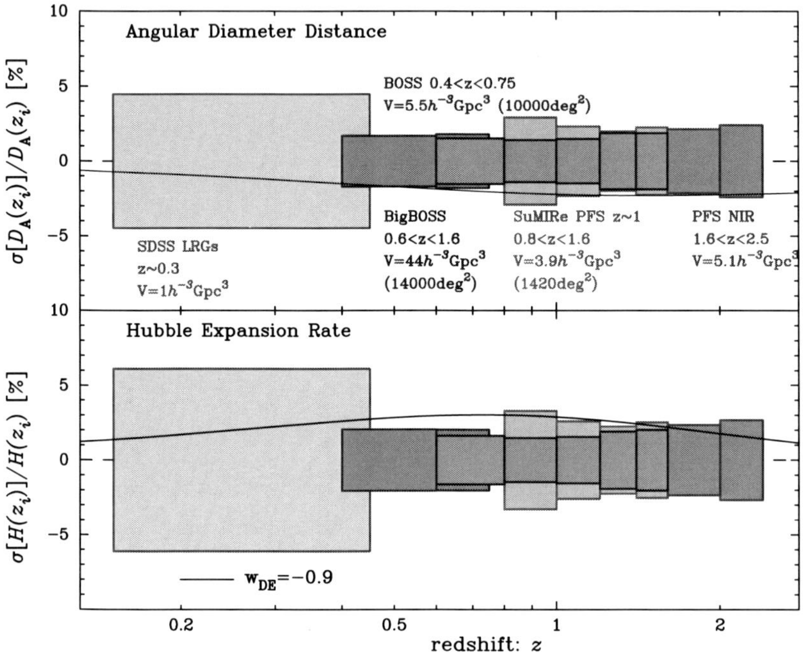

Fig. 4. The expected fractional errors of determining the angular diameter distance and the Hubble expansion rate in each redshift bin. The PFS prediction can be compared with the precisions of the existing and ongoing SDSS/BOSS surveys.

The PFS collaboration has grown into a large, multi-national collaboration, currently including astronomers from Japan, Caltech/JPL, Princeton and Johns Hopkins Universities, the Laboratoire d'Astrophysique deMarseille, Academia Sinica Institute of Astronomy & Astrophysics (ASIAA) Taiwan, the University of São Paulo and the Laboratorio Nacional de Astrofísica in Brazil. PFS is expected to receive first light in 2017.

2. XMASS

The XMASS experiment[3] aims at the direct detection of the cold dark matter in the Galactic halo using liquid Xenon as the target material. Liquid Xenon has a great scintillation light yield of 42,000 photons per energy deposition of MeV, which is comparable with NaI(Tl) scintillator. This enables the detection of low-energy signals of nuclear recoils caused by elastic scatterings with dark matter particles. Because Xenon has a high atomic number ($Z = 54$) and its liquid state has high density ($\sim 3\mathrm{g/cm^3}$), the detector of liquid Xenon makes an effective target for the dark matter interaction and provides self-shielding against external gamma rays and neutrons by absorbing them in the outer region of liquid Xenon volume. Dark matter interacts throughout the detector and can be distinguished from background by requiring its interaction position to lie within a background-suppressed inner fiducial volume.

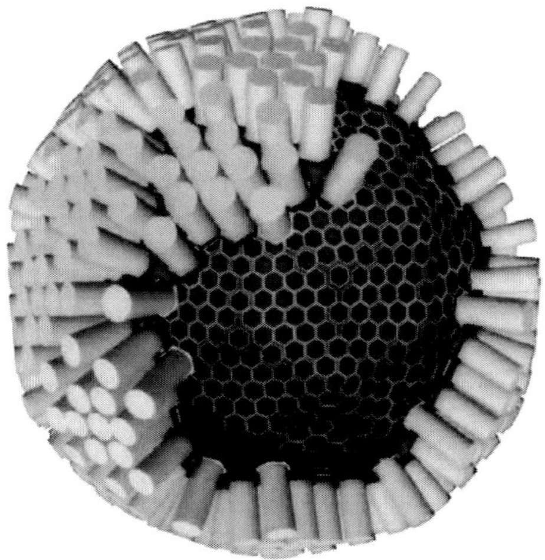

Fig. 5.   The XMASS phase-I detector. The 652 PMTs are arranged approximately in a spherical shape to view 857 kg of liquid xenon.

The XMASS phase-I detector contains a volume of 857-kg liquid Xenon viewed by 642 quartz-window photomultiplier tubes (PMTs). These PMTs are opted for 175-nm scintillation light and manufactured using the low-radioactivity material with the residual activity 100 times lower than that of regular PMTs. The fiducial mass of the detector is about 100 kg. Figures 5 and 6 show the schematic view and a photograph of the detector, respectively. The XMASS phase-I detector is immersed in the water tank equipped with 72 twenty-inch PMTs providing more than 4 meter thick water shield. It results in further reduction of external gamma rays (10^{-3} reduction with 2 m thick water) and fast neutrons. The expected sensitivity to

Fig. 6.   Photograph of XMASS phase-I detector.

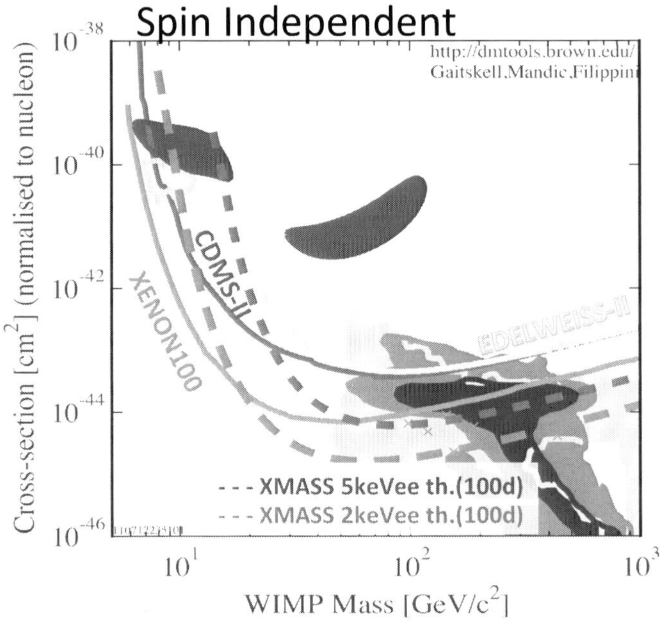

Fig. 7.   The sensitivity to weakly-interacting massive particles (WIPMs) dark matter expected from 1 year exposure with the XMASS phase-I detector.

weakly-interacting massive particles (WIPMs) dark matter is shown in Fig. 7. With 1 year exposure time XMASS expects to reach a sensitivity for spin-independent cross section of $\sigma \sim 2 \times 10^{-45}$ cm^2 for $50 - 100$ GeV/c^2 WIMPs. The XMASS collaboration currently includes 10 institutes and 41 collaborators. The XMASS phase-I detector had been completed and the data-taking is in progress.

3. Conclusion

Kavli IPMU has begun to embark on cosmology and particle astrophysics researches. Kavli IPMU and LeCosPA share the common physics goals. We are looking forward to working with people in LeCosPA.

We would like to thank the conference organizers for their superb hospitality.

References

1. Y. Komiyama *et al.*, Hyper Suprime-Cam: camera design, in Society of Photo-Optical Instrumentation Engineers (SPIE) Conference Series, ser. Society of Photo-Optical Instrumentation Engineers (SPIE) Conference Series, vol. 7735, Jul. 2010.
2. Richard Ellis, Masahiro Takada, Hiroaki Aihara, Nobuo Arimoto, Kevin Bundy, Masashi Chiba, Judith Cohen, Olivier Dore, Jenny E. Greene, James Gunn, Timothy Heckman, Chris Hirata, Paul Ho, Jean-Paul Kneib, Olivier Le Fevre, Hitoshi Murayama, Tohru Nagao, Masami Ouchi, Michael Seiffert, John Silverman, Laerte Sodre Jr, David Spergel, Michael A. Strauss, Hajime Sugai, Yasushi Suto, Hideki Takami, Rosemary Wyse, the PFS Team, "Extragalactic Science and Cosmology with the Subaru Prime Focus Spectrograph (PFS)," arXiv:1206.0737
3. Yoichiro Suzuki, "XMASS Experiment," Proceedings of Identification of Dark Matter 2008, 18-22 August, 2008, AlbaNova, Stockholm, Sweden.

TSINGHUA CENTER FOR ASTROPHYSICS AND THE DARK UNIVERSE

CHARLING TAO

CPPM/IN2P3/CNRS, Marseille, France
and
Tsinghua Center for Astrophysics (THCA), Tsinghua University, Beijing, China
E-mail: tao@cppm.in2p3.fr

This article summarizes the main activities of the Tsinghua Center for Astrophysics of Tsinghua University in Beijing, and the recent involvements on projects exploring the Dark Universe.

Keywords: Tsinghua Center for Astrophysics; Dark Universe; Dark Energy; Dark Matter; Gravitational Wave; High Energy astrophysics; Supernovae; CMB.

1. At the Beginning

The Tsinghua Center for Astrophysics (THCA) was founded in 2001 by Profs. Li Tipei and Shang Rencheng. A distinguishing characteristic of THCA is the engineering strength of Tsinghua University with expertise in nuclear physics, nuclear engineering, space and aeronautics, as well as electronics, computer, information technology, precision instruments and mechanics in the University departments.

The main research directions in THCA include high energy astrophysics and cosmology with space and ground observations in X-rays, gamma-rays, optical wavelengths, and more recently gravitational waves.

The first years of activity are only briefly mentionned in this presentation. They have been detailed in Prof. Zhang Shuang-Nan's presentation[1] at COSPA in 2009. I will develop the more recent involvements mainly in cosmology. For details, please refer to the center web site: http://www.thca.tsinghua.edu.cn/en.

2. High Energy Astrophysics in THCA

Details can be found under http://heat.tsinghua.edu.cn

2.1. *HXMT (Hard X-ray Modulation Telescope)*

HXMT jointly developed by the Chinese Academy of Sciences and THCA, is a high-energy X-ray telescope with electronics, anti-coincidence and collimator systems. HXMT should be China's first independently developed space astronomy satellite,

Fig. 1.   High precision ground test stand for HXMT, designed and built in Tsinghua.

using the original direct demodulation imaging method of Prof. Li to achieve wide-band X-ray (1-250 keV) high resolution imaging surveys, and study black hole binaries, and other celestial bodies. In April 2008, HXMT completed the project feasibility assessment. In October 2010, the State Council approved the HXMT project which should be financed through the 12th 5 year plan. Launch is foreseen for 2015. Figure 1 shows the ground test stand designed and built in Tsinghua.

2.2. *The gamma-ray burst polarization experiment (POLAR)*

POLAR is a dedicated polarization measurement experiment, aiming at the first reliable measurement of the polarization of the prompt gamma-ray emissions from all kinds of gamma-ray bursts, in order to understand the nature of the central engine of ultra-relativistic jets from which powerful gamma-ray emissions are produced. POLAR is a candidate payload of the space astronomy sub-system of China's Spacelab Mission.

2.3. *The Space Variable Objects Monitor (SVOM)*

The SVOM mission is a joint China-France mission which is dedicated to discover and make rapid multi-wavelength observations of gamma-ray bursts (GRBs) and their afterglow emis-sions, in order to understand the physics of stellar collapse, black hole formation, neutron star merging and ultra-relativistic jets, as well as to use GRBs to probe the evolution of the Universe.

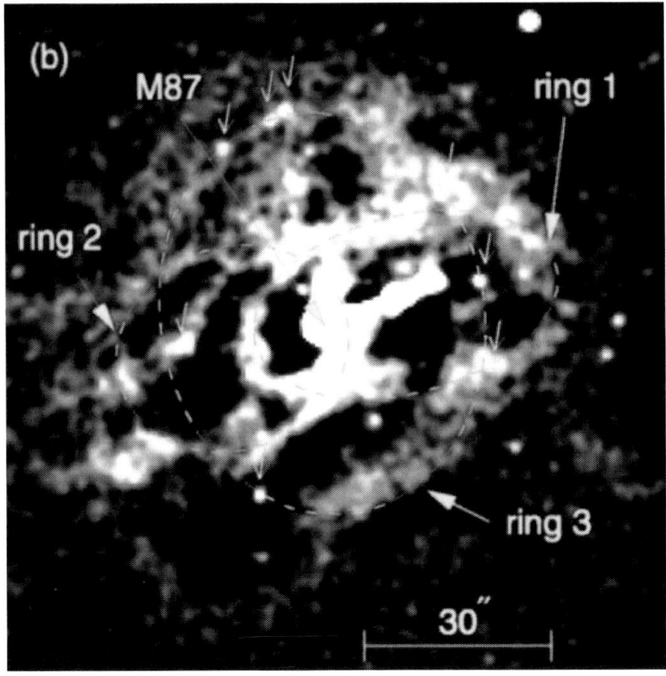

Fig. 2. A three ring structure around the jet of the M87 galaxy was discovered, cf http://heasarc.gsfc.nasa.gov/docs/objects/heapow/archive/active_galaxies/m87_rings.html.

2.4. *THCA developments of innovative techniques*

THCA is also involved in developing innovative techniques of X-ray polarization measurements and sub-arcsecond angular resolution X-ray imaging, for the next generation space astronomy missions.

2.5. *THCA analysis of X-ray data*

A number of interesting results have been obtained by THCA researchers (faculty, postdocs and students, lead by Feng Hua). Two examples:

1. The discovery of a three ring structure around the jet of the M87 galaxy was reported by HEASARC picture of the week[2] (cf Figure 2).

2. With three joint observations of the M82 galaxy using NASA's Chandra telescope and the European XMM-Newton telescope, strong evidence was found in 2010 for two candidate intermediate-mass black holes[3,4] (NASA news in Figure 3).

3. Optical Astrophysics in THCA

THCA faculty and students are involved in data analysis and improving observation techniques.

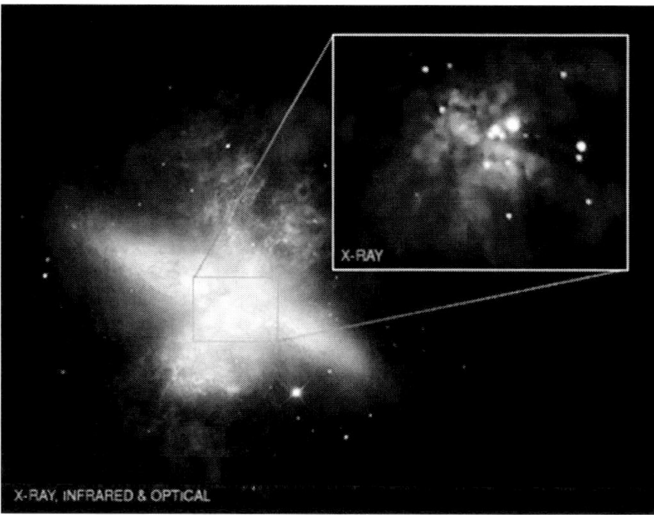

Fig. 3. Evidence for 2 intermediate-mass black holes in M82 using Chandra and XMM telescopes, cf http://chandra.harvard.edu/press/10_releases/press_042910.html.

3.1. *40-cm telescope*

A 40-cm small telescope has been installed in 2001 in Tsinghua Observatory (cf Fig. 4), for pedagogical purpose. This allows practice of observation on campus for students and helps popularize astronomy.

3.2. *80-cm telescope*

An 80-cm reflector telescope was installed in 2003 in Hebei Xinglong (cf Fig. 5), an ongoing fruitful collaboration with the National Astronomical Observatory. Tsinghua University - National Observatory Telescope (TNT), began observations in 2004. The telescope is used mainly for transient searches (supernovae, gamma-ray burst afterglow, active galactic nuclei, etc...).

3.3. *Spectrographs*

Tsinghua University department of precision instruments is now collaborating with the Thirty Meter Telescope (TMT) project, on the IRIS spectrographs. This expertise will help us design in THCA, new spectrographs for our future astrophysics projects.

3.4. *AST3 and KDUST in Dome A*

THCA is actively involved in the construction of the National Antarctic astronomical observatory. We contribute to the 50 cm (effective aperture) telescopes of the Schmidt telescope array (AST3). As a member of the Antarctic Astronomy Center,

Fig. 4.   A view of Tsinghua Observatory

Fig. 5.   The TNT telescope at Xinlong, Hebei province

THCA participates also in the common management of the telescope. The center will use the Antarctica telescopes to carry out time-domain astronomy (in particular supernovae, for which THCA is coordinator). The project is part of an international collaboration where US and Australian scientists successfully measured with Chinese astronomers the site parameters relevant for astronomical observations. The

Fig. 6. On the left is the first AST3 during tests in Xuyu Observatory, on the right, the 4 astronomers arriving in Dome A to install the first AST3 telescope.

first of the three foreseen AST3 telescopes with an i-band optical filter, was successfully installed in Dome A, Antarctica during the winter 2001-2012 traverse and is giving its first light in Antarctica. Figures 6 are showing the AST3 during testing, and the installation team in Dome A. News can be followed on http://aag.bao.ac.cn.

The Antarctic Observatory medium-term plan is to install a 2.5 m optical / infrared wide field of view survey telescope (KDUST), to study Dark Energy. THCA will also actively promote and participate in the project.

4. Supernova Research in THCA

Supernova (SN) science is now well known for its cosmology impact: The 2011 Nobel Prize underlined the fundamental contribution of SN Type Ia (SNIa) surveys to the discovery of the mysterious Dark Energy. SNIa are not perfect standard candles, but they can be "standardised" enough to-day to provide a measurement of the equation of state of Dark Energy, with a Hubble diagramme (distances as a function of redshift).

I was a member of SNLS,[5] a SNIa systematic rolling search, which measurements confirmed (with more than 500 SNIa observed today), the reality of the acceleration of the Universe hinted by the 2 pioneering teams. The "2-sigma" effect back in 1998 is now a clear convincing signal.

Conley *et al.*[6] combined best SNIa distance estimates to set the best constraints to-date on the Dark Energy equation of state, as shown in Figure 7, assuming the Universe is flat. More precise results can be obtained with multiprobe combinations (CMB, Baryonic oscillations, weak gravtitional lensing, ...) But the single best probe is still SNIa.

SNLS reached the *"systematics floor"*, with existing methodology, which means that in order to improve on cosmology with SNIa, we need to improve on calibrations, understanding of SNIa evolution and host galaxy extinctions, SNIa standardization techniques, and distance measurement methods.

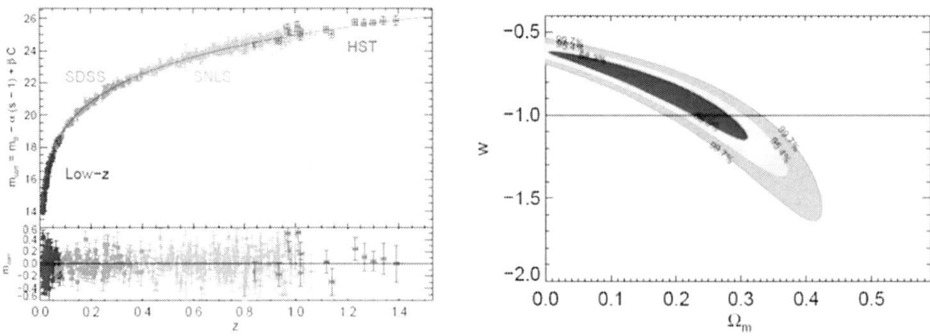

Fig. 7.   Hubble diagramme (left) and constraints on the DE equation of state (right) (from Conley *et al.*[6]).

THCA has joined the SNFactory effort to understand better nearby SNIa and their environment in order to test the diversity of SNIa and find new ideas for improving SNIa constraints on cosmological parameters. SNFactory aims at providing the reference low z anchor and database for SNIa studies.

Wang Xiaofeng proposed a type of correction based on the color parameter;[7] He found an average extragalactic dust absorption coefficient Rv= AV/[(E(B-V)]=2.3 (AV for the amount of extinction, E(B-V) is the color parameter), which is different from the Milky Way dust (Rv = 3.1), for the sample of SNIa he used. This result is disputed, the latest results with SNFactory data by Nicolas Chotard (now a postdoc in THCA) *et al.* indicates an intermediate value, Rv = 2.8 ± 0.3, compatible with both the Milky Way and the Wang *et al.* values.[8]

Wang Xiaofeng also proposed a classification of SNIa based on their expansion velocity.[9] This spectral classification reveals differences between the two classes of SNIa and the intrinsic dispersion within each class is reduced by roughly 50% (ie from 9% to 6%). This study also helps to understand the differences in the explosion physics of the progenitor star. The normal might correspond to double degenerate explosions, and the high velocity sample would be due to single degenerate explosions. But statistics for the high velocity sample is still low.

I am co-lead of the supernova and transient working group of the Euclid project,[10] which has been recently selected by the European Space Agency (ESA) on October 4th 2011 as the third M-class mission, launched in 2019. Euclid main science requirement is to study Dark Energy with weak gravitational lensing and galaxy clustering. I am also a member of LSST and and active in Antarctica (AST3, KDUST) projects for SN research and other dark energy studies, and hope to develop these activities with new THCA members.

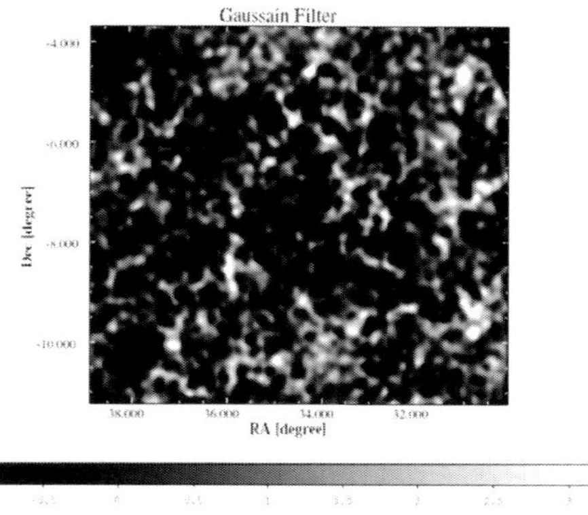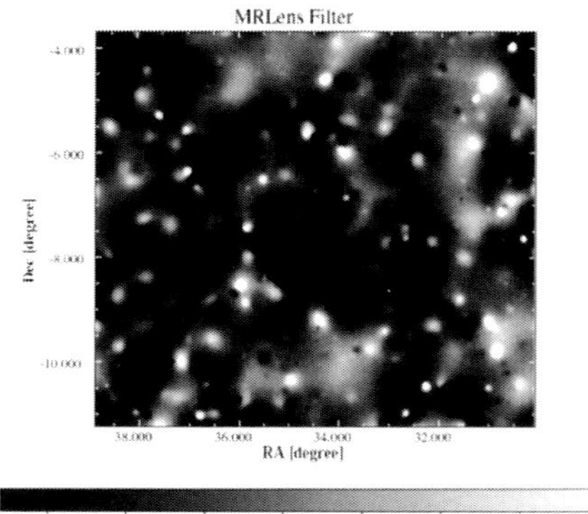

Fig. 8. Weak Lensing detected cluster maps of W1 from Shan *et al.*[12]

5. Studies on Gravitational Lensing in THCA

A very important and promising probe for constraining Dark Energy and other cosmological parameters in the next years is through gravitational lensing (weak, strong and flexion), which give estimates of total masses, in contrast to the more biased mass estimates from optical or IR and X ray surveys. The first CFHT and COSMOS results with cosmic shear are promising, but a number of systematic issues need to be solved. The work of Sun Lei (my former student), shows that the impact of catastrophic redshifts can be limited by sufficient spectroscopic calibrations of the photometric redshifts.[11] Many other systematics are/will limit the lensing results, and we are studying these effects.

Gravitation lensing is a powerful tool for other science: Shan HuanYuan, a postdoc in THCA, was the first to find clusters with weak lensing (cf Fig. 8 for a map) from the public CFHT W1, 64 sq deg data.[12]

We are extending this work to the whole CFHT published maps (171 sq deg) and hope that the extended CFHT-S3 proposal with 8000 sq deg will soon be approved and offer us independent interesting cluster count constraints on cosmological parameters. In the future, EUCLID, LSST, KDUST, China 2m space telescope should provide data for continuing this study.

Gravitational lensing can also provide the matter power spectrum at small scales, which should be different for cold and warm DM of keV scale. Whether DM is cold or warm is a hot topic, which has important incidence on the many direct and indirect searches for DM (cf my review[13]). We are currently studying how trustworthy present evidence for warm dark matter is, and if we can confidently continue and develop our program to search for GeV Cold DM.

6. Dark Matter Searches in THCA

The possibility of an underground tunnel under Jinping Mountain was the culminating point of a >20 years search for a good location of an underground laboratory in China. Thanks to the Ertan hydroelectric power company, China's first extremely deep underground laboratory, the China Jinping Underground Laboratory (CJPL), was opened and put into operation in Yalongjiang Jinping Hydropower Station in Sichuan province on December 12, 2010. CJPL is located 2,400 meters below the surface and is the deepest lab in the world covered by rocks.[14]

The laboratory will host two DM experiments (cf presentations in these proceedings of H. Wong[15] on CDEX[16] with sub-keV Germanium detectors, and K. Ni[17] on PANDAX with liquid Xenon). THCA is involved in the background neutron measurements for the Jinping site, and is preparing next generation DM experiments, with directional TPC.

7. Reanalysis of WMAP Data

Prof. Li Tipei and Dr. Liu Hao discovered a serious systematic error in the results of the WMAP satellite microwave background temperature maps.[18] An independent WMAP data analysis software system was developed and published.[19,20] There are significant differences with the WMAP official temperature map. Errors in the WMAP data processing have been identified and corrections proposed.[21] Independent checks support Li and Liu's conclusions, which imply an unexpected very small quadrupole contribution. The Planck mission has inished taking data, and should publish its results next year. Should the zero quadrupole result prove true, it certainly would question the present inflationary universe, but the Doppler dipole moment from the motion of the antenna relative to the CMB is a difficult systematic effect to subtract.[22] Liu and Li[23] predict that scan-induced anisotropies will also produce an artificially aligned quadrupole, with the scan strategy of the Planck mission. The scan-induced anisotropy is a common problem for all sweep missions and, like foreground emissions, has to be removed from observed maps. Without doing so, CMB maps from COBE, WMAP, and Planck are not reliable for studying the CMB anisotropy at the largest scales. The results were selected as cover story of the Royal astronomical association of England *"News and Reviews on Astronomy and Geophysics"* of Oct. 2010 (cf Fig. 9).

8. Gravitational Waves in THCA

The Laser Interferometer Gravitational Wave Observatory (LIGO) is one of the world's highest precision gravitational wave detection collaboration. Tsinghua University LIGO group (http://ligo.org.cn) lead by Prof. Cao Junwei, has mainly focused on gravitational wave data analysis using advanced computing technologies. The group works on the pipeline for trigger analysis, data quality and veto analysis, and real-time computing infrastructure for background burst identification and

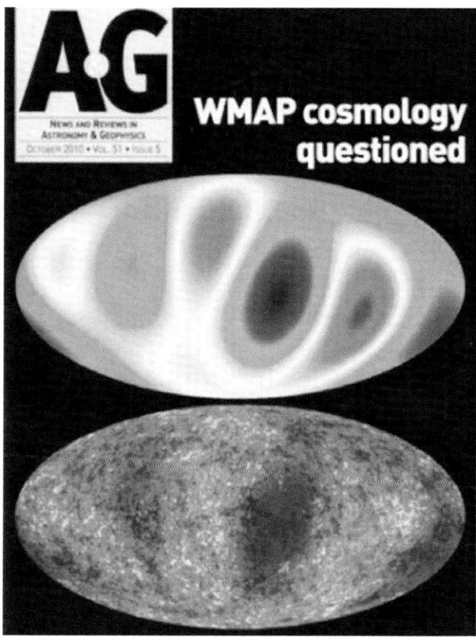

Fig. 9. Oct. 2010 cover of the journal of the Royal Astronomical association of England "News and Reviews on Astronomy & Geophysics" with an article discussing Li and Liu questioning of the WMAP results.

rejection. It is also exploring new methods to enable LIGO applications using grid, cloud, virtualization and GPU technologies. The quality of the work has been recognized by the international collaboration, the board meeting in September 2010 accepted Tsinghua University as a member of the LIGO scientific collaboration. The group is also active in the preparation for future gravitational wave projects and member of the China Gravitational Wave Working Group (CGWG).

9. Conclusions

It has been a great pleasure to visit LeCosPA and learn from its experience. I hope THCA and LeCosPA will find common projects to work on.

THCA pays special attention to interdisciplinary researches across the boundaries of astronomy, physics, cosmology, instrumentation, computational science, nuclear engineering, space and aeronautics engineering. We welcome academic visitors, excellent applicants for faculty positions at all levels, and excellent students (with training in the fields of astronomy, astrophysics, physics, and engineering in many disciplines) for graduate studies. For further information about THCA, especially for faculty opening applications and graduate studies at THCA, please contact THCA's academic secretary Dr. Guo-Qing Liu at liuguoqing@tsinghua.edu.cn.

Acknowledgments

Many thanks to the organizers of the LeCosPA symposium, in particular to Pisin Chen for his invitation and his wonderful hospitality.

THCA research is supported by the Chinese Ministry of Education, the National Natural Science Foundation of China, the Ministry of Science and Technology, the Chinese Academy of Sciences and China's National Space Agency, and Tsinghua University. We are especially grateful to IHEP and NAOC for long term supports of various foms including academic collaborations and providing direct funds.

Li Shiyu has kindly checked the layout and references numbering. Thanks for the rereading and for the latex version to her and to Dr. Liu Guoqing, who manages THCA with Prof. Shang, and simplifies so much my daily life in Tsinghua.

I thank my French collaborators for their patience while I am often teaching and working in Tsinghua, the CPPM laboratory director Eric Kajfasz, IN2P3 director for astroparticle physics and cosmology Stavros Katzanevas, for their understanding of the importance of the collaboration with China, and the two international sino-french laboratories in astrophysics (Origins) and particle physics (FCPPL), which are supporting THCA exchanges with France.

Last but not least, I want to add that my respect and admiration for Prof. Li Tipei, and the reputation of Tsinghua students, are the main reasons why I accepted to join his efforts to develop THCA.

Appendix A. THCA Members and Research Interests

THCA faculty and students are coming from the Department of Physics, Department of Engineering Physics, and the Information Technology Research Institute. The center currently has 5 Professors, 4 Associate Professors (/Research Associates), 1 tenure track assistant professor, 6 other staff.

- Li Tipei, professor, founding Director THCA, academician: high-energy astrophysics, cosmology
- Shang Rencheng, Deputy Director and co-founder of THCA: nuclear physics, space astronomical instruments
- Feng Hua, Research Associate, THCA Deputy Director: high-energy astrophysics, space astronomical instruments
- Lou Yuqing, Yangtze Distinguished Professor: Theoretical Astrophysics (celestial magnetic fluid, accretion and outflow)
- Cao Junwei, professor, IT, gravitational wave astronomy
- Zhang Youhong, associate professor: High Energy Astrophysics
- Wang Xiaofeng, tenure track associate professor, Optical Astronomy (supernova)
- Hu Jian, tenure track assistant professor, Theoretical Astrophysics
- Zhou Jianfeng, research associate: Virtual Observatory

- Chen Yibao, lecturer, computing
- Charling Tao, Director THCA, astroparticle and cosmology

References

1. S. N. Zhang, AAPPS Bulletin, Vol. 19, No. 2 (2009)
2. H. Feng, S. N. Zhang, Y. Q. Lou & T. P. Li, X-ray three-ring structure around the M87 galaxy in the core region of the Virgo cluster, Astrophys. J., 607, L95(2004)
3. H. Feng & P. Kaaret, Identification of the X-ray Thermal Dominant State in an Ultraluminous X-ray Source in M82Astrophys. J., 712: L169-L173 (2010)
4. H. Feng, F. Rao & P. Kaaret, Discovery of Millihertz X-Ray Oscillations in a Transient Ultraluminous X-Ray Source in M82, Astrophys. J., 710: L137-L141 (2010)
5. Astier et al., The Supernova Legacy Survey: measurement of Ω_M, Ω_Λ and w from the first year data set, Astronomy and Astrophysics, Volume 447, Issue 1, pp.31-48 (2006)
6. Conley et al. Supernova Constraints and Systematic Uncertainties from the First Three Years of the Supernova Legacy Survey, the Astrophysical Journal Supplement, Volume 192, Issue 1, article id. 1 (2011)
7. X. Wang et al., A Novel Color Parameter as a Luminosity Calibrator for Type Ia Supernovae, Astrophy. J., 620: L87-L90 (2005)
8. N. Chotard et al., The reddening law of Type Ia Supernovae: separating intrinsic variability from dust using equivalent widths, Astronomy & Astrophysics, 529,id.L4 (2011)
9. X. Wang et al. Improved Distances to Type Ia Supernovae with Two Spectroscopic Subclasses., Astrophys. J., 699: L139-L143 (2009)
10. R. Laureijs et al., Euclid Definition Study Report,arXiv:1110.3193 (2011)
11. L. Sun, Z. Fan, C.Tao et al., Catastrophic Photo-z Errors and the Dark Energy Parameter Estimates with Cosmic Shear, Astrophysical Journal, Volume 699, Issue 2, pp. 958-967 (2009).
12. H. Y. Shan, J. P. Kneib, C. Tao et al., Weak Lensing Measurement of Galaxy Clusters in the CFHTLS-Wide Survey, The Astrophysical Journal, Volume 748, Issue 1, article id. 56 (2012)
13. C.Tao, Astrophysical constraints on Dark Matter, EAS Publications Series, Volume 53, pp.97-104 (2012)
14. K. J. Kang et al., Status and prospects of a deep underground laboratory in China, Journal of Physics: Conference Series, Volume 203, Issue 1, pp. 012028 (2010).
15. Wong, H., these proceedings (2012)
16. Q. Yue and H. T. Wong, Dark Matter Search with sub-keV Germanium Detectors at the China Jinping Underground Laboratory, arxiv 1201.5373 (2012)
17. K. X. Ni, these proceedings (2012)
18. H. Liu & T. P. Li, Systematic distortion in CMB maps, Sci China G: Phy Mech Astron, 52: 804-808 (2009)
19. H. Liu & T. P. Li, Improved CMB map from WMAP data, arXiv:0907.2731 (2009)
20. H. Liu & T. P. Li, Inconsistence between WMAP data and released map, Chinese Sci Bull, 55:907-909 (2010)
21. Liu H. & Li T.P., Pseudo-dipole signal removal from WMAP data, Chinese Sci Bull 55(3): 1-5 (2010)
22. Liu,H, Xiong,S.L. & Li,T.P., Monthly Notices of the Royal Astronomical Society: Letters, Volume 413, Issue 1, pp. L96-L100.
23. Liu H. & Li T.P., Observational Scan-induced Artificial Cosmic Microwave Background Anisotropy, Astrophysical Journal, 732, 125 (2011);

Section II

Gravity and Spacetime

SPACE, TIME, MATTER: 1918–2012

GABRIELE VENEZIANO

Collège de France, 11 place M. Berthelot, 75005 Paris, France
and
Theory Division, CERN, CH-1211 Geneva 23, Switzerland
E-mail: gabriele.veneziano@cern.ch

Almost a century has elapsed since Hermann Weyl wrote his famous "Space, Time, Matter" book. After recalling some amazingly premonitory writings by him and Wolfgang Pauli in the fifties, I will try to asses the present status of the problematics they were so much concerned with.

Keywords: Elementary particles; gravity and cosmology; unification of all forces; quantum gravity; string theory.

1. Hermann Weyl and Wolfgang Pauli Facing Three Revolutions

At the beginning of last century three revolutions shook as many scientific beliefs:

- The belief in absolute determinism when Max Planck, in 1900, introduced $\hbar$ and started the Quantum Revolution;
- The belief in absolute time when Albert Einstein, in 1905, building on the invariance of the speed of light, c, formulated Special Relativity;
- The belief in an absolute geometry of space-time when Albert Einstein again, in 1915, starting from the Galilean universality of free-fall, arrived at a geometric theory of gravity, General Relativity, in which the gravitational constant, G, controls the amount by which matter deforms the geometry of space-time.

The best physicists of the time were deeply impressed by these amazing revolutions and wrote vividly about them. In 1918 Hermann Weyl wrote

Raum-Zeit-Materie

while in 1950 he added a preface to the American printing of its 4th 1921 edition.

In 1921 Wolfgang Pauli wrote

Relativitätstheorie

and in 1956 he reedited the original text adding several interesting notes.

Both Weyl and Pauli were expressing concern about the lack of a theory of matter exhibiting the same compelling and simple beauty as Maxwell's or Einstein's.

Here are some excerpts from their writings in fifties.

From H. Weyl's 1950 preface

After stressing the importance of (a modified version of) his gauge invariance principle and the need for a "unitary[a] theory" encompassing electromagnetic, gravitational and "electronic" fields, he writes, with amazing foresight:

*"Ultimately fields for the other elementary particles should be also included unless quantum physics succeeds in interpreting them all as **different quantum states of one particle**"*

He acknowledges the success of Dirac's theory of the electron, but, alluding to the infinities encountered in quantum-relativistic theories, he warns:

" Difficulties of the gravest kind turn up when one passes from one electron or photon to the interaction among an undetermined number of such particles."

He then concludes:

"A solution is not yet in sight: may need deep modifications of the foundations of Quantum Mechanics where the elementary electric charge, e, is accounted for in a fundamental way like c and ℏ are in Relativity and Quantum Mechanics."

From W. Pauli's 1956 notes

Not so many on the Special and General Relativity parts. Several notes, instead, on the last part (Chapt. 8) devoted to elementary particles. He is very sharp and critical about most proposals, including Einstein's, entirely based on classical concepts. The classical theory, in his opinion, is doomed. He is quite positive (and incidentally very clear!) about the Kaluza-Klein (KK) idea. But concludes:

*"The question of whether the KK formalism will have a bearing on physics leads us to the more general and yet unresolved problem of realizing a **synthesis between General Relativity and Quantum Mechanics**.*

He then concludes with:

*"**New elements**, extraneous to the concept of a field as a continuum, **should be added** in order to arrive at a satisfactory solution of the problem of matter.*

These writings, as we shall see, were much ahead of their times: the issues they raised remain at the heart of theoretical physics even today.

2. What Has Happened Since?

While not much has happened as far as *space* and *time* are concerned[b], giant steps along the lines advocated by Hermann Weyl and Wolfgang Pauli have been carried out in what concerns *matter*.

---

[a]Read "unified" in modern terminology.

[b]Unless the OPERA anomaly is a real one!

2.1. *The long road to the Standard Model of Nature*

The first attempts to combine Special Relativity and Quantum Mechanics took place in the 30s; eventually, in the late forties, these led to Quantum-Field-Theory (QFT) and, in particular to Quantum Electro-Dynamics (QED). But it took another 30-40 years of hard experimental and theoretical work before it was realized that *all* non-gravitational interactions can be described by a special class of QFTs. In spite of their obvious phenomenological differences, they correspond to different realizations of one and the same deeper structure, that of a gauge theory.

By the mid seventies physicists had formulated what we may call a Standard Model of Nature (SMN), one so far not contradicted by experiments. It is based on two pillars:

- **General Relativity** for the gravitational interaction;
- A **Gauge Theory** for the non gravitational interactions.

Let us then ask ourselves: is there a message that Nature wants to convey through the SMN paradigm?

2.2. *Two questions, two (personal) answers*

- Q_1: Why Gauge Theories?

  A_1: A gauge theory is the way to describe massless $J = 1$ particles, such as the photon.

  A massless $J = 1$ particle (or an electromagnetic wave) has 2 physical polarizations, while a massive one has 3. Gauge invariance is a (local) symmetry that allows to remove ("gauge away" in the theorist's jargon) the unphysical polarization of a $J = 1$ massless particle while keeping Lorentz invariance explicit. Hence:

  Message No.1: Nature likes $J = 1, m = 0$ particles and is therefore well described by a gauge theory.

- Q_2: Why General Relativity?

  A_2: A massless $J = 2$ particle has also 2 physical polarizations, while a massive one has 5. General covariance is a (local) symmetry that allows to remove the unphysical polarizations of a $J = 2$ massless particle while retaining explicit Lorentz invariance. Furthermore, interactions mediated by a massless $J = 2$ particle necessarily acquire a geometric meaning leading to an emergent curved space-time. Hence:

  Message No. 2: Nature likes $J = 2$ massless particles and is therefore well-described by General Relativity!

The question still remains, of course, of why Nature likes $m = 0, J = 1, 2$ particles. Before attempting an answer, a little reminder of the successes (and puzzles) of our present Standard Model of Nature.

3. Successes of the Standard Model of Nature (SMN)

3.1. *The Standard Model of Elementary Particles (SMEP)*

The SMEP has been very widely tested in accelerator experiments (most recently at LEP, HERA, the Tevatron, LHC). Its quantum-relativistic nature manifests itself through real and virtual particle production. Taking these effects into account is essential for agreement between theory and experiment.

The (updated) Standard Model of non-gravitational interactions can be written down in just 4 lines

$$L_{SM} = L_{Gauge} + L_{Yukawa} + L_{Higgs} + L_{mass}$$

$$L_{Gauge} = -\frac{1}{4}\sum_a F^a_{\mu\nu}F^a_{\mu\nu} + \sum_{i=1}^{3} i\bar{\Psi}_i\gamma^\mu D_\mu\Psi_i + D_\mu\Phi^*D^\mu\Phi$$

$$L_{Yukawa} = -\sum_{i,j=1}^{3} \lambda_{ij}^{(Y)}\Phi\Psi_{\alpha i}\Psi^c_{\beta j}\epsilon_{\alpha\beta} + c.c.$$

$$L_{Higgs} = -\mu^2\Phi^*\Phi - \lambda(\Phi^*\Phi)^2$$

$$L_{mass} = -\frac{1}{2}\sum_{i,j=1}^{3} M_{ij}\,\nu^c_{\alpha i}\nu^c_{\beta j}\epsilon_{\alpha\beta} + c.c. \tag{1}$$

It is basically the same Lagrangian we had in the seventies with just one interesting extension (last term in (1)) due to the discovery that neutrinos have mass. The SMEP can easily incorporate neutrino masses and mixing through a very mild extension, e.g. by introducing some highly massive "sterile" neutrinos that have no gauge interactions.

In Fig. 1 we recall a well-known plot of LEP's precision data with a best fit to a large number of observables. The quality of the fit (measured through a standard χ^2 parameter) depends of the yet unknown mass of the Brout-Englert-Higgs (higgs for short) boson with a marked preference for a mass close to LEP direct search's lower bound of 114 GeV.

The current situation at the LHC (see Fig. 2) has now restricted the possible mass of the Higgs boson to a narrow window which, happily enough, is very close to the above lower bound, with even some (so far inconclusive) evidence of "activity" around 125 GeV. The good news is that we will know soon!

3.2. *The Standard Model of Gravity and Cosmology (SMGC)*

Tests of the Equivalence Principle (on which the whole construction of General Relativity is based) have reached incredible precision (10^{-13} via universality of free-fall, with 10^{-15} possibly within reach with forthcoming satellites). Corrections to Newtonian gravity have also been better and better tested excluding deviations from GR at the level of 1/1000.

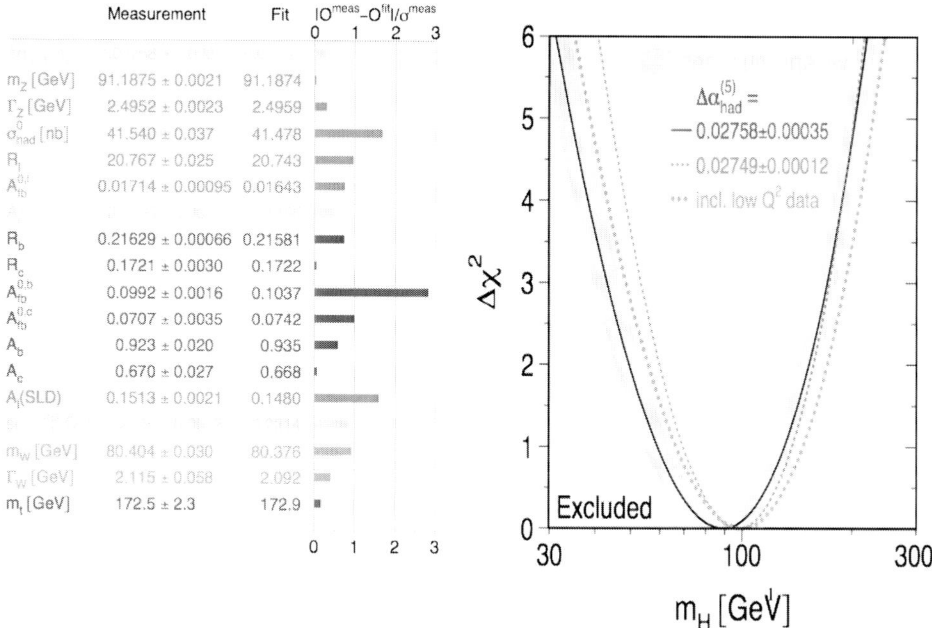

Fig. 1. Summary of precision tests of the SMEP after LEP2. On the right the dependence of the quality of the fit upon the unknown higgs-boson mass and the lower limit based on direct searches.

But General Relativity also makes entirely new predictions with respect to Newtonian gravity:

- Existence of black holes: for these there is, by now, overwhelming evidence. It looks that, in particular, the center of galaxies (including our own) is occupied by gigantic black holes that keep accreeting surrounding matter (active galactic nuclei or AGNs).
- Existence of gravitational waves: for these we only have, so far, indirect evidence from the behaviour of binary systems (see Fig. 3) such as a pair of pulsars whose periods can be monitored to high precision and follow a time evolution in agreement with GR's prediction of energy loss by gravitational-wave emission. The direct detection of gravitational waves is an ongoing process, particularly now at earth-based interferometers (LIGO, VIRGO) but with a plan to put one day a gigantic interferometer (LISA?) in space.

Finally, we now have, since the eighties, a standard model of cosmology (SMC), the so called concordance model based on the inflationary paradigm and on the existence of two new forms of matter/energy in the Universe: dark matter and dark energy. Three independent sets of data (see Fig. 4), from CMB anisotropies

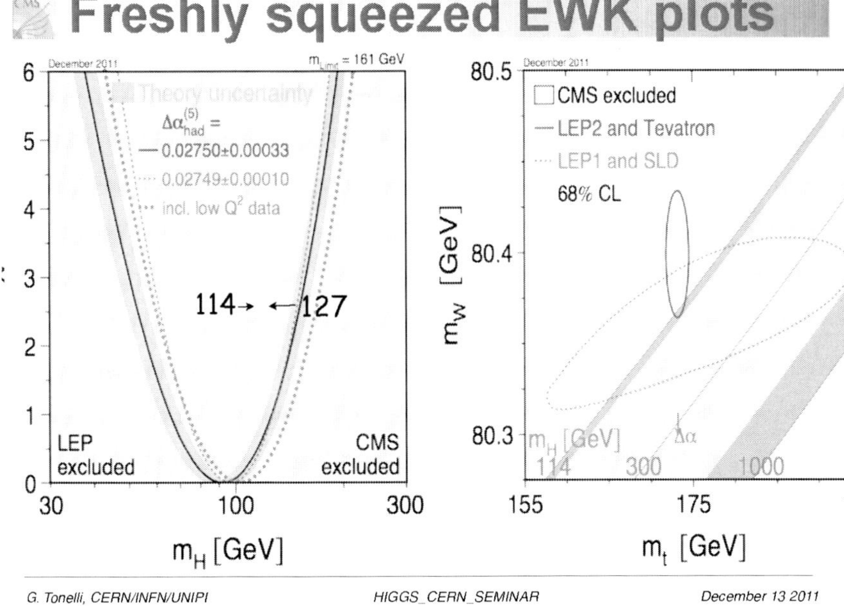

Fig. 2. The small window left for the higgs-boson to hide after the 2011 LHC run. The plots are from the CMS collaboration. Those from ATLAS are very similar.

(WMAP), large scale structure (LSS) and type Ia supernovae, all converge towards the conclusion that only about 4% of the energy of the universe is made by ordinary particles. About 23% is made of an unknown form of (cold) dark matter and 73% of a mysterious dark energy responsible for accelerating the expansion of the universe in our recent past (i.e. since a redshift of order 1).

The (updated) Standard Model of gravitational interactions in 2 lines

$$L_{SMGC} = L_{EH} + L_{CC}$$
$$L_{EH} = -\frac{1}{16\pi G_N}\sqrt{-g}\, R(g)$$
$$L_{CC} = \frac{1}{8\pi G_N}\sqrt{-g}\, \Lambda \;,\; \text{with}\; G_N\Lambda \sim 10^{-120}\,. \tag{2}$$

Here too the original model has been extended, this time through a (tiny) cosmological-constant term.

In spite of their respective successes both the SMEP and the SMGC suffer from some phenomenological puzzles as well as from more conceptual, theoretical problems. We shall briefly discuss them in turn.

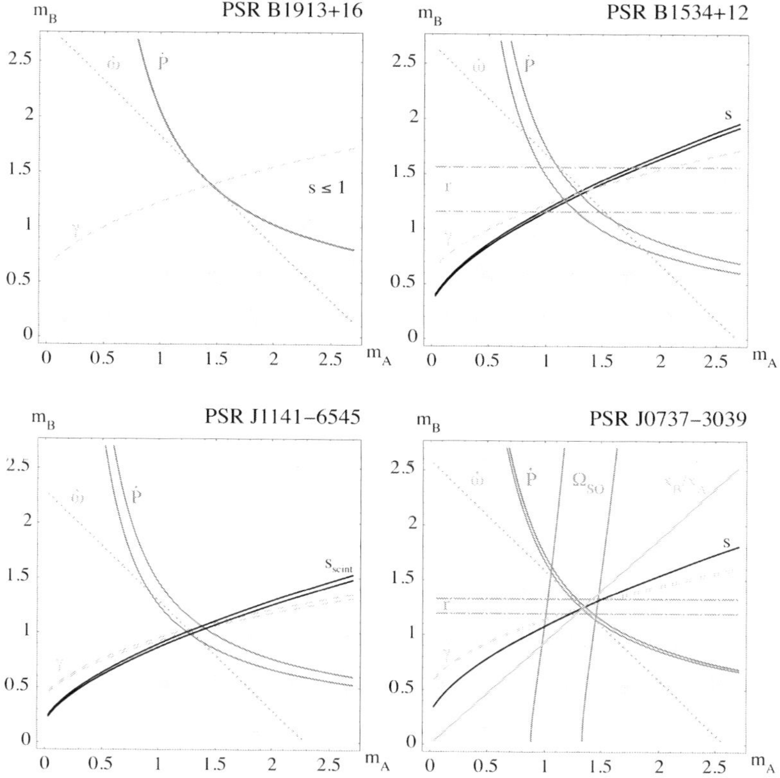

Fig. 3. Tests of General Relativity on various binary systems. The masses of the two companions are unknown but there can as many as seven different observables to be fitted.

4. Phenomenological Puzzles

4.1. *Particle physics*

Here is a long, yet incomplete, list of questions:

- Why the gauge group $G = SU(3) \otimes SU(2) \otimes U(1)$?
- Why do the fermions belong to such a bizarre, highly reducible representation of G?
- Why 3 families? Who ordered them (Cf. I. Rabi's question about the muon)?
- Why such an enormous hierarchy of fermion masses?
- Can we understand the mixings in the quark and lepton (neutrino) sectors? Why are they so different?
- What's the true mechanism for the breaking of G?
- If it is the Higgs *et al.* mechanism: what keeps the higgs-boson "light"?
- If it's supersymmetry, why did we see no signs of it yet?

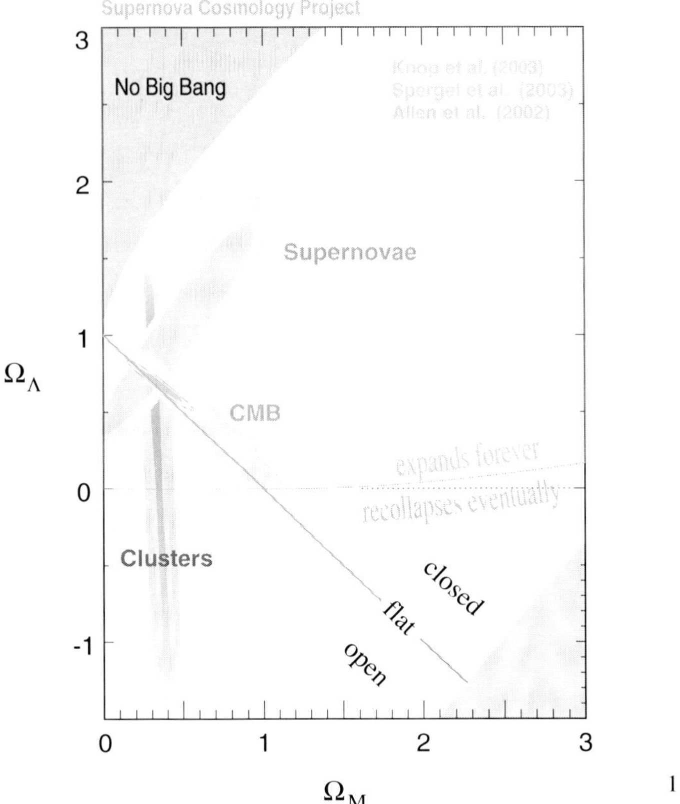

Fig. 4.   The concordance model as it emerges from three different sets of data..

- Why no strong CP violation?
- If it's Peccei-Quinn symmetry breaking where is the axion?
- ...

4.2. *Gravitation and Cosmology*

Similarly we may ask:

- Has there been a big bang, a beginning of time?
- What provided the initial (non vanishing, yet small) entropy?
- Was the big-bang fine-tuned (homogeneity/flatness problems)?
- If inflation is the answer: Why was the inflaton initially displaced from its potential's minimum?
- Why was it already fairly homogeneous?
- What's the nature of Dark Matter?

- What's the nature of Dark Energy? Why is Ω_Λ of $O(1)$ *today?*
- What's the origin of the matter-antimatter asymmetry?
- …

5. Conceptual/Theoretical Problems

In spite of the claimed common origin of gauge and gravity from massless spinning particles the SMN is "limping". The two legs it is resting on are uneven. In particular, the General Relativity side should be elevated to a full quantum theory. I see two main reasons (among others) to be unhappy about leaving gravity classical, as some people have advocated:

1. In order to avoid the singularities that are known to be ubiquitous in the classical theory;

2. In order not to give up an appealing quantum origin of the large-scale structure.

Unfortunately there are "Quantum Headaches" blocking the way towards quantization of gravity.

Relativistic QM (i.e. QFT) reintroduces an UV problem that had been solved by its non-relativistic predecessor (solving a UV problem had been Planck's original motivation for introducing $\hbar$). The already mentioned virtual pair creation allowed by Special Relativity plus Quantum Mechanics leads to infinities since virtual particles of arbitrarily high energy are copiously produced in any local Quantum Field Theory.

This is already true for Gauge Theories. It is simply worse for quantum gravity since the gravitational interaction grows with energy.

We have a recipe, known as renormalization, for handling the UV infinities of gauge theories and still get a (partially) predictive theory. Is this good enough? It was not, apparently, for Hermann Weyl!

Attempts to do the same for General Relativity have failed so far. The only way to make sense of quantum gravity appears to be to soften General Relativity below a certain length scale. Like Fermi's theory with respect to the SMEP, General Relativity would then just be a large-distance approximation to a better theory.

Another problem is what I would call "The missing quantum corrections". Radiative corrections to marginal and irrelevant (at large distance) operators have been seen in precision experiments. e.g.

- Running of gauge couplings;
- Effective 4-fermi interactions;
- Flavour anomalies.

By contrast, radiative corrections to relevant operators have not been seen. Best known examples:

- Scalar masses (what keeps the higgs-boson light?);
- Cosmological constant (what keeps Λ small?).

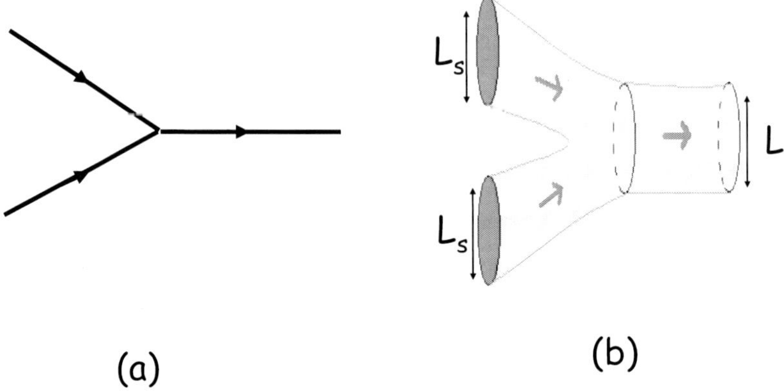

Fig. 5.   (a) In quantum field theory interactions are localized at points in space-time; (b) In string theory they are smeared over spece-time regions of size L_s.

There is a well known IR-UV connection saying that the most relevant (in the IR) operators receive quantum contributions from the deep UV, and vice-versa. The mystery of those missing quantum corrections may be telling us that the SMEP and the SMGC are not the full story: they are just effective theories in search of an *ultraviolet completion* that might explain away the huge fine-tunings needed in Quantum Field Theories. In the rest of the talk I will try to answer the following question about such a completion: Could it be String Theory?

6. Quantum String Theory and Its Miracles

The paradigm underlying string theory can be simply formulated as follows:

Every elementary particle, previously seen as point-like, is nothing but a vibrating string satisfying the laws of special relativity and quantum mechanics.

The 3 ingredients are essential: indeed, Strings + Special Relativity + Quantum Mechanics represent a magic cocktail leading us to a grand synthesis!

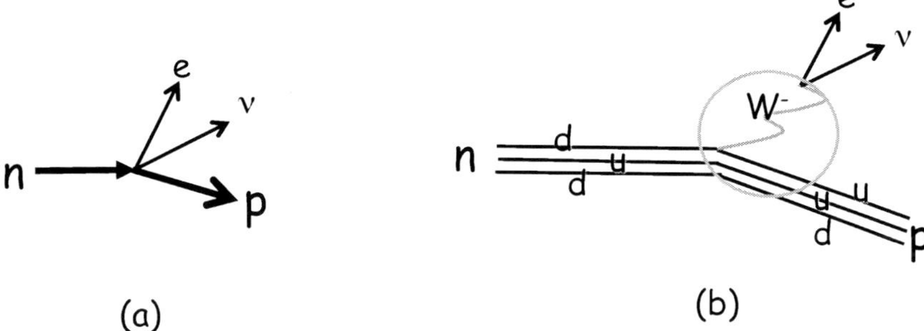

Fig. 6.   (a) In Fermi's theory four fermion interact at a single point. (b) In the SMEP the 4-point vertex is "resolved" into two separated 3-point vertices.

The following are two major miracles of quantum[c] strings:

- **Finite Size**

  Classical string theory is scale free. For this reason classical strings have no characteristic size. The string tension, T, is *not* a fundamental length: it is just a classical conversion factor from string length to string mass and vice versa. On the contrary, quantum strings have a characteristic size determined by Quantum Mechanics (in units where $c = 1$):

  $$L_s = \sqrt{\frac{\hbar}{T}} \tag{3}$$

  Note the analogy with classical gravity in $D = 4$ where G has dimensions of an inverse tension but does *not* provide a unit of length, while, at the quantum level, one can construct a fundamental (Planck) length as

  $$L_P = \sqrt{\hbar G} \tag{4}$$

---

[c]Classical strings are completely useless for giving interesting classical field theories

72

Without QM strings become lighter and lighter as they shrink and their mass goes to zero as one squeezes them to a point. With QM strings are lightest when their size is $O(L_s)$ just like a oscillator which, in its ground state has a well defined (and finite!) Δx. In string theory there is an interesting symmetry between x (actually $x' = \partial_\sigma x$) and p making the ground state to be quite symmetric as well (if momenta are converted into lengths using the string tension).

The consequences of this minimal size of strings are profound: they are no-doubt at the basis of the good UV behaviour of string loops. When we compare a Feynman diagram of QFT with a string diagram (Fig. 5), we see that, in the former, interactions occur at precisely defined points in space and time whereas, in the latter, they are smeared over regions of order L_s. One can even note an analogy with comparing the old (1934) Fermi theory of weak interactions with the SMEP (Fig. 6). In the former 4 fermions (a proton, a neutron, an electron, and an antineutrino) all interact at one space-time point. In the SMEP the interaction is mediated by a (virtual) W-boson that replaces a single 4-Fermi interaction by two local fermion-fermion-boson vertices separated by the W-boson propagator. This is enough to turn a non-renormalizable theory (Fermi's) into a renormalizable one.

- **J without M**

  A classical string cannot have angular momentum without having a finite length, hence a finite mass. One can prove a strict inequality between mass and angular momentum that holds true for a any classical motion:

$$\frac{M^2}{2\pi T} \geq J \tag{5}$$

with the equality sign holding for a rigid, rotating rod. A quantum string, instead, can have up to two units of angular momentum without gaining mass. Indeed the above classical inequality is slightly modifies to:

$$\frac{M^2}{2\pi T} \geq J + \frac{1}{2}\hbar \sum_1^\infty n = J - \alpha_0 \hbar \ , \quad \alpha_0 = 0, \frac{1}{2}, 1, \frac{3}{2}, 2. \tag{6}$$

where the divergent sum has been regularized in the only possible consistent way.

Thus, QST predicts, among others, the existence of the massless $J = 1$ and $J = 2$ states[d] that the SMN badly needs. Furthermore, the common (stringy) origin of gauge and gravity carriers implies a quantitative unification of all forces (recall H. Weyl's quest for a "unitary theory"). Indeed:

- $m = 0, J = 1$: provides the photon and other carriers of non-gravitational interactions;

---

[d]The inevitability of massless spinning states was one reason for abandoning the old string theory in favour of QCD, showing that string theory *can* be falsified by large-distance experiments!

$- \; m = 0, J = 2$: provides the graviton, the carrier of gravity.

This amazing property of quantum strings may well provide answers to the two questions we asked at the beginning: Why does Nature like $J = 1$ massless particles? Why does it also like $J = 2$ massless particles? and explain why it is well described by a Gauge Theory plus General Relativity.

To summarize: String Theory contains all the ingredients needed to provide a unified and finite theory of elementary particles, and of their gauge and gravitational interactions, not just compatible with, but based on, Quantum Mechanics! The quantitative unification of all forces occurs at the string-energy scale $M_s \sim \hbar l_s^{-1}$. Equating gravitational and gauge interactions at $E = M_s c^2$ implies:

$$\frac{M_s}{M_P} \sim \sqrt{\alpha_g} \;\; ; \;\; \text{closed strings}$$

$$\frac{M_s}{M_P} \sim \alpha_g \;\; ; \;\; \text{open strings} \tag{7}$$

Consequently, the string energy scale M_s should be somewhere between the GUT (10^{16} GeV) and the Planck (10^{19} GeV) scale, a very large scale indeed if compared with the reach of accelerators! Does it mean that string theory will never be tested? My answer is a definite *no* for at least three reasons:

- First of all such energies have existed, most likely, in the very early Universe. This most likely left some observable relics of a string-dominated early epoch like it is the case for conventional inflation.
- A generic prediction of string theory is the existence of (most likely 6) extra dimensions of space which must be tiny but could nevertheless be felt through their "zero-modes" (gauge bosons, fermions...). T-duality (a symmetry related to the already mentioned interchange $x' \leftrightarrow p$) suggests that the sizes of extra dimensions should be dynamically fixed at the string scale L_s, but models can be conceived where such sizes are as large as a fraction of a millimiter (provided the SM particles cannot move in them) leading to interesting experimental consequences (short distance modifications of gravity, weird LHC events, ...).
- Another characteristic property of string theory is the absence of free dimensionless parameters. They are replaced by (scalar) fields whose values provide the "Constants of Nature", e.g. the fine-structure constant (Cf. Weyl's remark about e!). Are all these dynamically determined? While today these constants look to be space and time-independent, their variations may have played an important role in early cosmology. The possibility that they are slightly varying even today is all but excluded. Furthermore, the abovementioned scalar fields, if light, could provide new long-range forces and induce violations of the EP.

  The conclusion is that we do not have so far tests of string theory simply because of our present inability to solve the theory non-perturbatively.

If taken at tree-level, string theory is already ruled out by large-distance experiments!

7. Conclusions

- Almost a century after Hermann Weyls book, we have made enormous progress in describing matter. Since the mid seventies we have a "Standard Model" of elementary particles, an achievement second only to the discoveries of relativity and quantum mechanics over a 100 years ago.
- The SMEP unifies conceptually three of the four known interactions. They are all described by a quantum-relativistic gauge theory.
- Not so much has happened, in comparison, to our theory of gravity, still based on Einstein's general relativity but:
    - Its experimental confirmations have become sharper and sharper.
    - We also have a much more satisfactory standard model for cosmology.
- Meanwhile, a number of phenomenological puzzles have been uncovered as experimental data in astrophysics and cosmology have caught up in precision with those coming from accelerators. In particular, 96% of the present energy in our observable universe is "dark".
- I am sure that the next few decades will witness much further progress in these areas thanks to the joint efforts of particle, astro- and cosmophysicists. Their ambitious experimental and theoretical programs will greatly benefit from stronger and stronger interconnections among these different areas of fundamental physics.
- In this effort, LeCosPA, by its very acronym, is bound to play a major role!

At a more theoretical level:

- The task of unifying GR and the SM in a fully quantum setup has proven tougher than anticipated. An ultraviolet completion of the Standard Model of Nature looks necessary.
- Since the mid eighties there are hopes that a fully consistent framework for describing all particles and their mutual interactions can be provided by *quantum* string theory.
- It is still too early to say whether string theory will be able to successfully address the deepest remaining puzzles in particle physics, astrophysics and cosmology or whether it will make the unhappy end of its hadronic predecessor.

Even if the latter should turn out to be the case, String Theory had many spinoffs:

- In many branches of mathematics;
- In providing an "existence proof" that gravity and quantum mechanics can be reconciled;

Thank you & many wishes to LeCosPA

Dragon from the Lantern Fest, Taipei, Feb. 6, 2012.

- In providing examples of black holes whose thermodynamic (Bekenstein-Hawking) entropy can be given a Statistical Mechanics interpretation;
- In providing new techniques to perform perturbative calculations in gauge theories;
- In giving new handles on gauge theories at strong coupling through a dual weak-coupling gravity description (Cf. the AdS/CFT calculation of shear viscosity and other transport coefficients which are relevant in heavy ion collisions).
- In providing a dual description of solid state problems?
- In suggesting new cosmological scenarios without a beginning of time.

I wonder what Hermann Weyl, Wolfgang Pauli (and Albert Einstein!) might have thought about it ...

GENERAL RELATIVITY WITHOUT PARADIGM OF SPACE-TIME COVARIANCE: SENSIBLE QUANTUM GRAVITY AND RESOLUTION OF THE "PROBLEM OF TIME"

CHOPIN SOO

Department of Physics, National Cheng Kung University, Tainan 701, Taiwan
E-mail: cpsoo@mail.ncku.edu.tw

HOI-LAI YU*

Institute of Physics, Academia Sinica, Taipei 115, Taiwan
** E-mail: hlyu@phys.sinica.edu.tw*

Covariance of space and time in General Relativity (GR) entails a number of technical and conceptual difficulties. Remarkably, these can be resolved by a paradigm shift from full 4-dimensional general coordinate invariance to invariance only with respect to spatial diffeomorphisms. The framework for a theory of gravity with this paradigm shift, from quantum to classical regimes, is presented; GR is contained as a special case. Appositely formulated as a master constraint, the Hamiltonian constraint now determines only dynamics; and is relieved of its dual role of generating symmetry transformations. The Dirac algebra, in which 4-dimensional diffeomorphism symmetry is only realized on-shell, is replaced by the master constraint algebra which possesses only spatial diffeomorphism gauge symmetry, both on- and off-shell. Decomposition of the spatial metric into unimodular and determinant, q, factors results in mutually commuting pairs of canonical variables. The classical content of GR can be captured with a Hamiltonian constraint linear in the trace of the momentum. This implies a theory of quantum gravity can be described by a Schrodinger equation first order in intrinsic time $\ln q$ accompanied with positive semi-definite probability density. The semi-classical Hamilton-Jacobi equation is also first order in intrinsic time, with the implication of being complete; and gauge-invariant physical observables can be constructed from integration constants of its complete integral solution. Classical space-time, with direct correlation of its proper times and intrinsic time intervals, emerges from constructive interference; and the physical content of GR can be regained from a theory with a true Hamiltonian generating intrinsic time translations, but with only spatial diffeomorphism symmetry. The framework also prompts natural extensions towards a well-behaved quantum theory of gravity.

Keywords: Paradigm shift; General relativity; Quantum Gravity; Intrinsic time.

1. Introduction and Overview

A central thesis of Einstein's General Relativity (GR) is the covariance of space and time encapsulated in the invariance of the theory with respect to general coordinate transformations. While the assumption of full 4-dimensional diffeomorphism symmetry has been crucial to the birth and subsequent development of the theory, it also entails a number of technical, conceptual and interpretational difficulties. Almost a century after its birth, a quantum theory of Einstein's GR remains incomplete, despite many recent advances. On the technical front, Einstein's theory fails to be renormalizable as a perturbative quantum field theory. At the conceptual level, the (super-)Hamiltonian constraint which is believed to dictate the dynamics of the theory has a dual role. It is also a generator of symmetry, with arbitrary lapse function as Lagrange multiplier. Full 4-dimensional diffeomorphism invariance with local Hamiltonian constraint and its consequent baggage of multi-fingered time, arbitrary lapse and gauged histories is hard to reconcile with the physical reality of of time.

Canonical quantum gravity has the guise of a formalism "frozen in time" as quantum states annihilated by the Hamiltonian constraint do not evolve in "time" (more precisely coordinate time x^0). In fact dx^o is merely one component of space-time displacement, with no invariant physical meaning in a theory with full space-time covariance. Yet a notion of time is needed to make sense of quantum mechanical interpretations. It is required both as the basis for the discussion of dynamics and evolution of quantum wave functions, and for the interpretation of probabilities with normalization of physical states at a particular instant of time. Even if a suitable degree of freedom in the theory is chosen as the intrinsic time (as many have advocated), a successful quantum theory of gravity with intrinsic time development will still need to reconcile a Klein-Gordon type Wheeler-DeWitt (WDW) equation[1,2] second-order-in-momenta with the positivity of "probabilities". Moreover, a resolution of the "problem of time"[3] in quantum gravity cannot be deemed complete if it fails to account for the intuitive physical reality of time and does not provide a satisfactory response to the basic question: what does intrinsic time development in quantum dynamics have to do with the passage of time measured by physical clocks in classical space-times?

At the deeper level of symmetries, the constraints of GR satisfy the Dirac algebra[4] which is the hallmark of space-time covariance and the embeddability of hypersurface deformations; and from which Einstein's theory of geometrodynamics can be uniquely regained. However, closer inspection reveals the Dirac algebra only corresponds to 4-dimensional general coordinate symmetry *on-shell*, modulo equations of motions (EOM). Off-shell, full 4-dimensional Lie derivatives involve time derivatives; these cannot be generated by constraints of a canonical theory which depend only on the spatial metric q_{ij} and its conjugate momentum $\tilde{\pi}^{ij}$. In Einstein's theory, the EOM, among other things, relate $\dot{q}_{ij}$ to $\tilde{\pi}^{ij}$ (through the extrinsic curvature) and the constraints then generate 4-dimensional diffeomorphisms on-shell. This is a clue that, without the help of EOM, a *quantum* theory of GR should not be able to, and need not, realize and enforce full 4-dimensional space-time covariance *off-shell*.

A key obstacle to the viability of GR as a perturbative quantum field theory lies in the deep conflict between unitarity and space-time general covariance: renormalizability of GR can be achieved with higher-order curvature terms, but space-time covariance requires time as well as spatial derivatives of the same (higher) order, thus compromising the stability and unitarity of the theory. By relinquishing full space-time covariance and by introducing higher-order spatial curvatures terms, Horava[5] achieved power-counting renormalizable theories of gravity. In loop quantum gravity, the master constraint program[6] for non-perturbative quantization of Einstein's theory seeks representations not of the Dirac algebra, but of the master constraint algebra which has the advantages of having structure constants rather than structure functions, and of spatial diffeomorphisms forming an ideal (thus allowing for the crucial decoupling of the equivalent quantum Hamiltonian constraint from spatial diffeomorphism generators). Reference 7 demonstrates that Horava gravity theories can be naturally realized as representations of canonical theories with first class master constraint algebra. This not only removes the canonical inconsistencies of projectable Horava theory, but also captures the essence of Horava gravity in retaining only spatial diffeomorphisms, generated by the (super-)momentum constraint $H_i = 0$, as the only local gauge symmetries. At the same time, the local Hamiltonian constraint $H = 0$ is equivalently enforced by the master constraint. The master constraint has been deployed as a mathematical device which can reproduce all the observables,[6] O, of Einstein's GR; in which O must commute both with H_i *and* H (equivalent to the extra requirement, above and beyond usual canonical rules, of $\{O, \{O, \mathrm{M}\}\}|_{\mathrm{M}=0} = 0$). In contradistinction, in this work a theory of gravity appositely formulated with master constraint marks a real paradigm shift in the symmetry, from full 4-dimensional general coordinate invariance to invariance only with respect to spatial diffeomorphisms. The immediate benefits and implications are very significant (detailed discussions will be presented in the next sections): enforcing $H(x) = 0$ through the master constraint $\mathrm{M} := \int_\Sigma \frac{[H(x)]^2}{\sqrt{q(x)}} = 0$ removes it from one of its dual role. It is no longer the generator of symmetry, and determines only dynamics. This can be seen from $\{f(q_{ij}, \tilde{\pi}^{ij}), m(t)\mathrm{M} + H_k[N^k]\}|_{\mathrm{M}=0 \Leftrightarrow H=0} \approx \{f, H_k[N^k]\} = \mathcal{L}_{\vec{N}} f$, which implies M generates no extra symmetry due to its quadratic dependence on H. This eliminates the extra freedom in multi-fingered time with arbitrary lapse and the consequent unphysical time development which are gauge histories. Instead, an intrinsic and more physical notion of time is to be found from a suitable degree of freedom in the theory.

A theory of geometrodynamics, with spatial metric and its conjugate momentum $(q_{ij}, \tilde{\pi}^{ij})$ as fundamental variables, is bequeathed with a number of remarkable features: positivity of the spatial metric guarantees the requirement of space-like separation between any two points on the initial value hypersurface (and allows, to quote Ref. 2, a notion of "simultaneity" and a common moment of a rudimentary "time"). In Arnowitt-Deser-Misner (ADM)[8] description of the space-time metric, these are labeled as constant-x^0 hypersurfaces. Quantum states however do not depend on x^0; the link between intrinsic and coordinate times is more subtle. Decomposition of the spatial metric into unimodular and determinant factors i.e. $q_{ij} = q^{\frac{1}{3}}\bar{q}_{ij}$, results in mutually commuting pairs of canonical variables, with $\ln q^{\frac{1}{3}}$ conjugate to the trace of the momentum π. For positive-definite spatial metrics, the generic ultra-local DeWitt supermetric[1] with deformation parameter λ has signature $(\text{sgn}[\frac{1}{3} - \lambda], +, +, +, +, +)$, and the single negative eigenvalue for $\lambda > \frac{1}{3}$ corresponds to the $\delta \ln q$ mode. It is not only the preeminent, but also apposite (as the discussion in next section will show), choice for the role of intrinsic time interval in quantum gravity. Although q is a scalar density and not a scalar - a criticism which has been used to disqualify it from the role of time variable - $\delta \ln q = \frac{\delta q}{q}$ is a scalar. Even in Galilean-Newtonian physics, it is *time interval*, and not absolute time, which is physical.

Based on the preceding concerns and observations, the framework for a theory of gravity, from quantum to classical regimes, will be presented; wherein several fortuitous circumstances collude to result in a compelling new formulation. The WDW Hamiltonian constraint of the ensuing theory factorizes into $(\beta \pi - \bar{H})(\beta \pi + \bar{H}) = 0$; and the vanishing of only one of the factors is sufficient for the purpose of capturing the classical content of GR (with $\beta^2 = \frac{1}{6}$). This fortuity, and the fact that π is precisely conjugate to $\ln q^{\frac{1}{3}}$, imply a theory of quantum gravity can be described by a Schrodinger equation *first order in intrinsic time* with the consequence of positive semi-definite probability density $\psi^\dagger \psi$. There is the additional promise of a well-behaved unitary quantum theory of gravity if the physical Hamiltonian generating intrinsic time development, $\bar{H}$, can be made positive semi-definite, and also renormalizable by including suitable higher-order spatial curvature terms, with Einstein's theory recovered in the limit of low curvatures. The Hamiltonian constraint is freed from its role of generator of symmetry transformations (and consequent unphysical "time"-development with arbitrary lapse) by the master constraint $\text{M} := \int \frac{(\beta \pi + \bar{H})^2}{\sqrt{q}} = 0$. The resultant first class algebra, $\{\text{M}, \text{M}\} = 0, \{H_i[N^i], \text{M}\} = 0, \{H_i[N^i], H_j[N'^j]\} = H_i[\mathcal{L}_{\vec{N}} N'^i]$, possesses only spatial diffeomorphism gauge symmetry, both on- and off-shell. The corresponding semi-classical Hamilton-Jacobi (HJ) equation is also first order in intrinsic time, with the implication of being complete;[11] and gauge-invariant physical observables can be identified with the integration constants of its complete integral solution. Classical space-time emerges from constructive interference of quantum wave functions.[9] The physical content of Einstein's EOM can remarkably be regained from a theory with $\frac{\bar{H}}{\beta}$ generating intrinsic time translations $\delta \ln q^{\frac{1}{3}}$, subject to constraints $\text{M} = H_i = 0$. Furthermore, the emergent classical space-times have ADM metrics with proper times directly correlated to intrinsic time by the explicit relation $d\tau^2 = [\frac{d \ln q}{(12\beta\kappa\bar{H}/\sqrt{q})}]^2$.

2. Theory of Gravity without Full Space-Time Covariance

2.1. *General framework, and quantum theory*

In an effort to solve the initial value problem, York[10] explored a conformal decomposition of the spatial metric q_{ij} on the Cauchy manifold Σ,

$$q_{ij} = \phi \bar{q}_{ij}, \tag{1}$$

of which $\phi = q^{\frac{1}{3}}$ (wherein $\det q_{ij} =: q$) with unimodular $\bar{q}_{ij}$ is a particular case. However, in focusing on the generic conformal factor ϕ, the "miracle" of $\phi = q^{\frac{1}{3}}$ was not fully revealed. Part of the miracle is that the decomposition results in a clean separation of mutually commuting canonical pairs, with symplectic potential,

$$\int \tilde{\pi}^{ij} \delta q_{ij} = \int \bar{\pi}^{ij} \delta \bar{q}_{ij} + \pi \delta \ln q^{\frac{1}{3}}, \tag{2}$$

confirming that $\ln q^{\frac{1}{3}}$ and $\bar{q}_{ij}$ are respectively conjugate to the trace $\pi = q_{ij}\tilde{\pi}^{ij}$ and traceless $\bar{\pi}^{ij} = q^{\frac{1}{3}}[\tilde{\pi}^{ij} - q^{ij}\frac{\pi}{3}]$ parts of the original momentum variable. Thus the only non-trivial Poisson brackets are

$$\{\bar{q}_{kl}(x), \bar{\pi}^{ij}(x')\} = P^{ij}_{kl}\,\delta(x, x'), \quad \{\ln q^{\frac{1}{3}}(x), \pi(x')\} = \delta(x, x'); \tag{3}$$

wherein the trace-free projector, $P^{ij}_{kl} := \frac{1}{2}(\delta^i_k \delta^j_l + \delta^i_l \delta^j_k) - \frac{1}{3}\bar{q}^{ij}\bar{q}_{kl}$ depends only on $\bar{q}_{ij}$. This clean separation carries over to the quantum theory, and permits a degree of freedom separate from the others to be identified as the carrier of temporal information. However, physical time intervals associated with $\delta \ln q^{\frac{1}{3}}$ can be consistently realized only when the dogma of full space-time covariance is relinquished.

Based on the signature of the supermetric, DeWitt proposed $\zeta \propto q^{\frac{1}{4}}$ (associated with the only negative mode of the supermetric) as the intrinsic time.[1] But an intrinsic clock is in fact not tied to 4-dimensional general covariance. Any generalized ultra-local DeWitt supermetric compatible with spatial diffeomorphism invariance, $G_{ijkl} = \frac{1}{2}(q_{ik}q_{jl} + q_{il}q_{jk}) - \frac{\lambda}{3\lambda - 1}q_{ij}q_{kl}$, has signature $[\text{sgn}(\frac{1}{3} - \lambda), +, +, +, +, +]$, and comes equipped with intrinsic temporal intervals $\delta \ln q^{\frac{1}{3}}$ provided $\lambda > \frac{1}{3}$ (Einstein's GR is the special case of $\lambda = 1$). Crucially it is this very same $\ln q^{\frac{1}{3}}$ (monotonic to q) which can be so cleanly separated from the rest of the degrees of freedom.

The Hamiltonian constraint of a theory of geometrodynamics quadratic in momenta is of the general form,

$$
\begin{aligned}
0 = \frac{\sqrt{q}}{2\kappa} H &= G_{ijkl}\tilde{\pi}^{ij}\tilde{\pi}^{kl} + V(q_{ij}) \\
&= -\frac{1}{3(3\lambda - 1)}\pi^2 + \bar{G}_{ijkl}\bar{\pi}^{ij}\bar{\pi}^{kl} + V[\bar{q}_{ij}, q] \\
&= -\beta^2\pi^2 + \bar{H}^2[\bar{\pi}^{ij}, \bar{q}_{ij}, q] \\
&= -(\beta\pi - \bar{H})(\beta\pi + \bar{H}).
\end{aligned}
\tag{4}
$$

In the above,

$$\bar{H}[\bar{\pi}^{ij}, \bar{q}_{ij}, q] = \sqrt{\bar{G}_{ijkl}\bar{\pi}^{ij}\bar{\pi}^{kl} + V[\bar{q}_{ij}, q]} = \sqrt{\frac{1}{2}(\bar{q}_{ik}\bar{q}_{jl} + \bar{q}_{il}\bar{q}_{jk})\bar{\pi}^{ij}\bar{\pi}^{kl} + V[\bar{q}_{ij}, q]}, \tag{5}$$

$\beta^2 := \frac{1}{3(3\lambda-1)}$, and $\bar{H}$ is independent of π and β. Einstein's GR (in which $\lambda = 1$ and $V[\bar{q}_{ij}, q] = -\frac{q}{(2\kappa)^2}(R - 2\Lambda_{eff}))$ is thus a particular realization of a wider class of theories, all of which factorizes miraculously as in the last step of Eq.(4).

Several crucial features should be emphasized:

(1) As written, (4) is a local constraint, in addition to those of spatial diffeomorphism invariance $H_i = 0$. This leads to a quandary: if the constraint algebra is first class, then H consistently generates multi-fingered time translation *symmetry* as well as "physically" evolving the theory with respect to "time" (as manifested by its dual roles in Einstein's theory with full general coordinate invariance). On the other hand if the algebra fails to close, H no longer generates symmetry, which relieves it from one of its dual roles to allow true physical time evolution. But as a *local* constraint, consistency demands final closure. So either further secondary constraints and/or restrictions on the Lagrange multipliers lead finally to closure, or the algebra is simply inconsistent.[5] All these latter cases imply true 4-dimensional diffeomorphism symmetry is absent. In fact, for the potential V of Einstein's GR, but with $\lambda > \frac{1}{3}$ (and $\lambda \neq 1$), secondary constraints lead to vanishing lapse function N,[7] indicating that only spatial diffeomorphism is intact.

(2)A master constraint formulation with $\mathrm{M} = \int H^2/\sqrt{q} = 0$ equivalently enforce the local constraint and physical content of Eq.(4). This permits H to determine dynamical evolution rather than generate symmetry. Moreover, consistency of a first class algebra which decouples M from H_i is attained, paving the road for quantization. The result is a theory with only spatial diffeomorphism invariance; with physical dynamics dictated by H, but encoded in M.

(3)On shell, M itself does not generate dynamical evolution, but only spatial diffeomorphisms; for instance, $\{q_{ij}, m(t)\mathrm{M} + H_k[N^k]\}|_{\mathrm{M}=0\Leftrightarrow H=0} \approx \{q_{ij}, H_k[N^k]\} = \mathcal{L}_{\vec{N}} q_{ij}$. Instead, true physical evolution can only be with respect to an intrinsic time extracted from the WDW constraint. As detailed above, $\ln q^{\frac{1}{3}}$ which is decoupled from all other degrees of freedom is the preeminent choice.

(4)The miraculous factorization $(\beta\pi - \bar{H})(\beta\pi + \bar{H}) = 0$ allows for the possibility that only $(\beta\pi + \bar{H}) = 0$ is all that is needed to recover the classical content of $H = 0$. This is a breakthrough:

(i)the semiclassical HJ equation is first order in intrinsic time, bearing in mind π is conjugate to $\ln q^{\frac{1}{3}}$, with its consequence of completeness.[11]

(ii)quantum gravity will now be dictated by a corresponding WDW equation which is a Schrodinger equation first order in intrinsic time (with the consequence of positive semi-definite probability density at any instant of intrinsic time). Thus the deep divide between a quantum mechanical interpretation (which requires *both* the notion of time and positive semi-definite probabilities) and the nature of usual Klein-Gordon type WDW equations second-order in intrinsic time (hence indefinite in "probabilities") no longer exists.

The starting point of a consistent quantum theory of gravity is a spatial diffeomorphism-invariant theory with $\mathrm{M}|\Psi\rangle = 0; \mathrm{M} := \int (\beta\pi + \bar{H})^2/\sqrt{q}$. Positive-semi-definite inner product for $|\Psi\rangle$ will equivalently imply[7]

$$(\beta\hat{\pi} + \hat{\bar{H}}[\hat{\bar{\pi}}^{ij}, \hat{\bar{q}}_{ij}, \hat{q}])|\Psi\rangle = 0,$$

$$\hat{H}_i|\Psi\rangle = 0. \tag{6}$$

In the metric representation, canonical momenta become functional differential operators, $\hat{\pi} = \frac{3\hbar}{i}\frac{\delta}{\delta \ln q}$, $\hat{\bar{\pi}}^{ij} = \frac{\hbar}{i}P^{ij}_{lk}\frac{\delta}{\delta \bar{q}_{lk}}$ which operate on $\Psi[\bar{q}_{ij}, q]$; and the Schrodinger equation, and HJ equation for semi-classical states $Ce^{\frac{iS}{\hbar}}$ are respectively,

$$i\hbar\beta\frac{\delta\Psi}{\delta \ln q^{\frac{1}{3}}} = \bar{H}[\hat{\bar{\pi}}^{ij}, q_{ij}]\Psi,$$

$$\beta\frac{\delta S}{\delta \ln q^{\frac{1}{3}}} + \bar{H}[\bar{\pi}^{ij} = P^{ij}_{kl}\frac{\delta S}{\delta \bar{q}_{kl}}; \bar{q}_{ij}, \ln q] = 0, \tag{7}$$

first order in intrinsic time derivative; and $\nabla_j \frac{\delta\Psi}{\delta q_{ij}} = 0$ enforces spatial diffeomorphism symmetry. Behold the appearance of a true Hamiltonian $\bar{H}$ generating time evolution w.r.t. $\frac{1}{\beta}\delta \ln q^{\frac{1}{3}}$ (or $\bar{H}/\beta$ w.r.t. $\ln q^{\frac{1}{3}}$). It is important to note that although $\ln q^{\frac{1}{3}}$ is not a scalar under spatial diffeomorphisms, $\delta \ln q^{\frac{1}{3}}$ is indeed a scalar apposite for the role of intrinsic time interval. How neatly too it separates canonically from the other degrees of freedom $\bar{q}_{ij}$.

2.2. *Emergence of classical spacetime*

The next task is to show the natural emergence of classical space-times (in particular, those described by Einstein's GR and its EOM) from Eq.(6) which has quantum dynamical evolution w.r.t. $\delta \ln q^{\frac{1}{3}}$. Many years ago Gerlach[9] demonstrated that classical space-time and its EOM can be recovered from the quantum theory through HJ theory and constructive interference. The first order HJ equation, which bridges quantum and classical regimes, has complete solution $\mathcal{S} = \mathcal{S}(^{(3)}\mathcal{G}; \alpha)$ which depends on 3-geometry $^{(3)}\mathcal{G}$ and integration constants (denoted generically here by α). Constructive interference with $\mathcal{S}(^{(3)}\mathcal{G}; \alpha + \delta\alpha) = \mathcal{S}(^{(3)}\mathcal{G}; \alpha)$; $\mathcal{S}(^{(3)}\mathcal{G} + \delta^{(3)}\mathcal{G}; \alpha + \delta\alpha) = \mathcal{S}(^{(3)}\mathcal{G} + \delta^{(3)}\mathcal{G}; \alpha)$ leads to[9]

$$\frac{\delta}{\delta\alpha}\Big[\int \frac{\delta\mathcal{S}(^{(3)}\mathcal{G}; \alpha)}{\delta q_{ij}}\delta q_{ij}\Big] = 0; \tag{8}$$

subject to constraints $\mathrm{M} = H_i = 0$. With the momenta identified with $\pi^{ij}(\alpha) := \frac{\delta\mathcal{S}(^{(3)}\mathcal{G};\alpha)}{\delta q_{ij}}$, and Lagrange multipliers δm and δN^i, the requirement of constructive interference for the emergence of classical EOM is equivalent to

$$0 = \frac{\delta}{\delta\alpha}\Big[\int (\pi^{ij}\delta q_{ij} + \delta N_i H^i) + \delta m\mathrm{M}\Big]$$

$$= \frac{\delta}{\delta\alpha}\Big[\int (\pi\delta \ln q^{\frac{1}{3}} + \bar{\pi}^{ij}\delta\bar{q}_{ij} + \frac{q^{ij}}{3}\delta N_i\nabla_j\pi + q^{-\frac{1}{3}}\delta N_i\nabla_j\bar{\pi}^{ij})\Big]. \tag{9}$$

Happily, for master constraint theories, there is no δm contribution (since $\mathrm{M} = 0 \Leftrightarrow H = 0$, and M being quadratic in H). From (7), $\pi = -\frac{\bar{H}}{\beta}$; integrating by parts and bearing in mind $\bar{H}$ is dependent on $\bar{\pi}^{ij}(\alpha)$, the resultant EOM is,

$$\frac{\delta \bar{q}_{ij}(x) - \pounds_{\vec{N}dt}\,\bar{q}_{ij}(x)}{\delta \ln q^{\frac{1}{3}}(y) - \pounds_{\vec{N}dt}\ln q^{\frac{1}{3}}(y)} = P_{ij}^{kl}\frac{\delta[\bar{H}(y)/\beta]}{\delta \bar{\pi}^{kl}(x)} = P_{ij}^{kl}\frac{\bar{G}_{klmn}\bar{\pi}^{mn}}{\beta \bar{H}}\delta(x-y) \qquad (10)$$

wherein $\pounds_{\vec{N}}$ denotes Lie derivative and $\delta \vec{N} =: \vec{N}dt$. Proceeding as in Ref. 9, the other half of Hamilton's equations

$$\frac{\delta \bar{\pi}^{ij}(x) - \pounds_{\vec{N}dt}\,\bar{\pi}^{ij}(x)}{\delta \ln q^{\frac{1}{3}}(y) - \pounds_{\vec{N}dt}\ln q^{\frac{1}{3}}(y)} = -\frac{\delta[\bar{H}(y)/\beta]}{\delta \bar{q}_{ij}(x)}, \qquad (11)$$

can be recovered. As predicted by quantum Schrodinger and semiclassical equations (7), $\bar{H}/\beta$ is the Hamiltonian for evolution w.r.t. intrinsic time $\ln q^{\frac{1}{3}}$.

Although the above derivation bear similarities to Gerlach's work,[9] important fundamental differences must be noted. In the Gerlach approach, $\delta N =: Ndt$ (associated with local constraint $H = 0$) will always contribute to the final EOM resulting in multi-fingered time with arbitrary lapse function. This is a consequence of 4-dimensional general coordinate invariance. In contradistinction, δm contribution does not arise for a master constraint theory. Not only is unphysical time development with arbitrary lapse function now evaded, but true evolution w.r.t. intrinsic time $\ln q^{\frac{1}{3}}$ is achieved. The "defect" that M does not generate dynamical evolution w.r.t. unphysical coordinate time is redeemed at a much deeper level through physical evolution w.r.t. intrinsic time. Through Eq.(10), the emergent ADM classical space-time has momentum related to the time derivative of the metric by

$$\frac{2\kappa}{\sqrt{q}}G_{ijkl}\tilde{\pi}^{kl} = \frac{1}{2N}\left(\frac{dq_{ij}}{dt} - \pounds_{\vec{N}}q_{ij}\right),\ Ndt := \frac{\delta \ln q^{\frac{1}{3}} - \pounds_{\vec{N}dt}\ln q^{\frac{1}{3}}}{(4\beta\kappa\bar{H}/\sqrt{q})}. \qquad (12)$$

In Einstein's GR the EOM with *arbitrary* lapse function N takes the form,

$$\frac{dq_{ij}}{dt} = \left\{q_{ij}, \int d^3x[NH + N_iH^i]\right\} = \frac{2N}{\sqrt{q}}(2\kappa)G_{ijkl}\tilde{\pi}^{kl} + \pounds_{\vec{N}}q_{ij}. \qquad (13)$$

This relates the extrinsic curvature to the momentum by $K_{ij} := \frac{1}{2N}\left(\frac{dq_{ij}}{dt} - \pounds_{\vec{N}}q_{ij}\right) = \frac{2\kappa}{\sqrt{q}}G_{ijkl}\tilde{\pi}^{kl}$. Contraction with $\frac{q^{ij}}{3}$ in fact yields

$$\frac{1}{3}Tr(K) = \frac{1}{2N}\left(\frac{d\ln q^{\frac{1}{3}}}{dt} - \pounds_{\vec{N}}\ln q^{\frac{1}{3}}\right) = -\frac{2\kappa}{\sqrt{q}}(\beta^2\pi)$$

$$= \frac{2\kappa}{\sqrt{q}}\beta\bar{H}, \qquad (14)$$

wherein the constraint (6), $(\beta\pi \mid \bar{H}) = 0$, has been made use of to arrive at the last step. Eq.(14) proves that the lapse function and intrinsic time are precisely related (a posteriori by the EOM) by the same formula as in Eq.(12). For a theory with full 4-dimensional diffeomorphism invariance (such as Einstein's GR with $\lambda = 1$ and consistent Dirac algebra of constraints), this relation is an identity which does not compromise the arbitrariness of N. However, it reveals (even in Einstein's GR) the physical meaning of the lapse function and its relation to the intrinsic time.

2.3. *Paradigm shift and resolution of "problem of time"*

Starting with only spatial diffeomorphism invariance and constructive interference, Eqs.(10) and (11), with physical evolution in intrinsic time generated by $\bar{H}$, are obtained. This relates the momentum to coordinate time derivative of the metric precisely as in Eq.(12). This means it is possible to interpret the emergent classical space-time (which can generically be described with ADM metric) from constructive interference to possess extrinsic curvature which corresponds precisely to the lapse function displayed in Eq.(12). However, only the freedom of spatial diffeomorphism invariance is realized, as the lapse is now completely described by the intrinsic time $\ln q^{\frac{1}{3}}$ and $\vec{N}$. All EOM w.r.t coordinate time t generated by $\int NH + N^i H_i$ in Einstein's GR can, remarkably, be recovered from evolution w.r.t. $\ln q^{\frac{1}{3}}$ and generated by $\bar{H}$ iff N assumes the form of (12). All the previous observations lead to the central revelation: full 4-dimensional space-time covariance (with its consequent baggage of multi-fingered time, arbitrary lapse and gauged histories) is a red herring which obfuscates the physical reality of time, and all that is necessary to consistently capture the classical physical content of even Einstein's GR is a theory invariant only w.r.t. spatial diffeomorphisms accompanied by a master constraint which enforces the dynamical content.

The paradigm shift from 4-dimensional general coordinate invariance to spatial diffeomorphism invariance leads to a *complete resolution of the problem of time*, from quantum to classical GR: classical spacetime, with consistent lapse function and ADM metric,

$$ds^2 = -\left(\frac{d\ln q^{\frac{1}{3}}(x,t) - dt \pounds_{\vec{N}} \ln q^{\frac{1}{3}}(x,t)}{[4\beta\kappa\bar{H}(x,t)/\sqrt{q}(x,t)]} \right)^2 + q^{\frac{1}{3}}\bar{q}_{ij}(x,t)(dx^i + N^i dt)(dx^j + N^j dt),$$

$$(15)$$

emerges from constructive interference of a spatial diffeomorphism invariant quantum theory with Schrodinger and HJ equations first order in intrinsic time development. Gratifying too is the correlation (for vanishing shifts) of classical proper time $d\tau$ and quantum intrinsic time $\ln q^{\frac{1}{3}}$ through $d\tau^2 = [\frac{d\ln q^{\frac{1}{3}}}{(4\beta\kappa\bar{H}/\sqrt{q})}]^2$. Physical reality of intrinsic time intervals cannot be denied. In particular, by Eqs.(13) and (14), proper time intervals measured by physical clocks in space-times which are solutions of Einstein's equations always agree with the result of Eq.(15).

2.4. *Improvements to the quantum theory*

The framework of the theory also prompts improvements to $\bar{H}$. Requirement of a real physical Hamiltonian $\bar{H}$ compatible with spatial diffeomorphism symmetry suggests supplementing the kinetic term with a quadratic form, i.e.

$$\bar{H} = \sqrt{\bar{G}_{ijkl}\bar{\pi}^{ij}\bar{\pi}^{kl} + [\frac{1}{2}(q_{ik}q_{jl} + q_{jk}q_{il}) + \gamma q_{ij}q_{kl}]\frac{\delta W}{\delta q_{ij}}\frac{\delta W}{\delta q_{kl}}}$$

$$= \sqrt{[\bar{q}_{ik}\bar{q}_{jl} + \gamma\bar{q}_{ij}\bar{q}_{kl}](\bar{\pi}^{ij}\bar{\pi}^{kl} + q^{\frac{2}{3}}\frac{\delta W}{\delta q_{ij}}\frac{\delta W}{\delta q_{kl}})} = \sqrt{[\bar{q}_{ik}\bar{q}_{jl} + \gamma\bar{q}_{ij}\bar{q}_{kl}]Q_+^{ij}Q_-^{kl}}. \quad (16)$$

$\bar{H}$ is then real if $\gamma > -\frac{1}{3}$. To lowest order for perturbative power-counting renormalizabilty, $W = \int \left[\sqrt{q}(aR - \Lambda) + g\tilde{\epsilon}^{ikj}(\Gamma_{im}^l\partial_j\Gamma_{kl}^m + \frac{2}{3}\Gamma_{im}^l\Gamma_{jn}^m\Gamma_{kl}^n)\right]$ (i.e. of the form of a 3-dimensional Einstien-Hilbert action with cosmological constant supplemented by a Chern-Simons action), and $Q_\pm^{ij} := \bar{\pi}^{ij} \pm iq^{\frac{1}{3}}\frac{\delta W}{\delta q_{ij}}$. The potential of the form of Einstein's theory with cosmological constant is recovered at low curvatures. The effective value of κ and cosmological constant can thus be determined as $\kappa = 8\pi c^3 G = \sqrt{\frac{1}{10\pi^2 a\Lambda(1+3\gamma)}}$ and $\Lambda_{eff} = \frac{3}{2}\kappa^2\Lambda^2(1 + 3\gamma) = \frac{3\Lambda}{20a\pi^2}$ respectively. The possibility of having a new parameter γ in the potential (different from λ in the supermetric) has been overlooked in previous works[5] in which resolution of the problem of time, evolution generated by physical Hamiltonian, consistency through master constraint theory, and first order Schrodinger and HJ equations were all absent. Furthermore, positivity of $\bar{H}$ (with $\gamma > -\frac{1}{3}$) is correlated with *real* κ and *positive* Λ_{eff}. There is also the intriguing feature that the lowest classical energy of the physical Hamiltonian $\bar{H}$ occurs when zero modes are present i.e. $\gamma \to -\frac{1}{3}$, leading, in this limit and for fixed κ, to $\Lambda_{eff} \to 0$. This, however, requires a thorough investigation of the renormalization group flow of γ and others parameters to deduce the exact behaviour of Λ_{eff} with physical energy scale, especially when matter and other forces are also taken into account.

3. Further Discussions

The inclusion of matter and other forces is rather straightforward as Standard Model fields do not couple to π; so the corresponding Hamiltonian of these fields can be appended to gravitational kinetic and potential energy terms of $\bar{H}$ in (16). With regard to the coupling of fermion fields to the vierbein and spin connections it should be pointed out although there is only spatial diffeomorphism invariance, Lorentz symmetry of the tangent space is intact, as the ADM metric $ds^2 = \eta_{AB}e_\mu^A e_\nu^B dx^\mu dx^\nu = -N^2 dt^2 + q^{\frac{1}{3}}\bar{q}_{ij}(x,t)(dx^i + N^i dt)(dx^j + N^j dt)$ is invariant under local Lorentz transformations $e_\mu'^A = \Lambda^A{}_B(x)e_\mu^B$ which do not affect metric components $g_{\mu\nu} = \eta_{AB}e_\mu^A e_\nu^B$; and a similar decomposition of the dreibein $e_i^a = e^{\frac{1}{3}}\bar{e}_i^a$ can be carried out without affecting the conjugate relation of the pair $(\pi, \ln q^{\frac{1}{3}} = \ln e^{\frac{2}{3}})$.

That there are two degrees of freedom can be ascertained in the following way: spatial diffeomorphism invariance constraints the physical momenta π_T^{ij} to be transverse ($\nabla_i\pi_T^{ij} = 0$) leaving 3 remaining degrees. These can be obtained through $\pi_T^{ij} = \pi^{ij} - \nabla^{(i}W^{j)}$, with W^i the solution for $\nabla_i\pi_T^{ij} = 0$. Having identified earlier that π is conjugate to $\ln q^{\frac{1}{3}}$ with $\delta \ln q^{\frac{1}{3}}$ as intrinsic time interval, further decomposition of $\pi^{ij} = q^{\frac{1}{3}}(\bar{\pi}_T^{ij} + \frac{\bar{q}^{ij}}{3}\pi_T)$ reveals that the symplectic potential reduces to

$$\int \pi^{ij}\delta q_{ij} = \int \pi_T^{ij}\delta q_{ij} - W^{(i}\nabla^{j)}\delta q_{ij},$$

$$= \int \left[q^{\frac{1}{3}}(\bar{\pi}_T^{ij} + \frac{\bar{q}^{ij}}{3}\pi_T)\delta\bar{q}_{ij} \right]\Bigg|_{q_{kl}^{phys.}}$$

$$= \int \left[\bar{\pi}_T^{ij}\delta\bar{q}_{ij} + \pi_T\delta \ln q^{\frac{1}{3}} \right]\Bigg|_{q_{kl}^{phys.}}; \tag{17}$$

wherein physical metric fluctuations are transverse, defined by $(\nabla^i \delta q_{ij})|_{q_{kl}^{phys.}} = 0$. This yields 2 physical canonical degrees of freedom $(\bar{q}_{ij}^{phys.}, \bar{\pi}_T^{ij})$, and an extra pair $((\ln q^{\frac{1}{3}})_{phys.}, \pi_T)$ to play the role of time and Hamiltonian (which, remarkably, is consistently tied to π_T by the dynamical equations (6) and (7)). For perturbations about any background $q_{ij}^* = q_{ij} - \delta q_{ij}$, the physical fluctuations $\delta\bar{q}_{ij}^{phys.} = (P_{ij}^{kl})^*\delta q_{kl}^{phys.}$ are the traceless $(q^{*ij}\delta\bar{q}_{ij} = 0)$ parts of the transverse modes $(\nabla^{*i}\delta q_{ij}^{phys.} = 0)$. This is just the correct accounting of perturbative graviton degrees of freedom.

Identification of a complete set of observables in theories with diffeomorphism invariance is often thought to be more than a challenging task. In this context, for a theory with HJ equation first-order-in-time, the solution is complete[11] in that it has as many integration constants (denoted earlier by α) as the number of degrees of freedom in the theory, plus an overall additive constant. These are all gauge invariant, and together with $\omega := \frac{\delta S}{\delta \alpha}$ (which express the coordinates in terms of time and the constants (α, ω)), provide general integrals of equations of motion which are well-suited to the role physical observables of diffeomorphism-invariant theories.

The symplectic 1-form $\int \pi\delta \ln q^{\frac{1}{3}} = \int \frac{2}{3}(\frac{\pi}{\sqrt{q}})\delta\sqrt{q}$ allows a different perspective. With the further restriction of $\nabla_i\pi = 0 \Leftrightarrow \frac{\pi}{\sqrt{q}} = T(t)$, York[10] interprets and deploys the scalar $\frac{\pi}{\sqrt{q}} = -\frac{TrK}{6\beta^2\kappa}$ as the "extrinsic time" variable. It follows that the Hamiltonian $\bar{H} = -\beta\pi$ is then proportional to $\sqrt{q}$, and the total energy to the volume. In our framework, York's restriction of spatially constant $\frac{\pi}{\sqrt{q}} = -\frac{\bar{H}}{\beta\sqrt{q}} = T(t)$ is a special case wherein, with vanishing shift vectors, $d\tau^2 = [d\ln q^{\frac{1}{12\beta^2\kappa T(t)}}]^2$. Although the extrinsic time variable is then invariant under spatial diffeomorphisms, it is however not invariant under 4-dimensional coordinate transformations which are supposedly symmetries of Einstein's theory. With the paradigm shift to just spatial diffeomorphism invariance, $\delta \ln q$ is well-suited to the role of physical time interval: it is a spatial diffeomorphism scalar with a gauge-invariant part which is spatially constant but can yet depend on t.

From the Schrodinger and HJ equations, the theory has the physical content of conjugate variables $(\bar{q}_{ij}, \bar{\pi}^{ij})$ subject to $H_i = 0$ evolving w.r.t. $\ln q^{\frac{1}{3\beta}}$ with effective

Hamiltonian $\bar{H}$. Thus proceeding from the action $\int [\bar{\pi}^{ij}\delta\bar{q}_{ij} - \bar{H}\delta\ln q^{\frac{1}{3\beta}}] - \int \delta N^i H_i$, and inverting for $\bar{\pi}^{ij}$ in terms of $\frac{\delta\bar{q}_{ij}}{\delta\ln q}$ from the EOM, yields the action functional as

$$S = -\int \sqrt{V}\sqrt{(\delta\ln q^{\frac{1}{3\beta}} - \pounds_{\delta\vec{N}}\ln q^{\frac{1}{3\beta}})^2 - \bar{G}^{ijkl}(\delta\bar{q}_{ij} - \pounds_{\delta\vec{N}}\bar{q}_{ij})\delta(\bar{q}_{kl} - \pounds_{\delta\vec{N}}\bar{q}_{kl})},$$

$$(18)$$

which is just the superspace proper time with $\sqrt{V}$ playing the role of "mass". This regains the generalized Baierlein-Sharp-Wheeler action[12] which has also been studied in Ref. 13 in a different situation.

In summary, the framework of a theory of gravity from quantum to classical regimes has been presented. The paradigm shift from full space-time covariance to spatial diffeomorphism invariance, together with master constraint formulation and clean decomposition of the canonical structure, permit physical dynamics and a logical resolution of the problem of time free from the burden of arbitrary lapse and gauged histories. The deep divide between quantum mechanics, which requires both the notion of time and positive semi-definite probabilities, and conventional canonical quantum gravity formulations with Klein-Gordon type WDW equations is overcome with a Schrodinger equation with positive semi-definite probability density at any instant of intrinsic time. Gauge invariant observables can be constructed from integrations constants furnished by the HJ equation which is also first order in intrinsic time. Classical space-time with direct correlation between its proper times and intrinsic time intervals emerges from constructive interference. The framework not only yields a physical Hamiltonian for Einstein's GR, but also prompts natural extensions and improvements towards a well-behaved quantum theory of gravity.

References

1. Bryce S. DeWitt, Phys. Rev. **160**, 1113 (1967).
2. J. A. Wheeler, *Superspace and the nature of quantum geometrodynamics*, in Battelle Rencontres, edited by C. M. DeWitt and J. A. Wheeler (W. A. Benjamin, New York, 1968).
3. See, for instance, C. Isham, *Canonical quantum gravity and the problem of time*, arXiv:9210011v1; K.V. Kuchar, Phys. Rev. **45**, 4443(1992), and references therein.
4. P. A. M. Dirac, *Lectures in Quantum Mechanics* (Yeshiva University Press, New York, 1964).
5. P. Horava, Phys. Rev. D **79**, 084008 (2009).
6. T. Thiemann, Class. Quantum Grav. **23**, 2211 (2006).
7. C. Soo, J. Yang and H. L. Yu, Phys. Lett. **B 701**, 275 (2011).
8. R. L. Arnowitt, S. Deser and C. W. Misner, Phys. Rev. **116**, 1322 (1959).
9. U. H. Gerlach, Phys. Rev. **177**, 1929 (1968).
10. J. W. York, Phys. Rev. Lett. **26**, 1656 (1971); ibid. **28**, 1082 (1972).
11. R. Courant and D. Hilbert, *Methods of Mathematical Physics* (Wiley Classics Edition, 1989).
12. R. F. Baierlein, D. H. Sharp and J. A. Wheeler, Phys. Rev. **126**, 1864 (1962).
13. J. Barbour, B. Foster and Niall O' Murchadha, Class. Quant. Grav. **19** 3217 (2002)

QUANTUM CORRECTIONS TO ENTROPIC GRAVITY

PISIN CHEN* and CHIAO-HSUAN WANG

Department of Physics, National Taiwan University, Taipei 10617, Taiwan
and
Leung Center for Cosmology and Particle Astrophysics,
National Taiwan University, Taipei 10617, Taiwan
** E-mail: chen@slac.stanford.edu*

The entropic gravity scenario recently proposed by Erik Verlinde reproduced Newton's law of purely classical gravity yet the key assumptions of this approach all have quantum mechanical origins. As is typical for emergent phenomena in physics, the underlying, more fundamental physics often reveals itself as corrections to the leading classical behavior. So one naturally wonders: where is $\hbar$ hiding in entropic gravity? To address this question, we first revisit the idea of holographic screen as well as entropy and its variation law in order to obtain a self-consistent approach to the problem. Next we argue that as the concept of minimal length has been invoked in the Bekenstein entropic derivation, the generalized uncertainty principle (GUP), which is a direct consequence of the minimal length, should be taken into consideration in the entropic interpretation of gravity. Indeed based on GUP it has been demonstrated that the black hole Bekenstein entropy area law must be modified not only in the strong but also in the weak gravity regime where in the weak gravity limit the GUP modified entropy exhibits a logarithmic correction. When applying it to the entropic interpretation, we demonstrate that the resulting gravity force law does include sub-leading order correction terms that depend on $\hbar$. Such deviation from the classical Newton's law may serve as a probe to the validity of entropic gravity.

Keywords: Quantum gravity; Entropic gravity; Generalized uncertainty principle; Entanglement entropy.

1. Introduction

The issue of how gravity and thermodynamics are correlated has been studied for decades, triggered by the seminal discovery by Bekenstein[1,2] on the area-law of black hole (BH) entropy and temperature. After Hawking's discovery of the BH evaporation and the interpretation of its temperature as the thermal temperature of blackbody radiation,[3] considerable efforts have been made to find the statistical interpretation of the proportionality of black hole entropy and its horizon area. See Ref. 4 and Ref. 5, for example, for a review. By now a well-accepted view is that the black hole entropy is associated with the external thermal state perceived by an observer outside the event horizon who has no access to the BH interior. Namely, the correlation between the degrees of freedom on opposite sides of the horizon results in a mixed state for observation from the outside, i.e., the 'entanglement entropy',[6,7] which depends upon the boundary properties and is referred to as the 'holographic entanglement entropy', see Ref. 8 for a review.

The inversion of the logic that describes gravity as an emergent phenomenon was first proposed by Zel'dovich[9] and Sakharov[10] in 1968, who suggested that gravity is induced by quantum field fluctuations. Invoking the area scaling property of entanglement entropy, Jacobson in 1995[11] used basic laws of thermodynamics to derive Einstein equations. In his perspective Einstein equations are now an equation of state rather than a fundamental theory. More ideas on emergent gravity have been recently proposed (See, for example, Refs. 12–15).

Similar to Jacobson's derivation of Einstein equations through thermodynamic, Verlinde treated gravity as an entropic force analogous to the restoring force of a stretched elastic polymer driven by the system's tendency towards the maximization of entropy,[15] and interestingly the Newton's law of gravitation was shown to arise. To arrive at the Newton's force law of gravity through the first law of thermodynamic $Fdx = TdS$, Verlinde first invoked the Compton wavelength of the test particle to find the change of entropy with respect to its displacement. He then invoked the holographic formula of entanglement entropy and the equipartition theorem to define the temperature experienced by the test particle. One cannot but notices that all these building blocks have quantum mechanical origin, or more specifically the presence of $\hbar$. Yet all the $\hbar$'s just get subtly cancelled and at the end a purely classical Newton's law has emerged. This is rather atypical for emergent phenomena in physics, where the underlying, more fundamental physics often reveals itself as corrections to the leading classical behavior. So one naturally wonders: where is $\hbar$ hiding in entropic gravity?

We here argue that as the concept of minimal length has been invoked in the Bekenstein entropic derivation, the generalized uncertainty principle (GUP), which is a direct consequence of the minimal length, should be taken into consideration in the entropic interpretation of gravity. It has been demonstrated that when GUP is invoked, the black hole Bekenstein entropy area law must be modified not only in the strong but also in the weak gravity regime[16] where in the weak gravity limit the GUP modified entropy exhibits a logarithmic correction. Such a log-correction is consistent with similar conclusions drawn from string theory, AdS/CFT correspondence, and loop quantum gravity considerations.[17–20] When applying it to the entropic derivations, we demonstrate that the resulting entropic gravity does include sub-leading order correction terms that depend on $\hbar$.

2. Entropic Gravity

In this section we briefly review how Verlinde arrives at classical Newton's law of gravity, and we shall clarify some points in his approach first. In Verlinde's picture, there is a spherical screen with radius R which centers at the massive source M and separates the universe into two components, one inside the sphere and the other stays outside. A particle of mass m is placed just outside the spherical screen, see Fig. 1. The spirit of this entropic gravity system is that for the test particle outside the sphere, it will interact thermodynamically with the screen on which

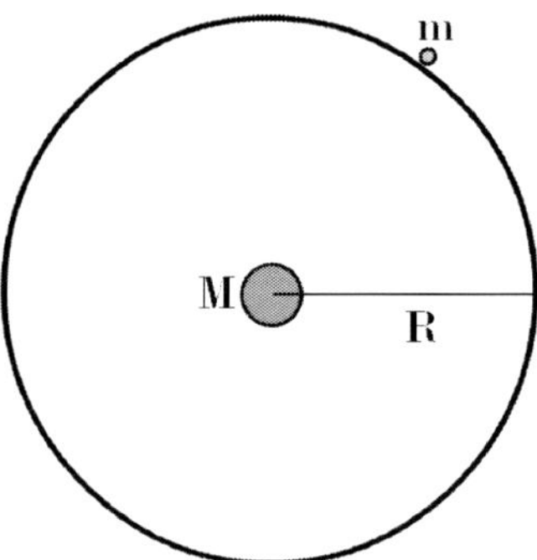

Fig. 1.   Verlinde's system: a massive source M is encoded by a spherical screen with radius R, and test particle m is placed just outside the screen.

the information of the massive source is registered. If the variation in the entropy occurs as the test particle moves, the test particle will then confronted a restoring force according to the first law of thermodynamics: $Fdx = TdS$. To find the form of this restoring force caused by the system's tendency toward the maximization of entropy, one first has to know how the entropy varies in response to the displacement of the test particle. If the temperature can also be determined, then putting these together one can arrive at the entropic force law.

Verlinde called the spherical screen in the system a 'holographic screen', on which the information content obeys 'holographic principle'. The appropriate terminology here should be 'the holographic formula for entanglement entropy' rather than 'the holographic principle' itself, please refer to Ref. 24 for detailed discussion on the distinctions. We therefore clarify the meaning of 'holographic screen' as the 'surface holding the holographic entanglement entropy property in the quantum gravity spacetime',

$$S_E = \frac{\mathrm{Area}(\Sigma)c^3 k_B}{4\hbar G} + \text{subleading terms.} \tag{1}$$

Here $\mathrm{Area}(\Sigma)$ is the area of the surface, and G is Newton's constant (see also Ref. 25 for a review). Under Verlinde's consideration, The number of degrees of freedom of the system is proportional to the area of this surface. The entropy on the screen is the entanglement entropy associated with the separation of the spacetime regions defined by this screen, and will follow the Bekenstein law $S_B = k_B c^3 A/4\hbar G$.

2.1. *Entropy variation law*

With the meaning of the holographic screen and the information content clarified we now proceed to see how the entanglement entropy changes as the test particle moves. We first review Verlinde's conjecture of the entropy variation law, in which we find some inconsistencies. Next we review a metric calculation of entropy variation law given by Fursaev.[26,27] Fursaev used two infinite surfaces as the holographic screens. We note, incidentally, that his direct calculation method was introduced[27] prior to Verlinde's conjecture We then present our calculation for spherical holographic screens following the same spirit as Fursaev. These three approaches to entropy variation law by Verlinde, Fursaev and us, respectively.

2.1.1. *Verlinde's approach*

Motivated by Bekenstein's argument "When a particle is one Compton wavelength from the horizon, it is considered to be part of the black hole," Verlinde proposed that the entropy on the screen decreases by $2\pi k_B$ when the test particle m moves one Compton wavelength away from the screen. Further assuming that the entropy varies linearly within small distance, Verlinde assumes that the variation of entropy associated with a small displacement Δx of the test particle m away from the screen is

$$\Delta S = -2\pi k_B \frac{\Delta x}{\lambda_m} = -2\pi k_B \frac{mc}{\hbar} \Delta x. \tag{2}$$

A crucial issue of this argument toward the entropy variation law has to do with the possible inconsistency in Verlinde's approach. There are two equations corresponding to the nature of entropy. One is the entropy variation law in Eq.(2) and the other is the Bekenstein law $S = A/4\,L_p^2$, which was implicitly used through the holographic formula of entanglement. A priori, these two formulas may or may not be compatible. This tension was also been pointed out by others.[28,29] While Verlinde conjectured the entropy variation law independently of the definition of entropy itself, we should point out that these two equations must be compatible since both of them are tightly related to the underlying form of entropy. In other words,

$$\Delta S = \frac{k_B c^3}{4\hbar G} \Delta A = -2\pi k_B \frac{mc}{\hbar} \Delta x \tag{3}$$

must hold under the entropic gravity framework.

Whereas Verlinde gave the argument associated with entropy without guaranteeing the compatibility, we will show that Eq.(3) can be calculated straight-forwardly in the weak gravity limit based on the knowledge of the spacetime.

2.1.2. *Fursaev's approach*

Fursaev studied the dynamics of the holographic surface by the displacement of test particle under the gravity in the weak field limit.[26] In his approach two infinite surfaces B_1 and B_2 are located at $z = z_1$ and $z = z_2$, respectively. These two surfaces separate the universe into two regions: a subspace between the two surfaces and its complement on the outside. The massive source M is placed between the spheres while a test particle is on the outside near one of the surfaces, see Fig. 2.

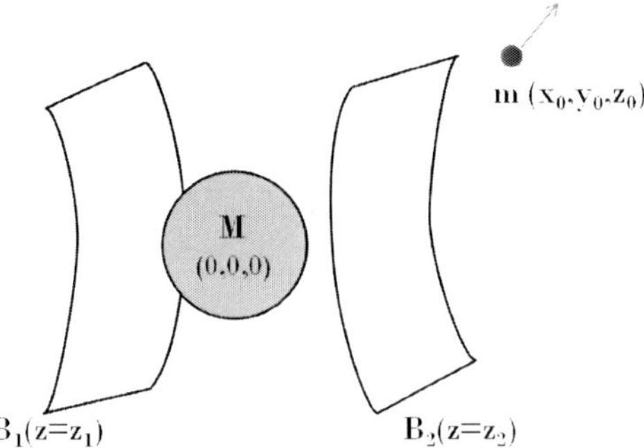

Fig. 2.   Fursaev's system: two infinite surface B_1 and B_2, with their z coordinates fixed, are placed around a massive source M. Outside the sphere is a test particle m whose displacement will affect the area of the surfaces.

In Fursaev's system there are a massive source M at $(x, y, z) = (0, 0, 0)$ and a test particle m at $(0, 0, r_0)$. The presence of the test particle should affect the above metric, which can be viewed as a back-reaction. The resultant metric in weak gravity limit is

$$ds^2 \simeq -\left(1 - \frac{2GM}{c^2\rho} - \frac{2Gm}{c^2\rho_0}\right) c^2 dt^2 + \left(1 + \frac{2GM}{c^2\rho} + \frac{2Gm}{c^2\rho_0}\right) \left(dx^2 + dy^2 + dz^2\right).$$
$$(4)$$

Here $\rho_0 = \sqrt{(x - x_0)^2 + (y - x_0)^2 + (z - z_0)^2}$. This modified metric makes intuitive sense based on the equivalence principle consideration.

A small segment of area on a infinite surface B_k, with $k = 1, 2$, is described as $da^2 = g_{xx} dx^2 g_{yy} dy^2$. The total area of surface B_k is therefore

$$A_k = \int\int dx dy \left(1 + \frac{2GM}{c^2\rho_k} + \frac{2Gm}{c^2\rho_{k,0}}\right).$$
$$(5)$$

Here $\rho_{k,0} = \sqrt{(x - x_0)^2 + (y - x_0)^2 + (z_k - z_0)^2}$ and $\rho_k = \sqrt{x^2 + y^2 + z_k^2}$.

When the distance between the test particle and the surface changes by an amount $\Delta r \approx \Delta z_0$, the surface area will correspondingly vary by an amount, to the leading order,

$$\Delta A_k = \frac{2Gm\Delta z_0}{c^2} \int\int dx dy \frac{\partial\left(1/\rho_{k,0}\right)}{\partial z_0} = -\frac{2Gm\Delta z_0}{c^2} \int\int dx dy \frac{(z_0 - z_k)}{\rho_{k,0}^3}$$
$$= -\frac{2Gm\Delta z_0}{c^2} \int_{u=0}^{\infty} \int_{\psi}^{2\pi} u\, du\, d\psi \frac{(z_0 - z_k)}{(u^2 + (z_k - z_0)^2)^{3/2}}$$
$$= -\frac{4\pi Gm\Delta z_0}{c^2} = -\frac{4\pi Gm\Delta r}{c^2},$$
$$(6)$$

where the change of variables with $x - x_0 = u\cos\psi$ and $y - y_0 = u\sin\psi$ have been made. The total variation of the area is then equal to

$$\Delta A = \Delta A_1 + \Delta A_2 = -\frac{8\pi Gm\Delta r}{c^2}.$$
$$(7)$$

With this, the entropy variation law Eq.(2) is successfully reproduced.

2.1.3. *Our approach*

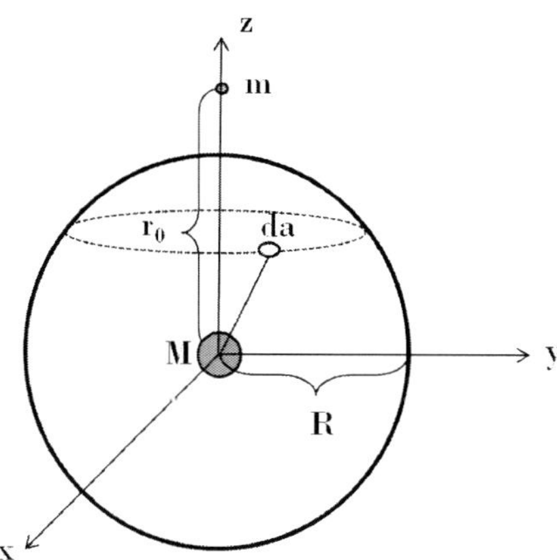

Fig. 3. Our system: a massive source M located at the origin is encoded by a spherical screen with radius R. A test particle m is placed outside the screen at a distance r_0 from the origin.

Although Fursaev has successfully reproduced the entropy variation law that is consistent with the Bekenstein law of entropy, his derivation is only valid for the special case of infinite surfaces. The more physically relevant geometry should be a sphere, on which one can introduce the uniform temperature more naturally. As discussed earlier although a 2-sphere has nonvanishing extrinsic curvature and is therefore not a minimal surface, as a special case it does satisfy the condition for the holographic entanglement entropy. We therefore follow Fursaev's approach but apply it to the variation of the surface area of a sphere, i.e., a spherical holographic screen, to recheck Verlinde's set up of the entropic gravity.

In the system of interest, there is a massive source M located at $(x, y, z) = (0, 0, 0)$ and a test particle m located at $(0, 0, r_0)$. A 2-sphere with radius $r = R$ surrounding M is a surface that possesses the holographic property of entanglement entropy, which partitions the universe into two complementary regions to which M and m separately belong (see Fig. 3). Similar to Fursaev's case, the area of the surface no longer equals to $4\pi R^2$ because of the slight warpage of the metric induced by the presence of the test particle. The metric in this system becomes

$$ds^2 \simeq -\left(1 - \frac{2GM}{\rho c^2} - \frac{2Gm}{c^2\sqrt{\rho_0{}^2 + \rho^2 - 2\rho_0\rho\cos\theta}}\right)c^2 dt^2$$

$$+ \left(1 + \frac{2GM}{\rho c^2} + \frac{2Gm}{c^2\sqrt{\rho_0{}^2 + \rho^2 - 2\rho_0\rho\cos\theta}}\right)\left(d\rho^2 + \rho^2 d\Omega^2\right), \tag{8}$$

where $\rho = R\left(1 - GM/Rc^2\right)$ and $\rho_0 = r_0\left(1 - GM/r_0c^2\right)$.

The surface area of the sphere is therefore

$$A = \int \rho^2 \sin\theta\, d\theta\, d\phi \left(1 - \frac{2GM}{c^2\rho} - \frac{2Gm}{c^2\sqrt{\rho_0{}^2 + \rho^2 - 2\rho_0\rho\cos\theta}}\right)$$

$$= 4\pi\rho^2 - \frac{8\pi GM\rho}{c^2} + A_m \;. \tag{9}$$

Here A_m is the correction of the surface area associated with the metric perturbation induced by the presence of the test particle m:

$$A_m = -\int_0^{2\pi} d\phi \int_{-1}^{1} d\cos\theta\, \frac{2Gm\rho^2}{c^2\sqrt{\rho_0{}^2 + \rho^2 - 2\rho_0\rho\cos\theta}}$$

$$= \frac{4\pi Gm\rho^2}{c^2}\frac{1}{\rho_0\rho}\sqrt{\rho_0{}^2 + \rho^2 - 2\rho_0\rho\cos\theta}\bigg|_{-1}^{1} = \left\{ \begin{array}{l} \frac{8\pi Gm\rho^2}{c^2\rho_0},\ \rho_0 > \rho\,, \\ \frac{8\pi Gm\rho}{c^2},\ \ \rho_0 < \rho\,. \end{array}\right. \tag{10}$$

Keeping the leading order in GmR/c^2 and GMR/c^2, we find

$$r_0 > R : A = 4\pi R^2 - \frac{8\pi GMR}{c^2} + \frac{8\pi GmR^2}{c^2 r_0}\;,$$

$$r_0 < R : A = 4\pi R^2 - \frac{8\pi GMR}{c^2} - \frac{8\pi GmR}{c^2}\;. \tag{11}$$

We see that while to the leading order the surface area A is equal to $4\pi R^2$, its correction induced by the presence of the test particle at r_0 is contributed by the $8\pi GmR^2/c^2 r_0$ term. (Here we assume that the test particle is outside the sphere.) Now we like to see how an infinitesimal displacement of the test particle m would further affect the surface area of the sphere. When the test particle makes a small displacement Δr_0 away from the sphere, the area will change by an amount

$$\frac{\partial A}{\partial r_0}\Delta r_0 = -\frac{8\pi GmR^2}{c^2 r_0{}^2}\Delta r_0\;. \tag{12}$$

Therefore if the entropy on the holographic screen follows the Bekenstein's law, then the entropy variation induced by the displacement of the test particle should be

$$\Delta S = k_B \frac{\Delta A}{4 l_p{}^2} = -\frac{2\pi k_B R^2}{r_0{}^2}\frac{mc}{\hbar}\Delta r_0\;. \tag{13}$$

When the test particle is just outside the sphere, that is, $R \approx r_0$, but with $R - r_0 \gg Gm/c^2$ to satisfy the weak field condition, the entropy variation on the sphere becomes

$$\Delta S = -2\pi k_B \frac{mc}{\hbar}\Delta r_0 = -2\pi k_B \frac{\Delta r_0}{\lambda_m}\;, \tag{14}$$

where $\lambda_m = \hbar/mc$ is the Compton wavelength of the test particle. We have thus obtained the entropy variation law suggested by Verlinde explicitly and consistently without invoking the ambiguous Compton wavelength argument.

2.1.4. *Temperature*

Once the entropy variation associated with the displacement of the test particle is established, the only remaining task is to define the temperature as the final step towards the entropic gravity force law. Here we suggest a heuristic derivation of the Hawking temperature for the Schwarzschild black hole in terms of its mass and entanglement entropy. In terms of black hole thermodynamics, the Hawking temperature can be viewed as the blackbody radiation temperature associated with its evaporation. In this regard, the averaged energy of a single photon is $2.7k_BT$ based on statistical mechanics. The degrees of freedom for a black hole is $N = S_B/k_B$ (see, for example, Ref.,[23] for such argument based on the holographic principle). We suppose that these degrees of freedom, N, are associated with the number of the blackbody photons and that the total energy of the blackbody radiation in turn takes up the entire rest mass energy, Mc^2, of the BH. Thus the temperature can be written as

$$T = \frac{Mc^2}{2S_B} , \tag{15}$$

up to a constant.

When the test particle makes a small displacement Δx relative to the screen, the entropy on the screen will change by an amount ΔS according to Eq.(2). The test particle will therefore experience a restoring force originated from the system's tendency to increase its entropy. This 'entropic force law' should thus follow the first law of thermodynamics $F\Delta x = T\Delta S$. Following the above assumptions and equating the area of the spherical screen to $4\pi R^2$ in the leading order approximation, we finally obtain the entropic force law that is identical to Newton's force law of gravity,

$$F = -\frac{GMm}{R^2} . \tag{16}$$

The minus sign in this force law indicates that the entropic force is oriented opposite to the direction of the displacement, just as in Newton's view of the gravitational force that is attractive between two massive sources.

While Newton's force law of gravitation seems to emerge elegantly through this entropic reasoning, we should emphasize again that both the entropy variation formula and the temperature formula involve an $\hbar$, which manifests their quantum origin. The complete cancellation between these two $\hbar$'s was due to the coincidence that both the number of bits, N, and the Bekenstein law are straight-forwardly proportional to the surface area of the holographic screen, which was fortuitous. We will argue in the next section that the entropy of entanglement is not exactly proportional to the area. As demonstrated in Ref. 16, the generalized uncertainty principle (GUP) implies a corrected formula for entanglement entropy not only in the strong gravity but also in the weak gravity regime.

3. Generalized Uncertainty Principle

In the derivation of the entropic gravity, the form of the entropy is a key ingredient. Extra care must therefore be taken in the determination of the BH entropy. With this in mind we emphasize that the holographic formulation of entanglement entropy is based on a cutoff length of the order of the Planck length. This introduction of the cutoff length implies the existence of a minimal length scale that is essential in the entropic interpretation of gravitational force. That is, the standard Heisenberg uncertainty principle, which is deduced under the Minkowski spacetime, must be modified, or generalized, when the spacetime cannot be reduced indefinitely but is subject to some minimal length scale.[30] Originally suggested in 1960s[31] based purely on the considerations of GR, GUP acquires additional theoretical support from string theory's perspective[32–36] since 1980s. One important implication of GUP is that the standard forms of Bekenstein entropy and Hawking temperature no longer hold as the size of a black hole approaches the Planck length.[16] A direct consequence of this GUP modified BH entropy is that the BH evaporation process will come to a stop when its Schwarzschild radius approaches the Planck length. As a result the Hawking evaporation should leave behind a BH remnant at Planck mass and size.

Based on GUP, it was found the black hole entropy is modified to a form different from the simple Bekenstein entropy expression. As the BH entropy is precisely the entanglement entropy on the BH horizon, we assert that under GUP the area-dependence of entanglement entropy is now expressed in the correct form as

$$S_{GUP} = 2\pi k_B \left\{ \frac{1}{\chi^2} \left(1 - \chi^2 + \sqrt{1 - \chi^2} \right) - \log \left[\frac{1}{\chi} \left(1 + \sqrt{1 - \chi^2} \right) \right] \right\}, \qquad (17)$$

here $\chi = M_p{}^2/M^2 = 16\pi L_p{}^2/A$. This expression will recover Bekenstein entropy as M_P/M goes to zero.

The correction to the semiclassical area law of black hole entropy has been extensively studied. For example a generic logarithmic term as the leading correction to black hole entropy has been found universal up to a coefficient of order unity based on string theory and loop quantum gravity considerations, see for example.[17–20] However a fundamental difference between the GUP and other approaches is that the GUP correction to the entanglement entropy as shown in Eq.(17) is an exact form, valid for both the UV and the IR limits. Therefore this GUP corrected form of entropy is also valid in the UV limit, which will be useful in our future work to extend our result to the strong gravity regime.

4. Quantum Effects in Entropic Gravity

In Verinde's entropic gravity scenario, the purely classical Newton's force law of gravitation is derived based on a quantum-mechanical and thermodynamical setup. To keep track of the underlying quantum dynamics, we now invoke generalized uncertainty principle to uncover the missing quantum contribution in entropic gravity.

Again we consider a spherical holographic screen, whose information content is defined by the GUP corrected entanglement entropy, encoding a massive source

M at the center and a test particle m placed just outside this spherical surface of radius R. The restoring force acting on the test particle m induced by the displacement from its (equilibrium) location will be derived based on the first law of thermodynamics.

First of all, the entropy variation law is directly affected by the GUP corrected form. Under GUP, the entropy varies with the surface area as

$$\Delta S = \frac{\partial S_{GUP}}{\partial A} \Delta A \, , \tag{18}$$

with $\Delta A = -8\pi G m/c^2$ as calculated before.

Next we determine the temperature on the screen. We see that the form of N is proposed on the prerequisite that the information of the space is proportional to the surface area of the screen. Under the GUP framework the entropy is no longer proportional to the area, so we compare $N = Ac^3/G\hbar$ with the Bekenstein entropy to arrive at the form for the number of bits on the screen as $N = S_{GUP}/4k_B$.

We again apply Eq.(15) to determine the temperature on the screen and find $T = 2Mc^2/Nk_B = Mc^2/2S_{GUP}$. Finally, using the first law of thermodynamics we arrive at the modified gravity force law:

$$F_{GUP} = F_N \frac{2((1+\eta) - 2\alpha(2+\eta))}{\eta(1+\eta)\left(-4\alpha + (1+\eta) + 4\alpha \log\left[2\sqrt{\alpha}(1+\eta)^{-1}\right]\right)} . \tag{19}$$

Here $F_N = GmM/R^2$ is Newton's gravitational force law, and we have introduced symbols $\eta = \sqrt{1 - 4G\hbar/c^3 R^2}$ and $\alpha = G\hbar/c^3 R^2$ to simplify the expression.

In the large distance limit where $R \gg L_p = \sqrt{G\hbar/c^3}$ and therefore $\alpha = G\hbar/c^3 R^2 \ll 1$, we can expand the force to the third order of α as

$$\begin{aligned}
F_{GUP} = F_N \{ &1 + \alpha[2 - \log\alpha] + \alpha^2[4 - 5\log\alpha + (\log\alpha)^2] \\
&+ \alpha^3[7 - 18\log\alpha + 8(\log\alpha)^2 - (\log\alpha)^3] + ...\} \, .
\end{aligned} \tag{20}$$

It is clear that this GUP-based force law recovers the classical Newton's gravitational force law in the infinite distance limit, while some subleading quantum corrections is present as long as α is finite. On the other hand these correction terms go to zero in the classical limit as $\hbar$ vanishes. These α-dependent terms, we conclude, are where $\hbar$ is hiding in entropic gravity.

5. Conclusion and Discussions

In this paper we raised the question about where $\hbar$ is hiding in entropic gravity. Through the reanalysis of the fundamental building blocks of entropic gravity, in particular the meaning of holographic screen and its associated entanglement entropy, we argued that the perfect cancellation of $\hbar$ among all the quantum mechanically originated inputs is broken if the more exact form of the BH entanglement entropy based on GUP is to replace the Bekenstein area law. Based on this we found, in the weak gravity limit, the hided $\hbar$'s in the form of logarithmic corrections to the classical Newton's law, in Eq.(20).

Before jump into Verlinde's entropic gravity formulation, we first review some features of gravity and the corresponding proposal of emergent gravity. The failure to incorporate GR into a fully quantum description and also other aspects of gravity triggered the conceptually different proposal from canonical formulation that gravity is a low energy emergent theory rather than a fundamental interaction. By describing gravity as a low energy collective effect, the hierarchy of gravity force scale and cosmological constant may also be resolved and the mysterious thermodynamics property of black holes may help to investigate the microscopic structure of spacetime.

In our attempt of seeking the missing $\hbar$'s, we reinvestigated all the fundamental assumptions in the existing derivations of entropic gravity. Within Verlinde's entropic gravity derivation, two ingredients involving entropy formula have been invoked without the guarantee of their mutual compatibility. We applied Fursaev's procedure to reproduce the leading order entropy variation in Verlinde's setup of spherical holographic screen.

While our approach manages to avoid the compatibility issue, there is a price to pay. In our alternative approach we have introduced the concept of spacetime metric and its deformation due to the presence of a massive object, which implicitly assumed the knowledge of GR. Yet the very attempt of entropic gravity is to deduce it from quantum mechanics and statistical physics alone without any prior knowledge of gravity. We are therefore at risk of a circular logic in our approach if gravity is to be interpreted as an emergent phenomenon. In this regard a more cogent and consistent argument without involving any gravity-related concept is needed towards an alternative entropy variation law, in order to assert the validity of the entropic framework of gravity as an emergent phenomena. By the same token, the existing derivations of entropic gravity also faces the similar issue since Newton's constant has been invoked as a fundamental constant from the outset instead of being a secondary, derived parameter of the theory as it should if gravity is to be an emergent phenomenon.

Under this light one can instead view our derivation of the entropic gravity not as an emergent phenomenon but as a means to deduce the 'quantum gravity force law' via the quantization of the information content on the surfaces in units of Planck area provided by GUP as well as the spacetime warpage effect in the presence of a massive particle provided by general relativity.

Although there are still rooms to improve in this line of approach to gravity, we have provided an exact form of quantum corrected entropic gravity force law based on the assumption of GUP as a fundamental input. Such quantum corrections, though minute, may serve as a probe to examine the concreteness of the entropic gravity interpretation in the the experimentally measurable scale of large distance and weak gravity limit.

Acknowledgements

We thank Debaprasad Maity, Taotao Qiu, Yen-Chin Ong , Chien-I Chiang, Nian-An Tung and Jo-Yu Kao for helpful and inspiring discussions. This research is supported by the Taiwan National Science Council (NSC) under Project No. NSC98-2811-M-002-501, No. NSC98-2119-M-002-001, and the US Department of Energy under Contract No. DE-AC03-76SF00515. We would also like to thank the NTU Leung Center for Cosmology and Particle Astrophysics for its support.

References

1. J. D. Bekenstein, *Phys. Rev. D.***7**, 2333 (1973).
2. J. D. Bekenstein, *Phys. Rev. D.***23**, 287 (1981).
3. S. W. Hawking, *Comm. Math. Phys.* **43**, 199 (1975).
4. M. Srednicki, *Phys. Rev. Lett.* **71**, 666 (1993).
5. J. Eisert, M. Cramer, and M.B. Plenio, *Rev. Mod. Phys.* **82**, 277 (2010).
6. D. Kabat, *Nucl. Phys. B* **453**, 281 (1995).
7. T. Jacobson, arXiv:gr-qc/9404039 (1994).
8. S. Ryu, and T. Takayanagi, *JHEP* **0608**, 045 (2006).
9. Ya. B. Zel'dovich, *Sov. Phys. Usp.* **11**, 381 (1968).
10. A.D. Sakharo, *Sov. Phys. Dokl.* **12**, 1040 (1968).
11. T. Jacobson, *Phys. Rev. Lett.* **75**, 1260 (1995).
12. G. E. Volovik, *Proc. Sci.*, QG-Ph 043(2007).
13. L. Sindoni, S. Liberati and F. Girelli, *AIP Conf. Proc.* **1196**, 258 (2009).
14. *Rept. Prog. Phys.* **73**, 046901 (2010).
15. E. P. Verlinde, *JHEP* **1104**, 029 (2011).
16. R. J. Adler, P. Chen, and D. I. Santiago, *Gen. Relativ. Gravit.* **33**, 2101 (2001).
17. S. N. Solodukhin, *Phys. Rev. D* **57**, 2410 (1998).
18. M, Cadoni, and M. Melis, *Entropy* **2010**, 12, 2244 (2010).
19. R. K. Kaul, and P. Majumda, *Phys. Rev. Lett.* **84**, 5255 (2000).
20. A, Ghosh, and P. Mitra, *Phys. Rev. D* **71**, 027502 (2005).
21. G. 't Hooft, arXiv:gr-qc/9310026v2 (1993).
22. L. Susskind, *J. Math. Phys.* **36**, 6377 (1995).
23. R. Bousso, *Rev. Mod. Phys.* **74**, 825 (2002).
24. P. Chen, C.-H. Wang, arXiv:gr-qc/1112.3078v1 (2011).
25. T. Nishioka, S. Ryu, and T. Takayanagi, *J. Phys. A: Math. Theor.* **42**, 504008 (2009).
26. D. V. Fursaev, arXiv:1006.2623 (2010).
27. D. V. Fursaev, *Phys. Rev. D* **77**, 124002 (2008).
28. L. Modesto and A. Randono, arXiv:1003.1998v1 (2010).
29. M. R. Setare and D. Momeni, arXiv:1004.2794v1 (2010).
30. S. Hossenfelder, Phys. Rev. D **73**, 105013(2006); S. Hossenfelder, M. Bleicher, S. Hofmann, J. Ruppert, S. Scherer, and H. Stcker, Phys. Lett. B **575**, 85-99 (2003).
31. C. A. Mead, *Phys. Rev.* **135**, B849 (1964); C. A. Mead **143**, 990 (1966).
32. G. Veneziano, *Europhys. Lett.* **2**, 199 (1986).
33. D. J. Gross and P. F. Mende,*Nucl. Phys. B* **303**, 407 (1988).
34. D. Amati, M. Ciafolini, and G. Veneziano, *Phys. Lett. B* **216**, 41 (1989).
35. K. Konishi, G. Paffuti and P. Provero, *Phys. Lett. B* **234**, 276 (1990).
36. E. Witten, *Phys. Today*, **Apr. 24** (1996).

BLACK HOLES AND THE GENERALIZED UNCERTAINTY PRINCIPLE

B. J. CARR

School of Physics and Astronomy, Queen Mary, University of London,
Mile End Road, London E1 4NS, England
E-mail: B.J.Carr@qmul.ac.uk

We propose a new way in which black holes connect macrophysics and microphysics. The Generalized Uncertainty Principle suggests corrections to the Uncertainty Principle as the energy increases towards the Planck value. It also provides a natural transition between the expressions for the Compton wavelength below the Planck mass and the black hole event horizon size above it. This suggests corrections to the the event horizon size as the black hole mass falls towards the Planck value, leading to the concept of a Generalized Event Horizon. Extrapolating this expression below the Planck mass suggests the existence of a new kind of black hole, whose size is of order its Compton wavelength. Recently it has been found that such a black hole solution is permitted by loop quantum gravity, its unusual properties deriving from the fact that it is hidden behind the throat of a wormhole. This has important implications for the formation and evaporation of black holes in the early Universe, especially if there are extra spatial dimensions.

Keywords: black holes; uncertainty principle; quantum evaporation; Planck relics.

1. Introduction

The crucial role of black holes in both macrophysics and astrophysics is summarized in Fig. 1. This shows the Cosmic Uroborus (the snake eating its own tail), with the various scales of structure in the universe indicated along the side. It can be regarded as a sort of "clock" in which the scale increases by a factor of 10 for each minute – from the Planck scale (10^{-33}cm) at the top left to the scale of the observable universe (10^{27}cm) at the top right. In between are quarks, protons, atoms and molecules (in the micro domain on the left), mountains, planets, stars and galaxies (in the macro domain on the right), and humans (at the bottom). The head meets the tail at the big bang because at the largest cosmological distances. one is peering back to an epoch when the universe was very small, so the very large meets the very small there. There might also be extra spatial dimensions at the top of the Uroborus, reflecting the higher dimensionality of the early universe.

The various types of black holes are also indicated in Fig. 1. These are labelled by their mass, this being proportional to their size. On the right are the well established astrophysical black holes: stellar remnants ($1M_\odot$), supermassive back holes in galactic nuclei ($10^6 M_\odot$) and quasars ($10^9 M_\odot$), and in some sense the universe it-

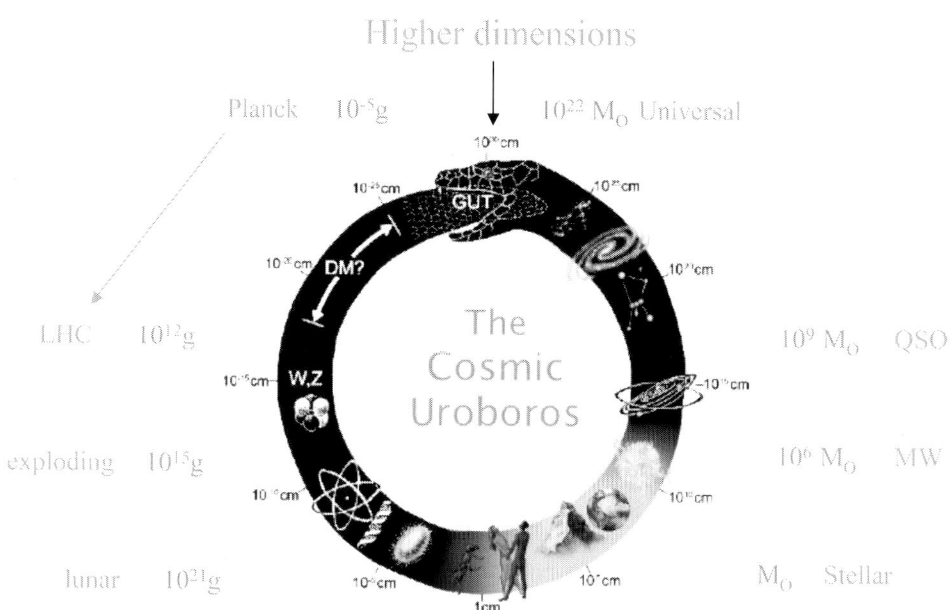

Fig. 1.   Black holes as a link between macro and micro physics.

self ($10^{22} M_\odot$). On the left are the more speculative "primordial" black holes. These formed in the compression of the big bang and could span the range from Planck relics (10^{-5}g) to the holes evaporating at the present epoch (10^{15}g) to the sort of lunar-mass holes (10^{27}g) which might provide the dark matter. If the extra spatial dimensions at the top of the Uroborus are "large" rather than just having the Planck scale, then the quantum gravity scale might be reduced to the LHC scale, in which case black holes might be produced in accelerators (10^{12}g). The remarkable symmetry between the black holes on the two sides of Fig. 1 illustrates the way in which they link the micro and macro domains.

In this talk, I will argue the black holes also link the macro and micro domains through what is termed the Generalized Uncertainty Principle (GUP). A key feature of the micro domain is the Uncertainty Principle, which implies that an object of mass M cannot be localized on a scale less than its Compton wavelength $R = \hbar/(Mc)$, while a key feature of the macro domain is the black hole, which arises when a mass M falls within the Schwarzschild radius $R = 2GM/c^2$. In the (M, R) diagram these two lines intersect at the Planck scales, $M_P \sim 10^{-5}$g and $R_P \sim 10^{-33}$cm.

The Compton and Schwarzschild lines presumably change their form as one approaches the Planck point due to quantum gravity effects. The GUP describes the modification to the Compton line as M increases towards M_P from *below* and has been the focus of considerable study in a series of papers by Pisin Chen and his collaborators.[1] The modification to the Schwarzschild line as M decreases towards M_P from *above* – which we term the Generalized Event Horizon (GEH) – has attracted less attention but the two ideas are closely related. This is because the generalized Compton scale asymptotes to the Schwarzschild form for $M \gg M_P$, while the GEH asymptotes to the Compton form for $M \ll M_P$. The possibility of unifying the Compton and Schwarzschild expressions in a single formula suggests some profound connection between the Uncertainty Principle and black holes. We describe this as the Black Hole Uncertainty Principle (BHUP) correspondence[2] and will argue that it is an explicit feature of loop quantum gravity (LGQ).

2. Generalized Uncertainty Principle/Generalized Event Horizon

The Heisenberg Uncertainty Principle (HUP) implies that the uncertainty in the position and momentum of a particle must satisfy

$$\Delta x > \hbar/(2\Delta p) \,, \tag{1}$$

where the factor of 2 must be included if one interprets the uncertainties as root-mean-squares.[3] It is well known that one can heuristically understand this result as reflecting the momentum transferred to the particle by the probing photon. Since the momentum of a particle of mass M is bounded by Mc, an immediate implication is that one cannot localize a particle of mass M on a scale less $\hbar/(2Mc)$. An important role is therefore played by the reduced Compton wavelength,

$$R_C = \hbar/(Mc) \,. \tag{2}$$

Formally, this can be obtained from the HUP with the substitution $\Delta x \to R$ and $\Delta p \to cM$ but without the factor of 2. In the (M, R) diagram of Fig. 2, the region corresponding to $R < R_C$ might be regarded as the "quantum domain" in the sense that the classical description breaks down there.

An object of mass M forms a black hole if it is compressed enough to form an event horizon. For a spherically symmetric object, general relativity implies that this corresponds to the Schwarzschild radius,

$$R_S = 2GM/c^2 \,. \tag{3}$$

The region $R < R_S$ might be regarded as the "relativistic domain" in the sense that there is no stable classical configuration in this part of Fig. 2. The boundaries given by Eqs. (2) and (3) intersect at around the Planck scales,

$$R_P = \sqrt{\hbar G/c^3} \sim 10^{-33}\text{cm}, \quad M_P = \sqrt{\hbar c/G} \sim 10^{-5}\text{g},$$

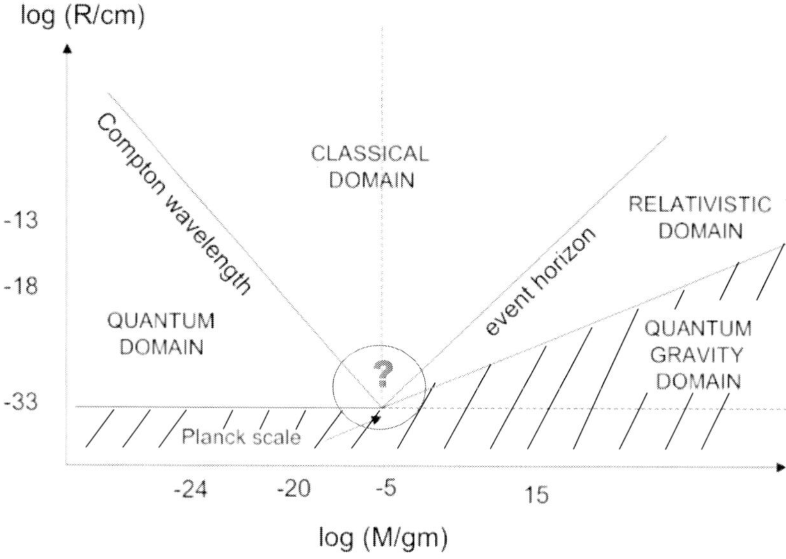

Fig. 2.   The division of the (M, R) diagram into the classical, quantum, relativistic and quantum gravity domains. The boundaries are specified by the Planck density, the Compton wavelength and the Schwarzschild radius. The GUP smoothes the transition between the last two.

and they divide the (M, R) diagram in Fig. 2 into three regimes (quantum, relativistic, classical). However, there are several other interesting lines in this diagram. The vertical line $M = M_P$ is often assumed to mark the division between elementary particles ($M < M_P$) and black holes ($M > M_P$), because one usually requires a black hole to be larger than its own Compton wavelength. The horizontal line $R = R_P$ is significant because a simple heuristic argument suggests that quantum fluctuations in the metric should become important below this.[4] Quantum gravity effects should also be important whenever the density exceeds the Planck value,

$$\rho_P = c^5/(G^2 \hbar) \sim 10^{94} \mathrm{g\,cm}^{-3} \,, \tag{4}$$

corresponding to the sorts of curvature singularities associated with the big bang or the centres of black holes. This implies

$$R < (3M/4\pi\rho_P)^{1/3} \sim (M/M_P)^{1/3} R_P \,, \tag{5}$$

which is well above the $R = R_P$ line in Fig. 2 for $M \gg M_P$. So one might regard the combination of this line and the $R = R_P$ line as specifying the boundary of the "quantum gravity" domain, as indicated by the shaded region in Fig. 2.

Although the Compton and Schwarzschild boundaries correspond to straight lines in the logarithmic plot of Fig. 2, this form presumably breaks down near the Planck point. Chen and colleagues[1] have discussed how the quantum boundary in Fig. 2 might be modified as one approaches the Planck point from the left. They argue that the Uncertainty Principle should take the GUP form

$$\Delta x > \hbar/\Delta p + \alpha R_P^2 (\Delta p/\hbar) , \tag{6}$$

where α is a dimensionless constant which depends on the particular model and the factor of 2 in the first term has been dropped. They offer a series of heuristic arguments for the second term in Eq. (6) but it essentially represents the gravitational effect of the probing photon rather than its momentum effect. The contrast between the HUP and GUP is indicated in Fig. 3.

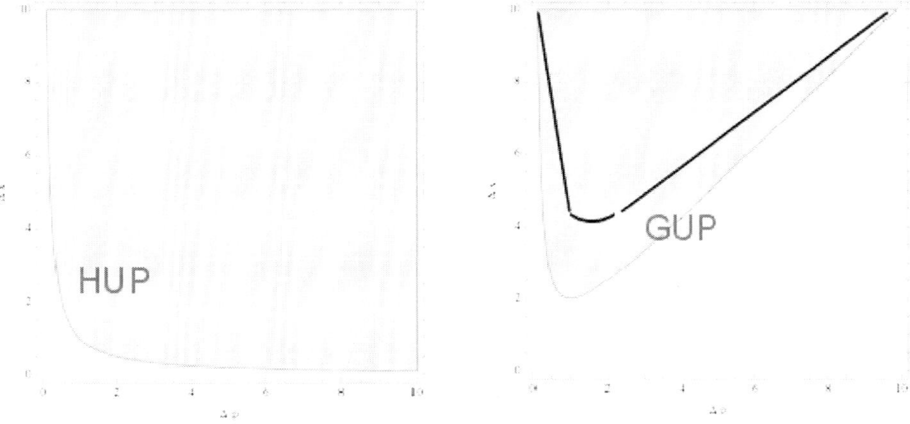

Fig. 3. Δx versus Δp for the HUP (left) and the GUP (right) in its linear (lower curve) and quadratic (upper curve) forms.

Variants of Eq. (6) can be found in other approaches to quantum gravity, such as non-commutative quantum mechanics or general minimum length considerations.[5] The GUP can also be derived in LQG because of polymer corrections in the structure of spacetime[6] and it is implicit in some approaches to the problem of quantum decoherence.[7] Finally, an expression resembling Eq. (6) arises in string theory,[8] although the second term cannot correspond to a black hole for $M \gg M_P$ because the string is too elongated to form an horizon.

The second term on the right of Eq. (6) is much smaller than the first term for $\Delta p \ll M_P c$. Since it can be written as $\alpha G(\Delta p)/c^3$, it roughly corresponds to the Schwarzschild radius for an object of mass $\Delta p/c$. Indeed, if we rewrite Eq. (6) using the same substitution $\Delta x \to R$ and $\Delta p \to cM$ as before, it becomes

$$R > \hbar/(Mc) + \alpha GM/c^2 \, . \tag{7}$$

The expression on the right might be regarded as a generalized Compton wavelength and is most naturally written in the form

$$R'_C = \frac{\hbar}{Mc} \left[1 + \alpha(M/M_P)^2\right] \, . \tag{8}$$

Here the second term can be regarded as a correction as one approaches the Planck point from the left, this being small for $M \ll M_P$.

One can also apply Eq. (8) for $M \gg M_P$ and it is striking that in this regime it asymptotes to the Schwarzschild form, apart from a numerical factor. This suggests that there is a different kind of positional uncertainty for an object larger than the Planck mass, related to the existence of black holes. This is not unreasonable since the Compton wavelength is below the Planck scale (and hence meaningless) here and also an outside observer cannot localize an object on a scale smaller than its Schwarzschild radius.

The GUP also has important implications for the black hole horizon size, as can be seen by examining what happens as one approaches the intersect point from the right. In this limit, it is natural to write Eq. (7) as

$$R > R'_S = \frac{\alpha GM}{c^2} \left[1 + \frac{1}{\alpha}(M_P/M)^2\right] \tag{9}$$

and this represents a small perturbation to the Schwarzschild radius for $M \gg M_P$ if one assumes $\alpha = 2$. Unfortunately, there is no reason for anticipating $\alpha = 2$ in the heuristic derivation of the GUP. Nor is it clear why a more precise calculation (within the context of a specific theory of quantum gravity) would yield this value.

This motivates an alternative approach in which the free constant in Eq. (7) is associated with the first term rather than the second. After all, the factor of 2 in the expression for the Schwarzschild radius is precise, whereas the coefficient associated with the Compton term is somewhat arbitrary. Thus one might rewrite Eq. (8) as

$$R'_C = \frac{\beta \hbar}{Mc} \left[1 + \frac{2}{\beta}(M/M_P)^2\right] \tag{10}$$

for some constant β, so that Eq. (9) becomes

$$R'_S = \frac{2GM}{c^2}\left[1 + \frac{\beta}{2}(M_P/M)^2\right].\tag{11}$$

This might then be regarded as defining a Generalized Event Horizon (GEH). Indeed, the mathematical equivalence of R'_C and R'_S underlies what we have termed the BHUP correspondence. The simplest scenario would have $\alpha = 2$ and $\beta = 1$ but there is no compelling reason for either expecting or demanding this.

An important caveat is that Eq. (6) assumes that the two uncertainties add linearly. On the other hand, since they are independent, it might be more natural to assume that they add quadratically:

$$\Delta x > \sqrt{(\hbar/\Delta p)^2 + (\alpha R_P^2 \Delta p/\hbar)^2}\,.\tag{12}$$

This corresponds to the upper curve on the right of Fig. 3. While the heuristic arguments indicate the form of the two uncertainty terms, they do not specify how one combines them. We refer to Eqs. (6) and (12) as the *linear* and *quadratic* forms of the GEP. The latter corresponds to a generalized Compton wavelength

$$R'_C = \sqrt{(\hbar/Mc)^2 + (\alpha GM/c^2)^2}\,.\tag{13}$$

As in the linear case, it might be more natural to use the unified expression

$$R'_C = R'_S = \sqrt{(\beta\hbar/Mc)^2 + (2GM/c^2)^2}\,.\tag{14}$$

leading to the approximations

$$R'_C \approx \frac{\beta\hbar}{Mc}\left[1 + \frac{2}{\beta^2}(M/M_P)^4\right]\tag{15}$$

for $M \ll M_P$ and

$$R'_S \approx \frac{2GM}{c^2}\left[1 + \frac{\beta^2}{8}(M_P/M)^4\right]\tag{16}$$

for $M \gg M_P$. These might be compared to the *exact* expressions in the linear case, given by Eqs. (10) and (11). Again, the simplest model would have $\alpha = 2$ and $\beta = 1$ but we will regard α and β as free parameters here. We will see below that a model inspired by LQG permits the existence of a new type of black hole whose horizon size has precisely the form (14) but without $\beta = 1$.

More generally, the BHUP correspondence might allow any unified expression for $R'_C(M)$ and $R'_S(M)$ which has the asymptotic behaviour

$$R'_C \equiv R'_S \approx \begin{cases} \beta\hbar/(Mc) & (M \ll M_P) \\ 2GM/c^2 & (M \gg M_P)\,. \end{cases}\tag{17}$$

One could envisage many unified expressions satisfying this condition. However, while both the linear and quadratic forms of the GUP have some physical basis, more general expressions would only be well motivated if some final theory of quantum gravity specifically predicted them.

3. Loop Black Holes

Loop Quantum Gravity (LQG) is based on a canonical quantization of the Einstein equations written in terms of the Ashtekar variables[9] . One important consequence of this is that the area is quantized, with the smallest possible value being

$$A_{\min} = 4\pi\sqrt{3}\gamma R_P^2 \,, \tag{18}$$

where γ is called the Immirzi parameter. This exact expression should not be taken too seriously, so we parametrize our ignorance with another constant ζ and use

$$a_o = A_{\min}/8\pi = \sqrt{3}\,\gamma\zeta R_P^2/2 \,. \tag{19}$$

The expected values of γ and ζ are of order 1 but the precise choice is not crucial. The remaining unknown constant is the dimensionless polymeric parameter δ. Together with a_0, this determines the deviation from classical theory.

One version of LQG, using the mini-superspace approximation, gives rise to cosmological solutions which resolve the initial singularity problem.[10] Another version gives a black hole solution, known as the loop black hole (LBH),[11] which has a self-duality property that removes the singularity and replaces it with another asymptotically flat region. The metric in this solution depends only on the combined dimensionless parameter $\epsilon \equiv \delta\gamma$, which must be small if quantum gravitational corrections are relevant only when the curvature is in the Planckian regime. More precisely, the metric can be expressed as

$$ds^2 = -G(r)c^2 dt^2 + \frac{dr^2}{F(r)} + H(r)d\Omega^{(2)}, \tag{20}$$

with $d\Omega^{(2)} = d\theta^2 + \sin^2\theta d\phi^2$ and

$$G(r) = \frac{(r - r_+)(r - r_-)(r + r_*)^2}{r^4 + a_o^2} \,,$$

$$F(r) = \frac{(r - r_+)(r - r_-)r^4}{(r + r_*)^2(r^4 + a_o^2)} \,,$$

$$H(r) = r^2 + \frac{a_o^2}{r^2} \,. \tag{21}$$

Here $r_+ = 2Gm/c^2$ and $r_- = 2GmP^2/c^2$ are the outer and inner horizons, respectively, and $r_* \equiv \sqrt{r_+ r_-} = 2GmP/c^2$, where m is the black hole mass and

$$P \equiv \frac{\sqrt{1 + \epsilon^2} - 1}{\sqrt{1 + \epsilon^2} + 1} \tag{22}$$

is the polymeric function. For $\epsilon \ll 1$, we have $P \approx \epsilon^2/4 \ll 1$, so $r_- \ll r_* \ll r_+$. Since $g_{\theta\theta}$ is not exactly r^2 in the above metric, r is only the usual radial coordinate asymptotically. In the limit $r \to \infty$ one has

$$G(r) \to 1 - \frac{2GM}{c^2 r}(1 - \epsilon^2) \, ,$$
$$F(r) \to 1 - \frac{2GM}{c^2 r} \, ,$$
$$H(r) \to r^2 \, , \tag{23}$$

so the deviations from the Schwarzschild solution are of order $GM\epsilon^2/(c^2 r)$. Here

$$M = m(1 + P)^2 \tag{24}$$

is the ADM mass, which is determined solely by the metric at flat asymptotic infinity and might be associated with the quantity M appearing in our earlier discussion.

The expression for $H(r)$ shows that the more physical radial coordinate is

$$R \equiv \sqrt{r^2 + \frac{a_o^2}{r^2}} \tag{25}$$

in the sense that this measures the proper circumferential distance. As r decreases from ∞ to 0, R first decreases from ∞ to a minimum value of $\sqrt{2a_0}$ at $r = \sqrt{a_0}$ and then increases again to ∞. In particular, the value of R associated with the event horizon is

$$R_{EH} = \sqrt{H(r_+)} = \sqrt{\left(\frac{2Gm}{c^2}\right)^2 + \left(\frac{a_o c^2}{2Gm}\right)^2} \, . \tag{26}$$

This is equivalent to Eq. (14), asymptoting to the Schwarzschild radius for $m \gg M_P$ and to the Compton wavelength for $m \ll M_P$ if we put $\beta = \sqrt{3}\gamma\zeta/4$.

The important physical implication of Eq. (25) is that central singularity of the Schwarzschild solution is replaced with another asymptotic region, so the collapsing matter bounces and the black hole becomes part of a wormhole. Equation (25) has three important cosequences: (1) it removes the singularity; (2) it permits the existence of black holes with $m \ll M_P$; and (3) it allows a unified expression for the Compton and Schwarzschild scales. Indeed, it seems remarkable that the purely geometrical condition (26) implies the quadratic version of the GUP given by Eq. (13).

4. GUP and Black Hole Thermodynamics

Let us first recall the link between black hole radiation and the HUP.[12] This arises because we can obtain the black hole temperature for $M \gg M_P$ by identifying Δx with the Schwarzschild radius and Δp with some multiple of the black hole temperature:

$$kT = \eta c \Delta p = \frac{\eta \hbar c}{\Delta x} = \frac{\eta \hbar c^3}{2GM} \, . \tag{27}$$

This gives the precise Hawking temperature if we take $\eta = 1/(4\pi)$. The second equality in Eq. (27) relates to the emitted particle and assumes that Δx and Δp

satisfy the HUP. The third equality relates to the black hole and assumes that Δx is the Schwarzschild radius. Both these assumptions require $M \gg M_P$ but the GUP and GEH suggest how they should be modified for $M \ll M_P$.

Adler *et al.*[1] calculate the modification required if Δp and Δx are related by the linear form of the GUP rather than the HUP. However, they still associate Δx with the Schwarzschild radius. In this case, using the α-formalism, Eq. (27) is replaced with

$$\frac{2GM}{c^2} = \frac{\hbar \eta c}{kT} + \frac{\alpha R_P^2 kT}{\hbar \eta c} , \tag{28}$$

which leads to a temperature

$$T = \frac{\eta M c^2}{\alpha k} \left(1 - \sqrt{1 - \frac{\alpha M_P^2}{M^2}} \right) . \tag{29}$$

This implies

$$T \approx \frac{\eta \hbar c^3}{2GkM} \left[1 + \frac{\alpha M_P^2}{4M^2} \right] \tag{30}$$

for $M \gg M_P$, which just represents a small perturbation to the standard Hawking temperature. However, the exact expression becomes complex when M falls below $\sqrt{\alpha}\, M_P$. Adler *et al.* infer that evaporation ceases at about the Planck mass, leading to stable relics. So the GUP stabilizes the ground state of a black hole just as the HUP stabilizes the ground state of a hydrogen atom.

The expression for the black hole temperature is different in the present (LQG) analysis for *two* reasons. First, the relationship between Δp and Δx is modified to the quadratic form. If one still associates Δx with the Schwarzschild radius, Eq. (27) is replaced with

$$\frac{2GM}{c^2} = \left[\left(\frac{\hbar \eta c}{kT} \right)^2 + \left(\frac{\alpha R_P^2 kT}{\hbar \eta c} \right)^2 \right]^{1/2} . \tag{31}$$

This leads to

$$T = \frac{\sqrt{2}\, \eta M c^2}{\alpha k} \left(1 - \sqrt{1 - \frac{\alpha^2}{4} \left(\frac{M_P}{M} \right)^4} \right)^{1/2} , \tag{32}$$

which implies

$$T \approx \frac{\eta \hbar c^3}{2GkM} \left[1 + \frac{\alpha^2}{32} \left(\frac{M_P}{M} \right)^4 \right] \tag{33}$$

for $M \gg M_P$. Therefore the deviation from the Hawking prediction is smaller than implied by Eq. (30) but the exact expression still goes complex for $M < \sqrt{\alpha/2}\, M_P$ (i.e. at a mass smaller by $\sqrt{2}$ than before).

However, there is a second discrepancy with Alder *et al.* in that the BHUP correspondence suggests Δx is given by Eq. (26) rather than $2GM/c^2$. This only has a small effect for $M \gg M_P$ but it makes a major *qualitative* difference for $M \ll M_P$ because Δx then scales as M^{-1} rather than M. This means that the black hole temperature no longer goes complex below the Planck mass, so one must consider the form of the temperature in the sub-Planckian regime.

We first calculate the temperature on the assumption that the GUP is given by Eq. (12) and the GEH by Eq. (14), i.e. we assume the parameters α and β are *independent*. In this case, Eq. (27) is replaced with

$$\left[\left(\frac{\hbar\eta c}{kT}\right)^2 + \left(\frac{\alpha R_P^2 kT}{\hbar\eta c}\right)^2\right]^{1/2} = \left[\left(\frac{\hbar\beta}{Mc}\right)^2 + \left(\frac{2GM}{c^2}\right)^2\right]^{1/2}. \tag{34}$$

This leads to Eq. (32) except that the last term becomes

$$\left(1 + \frac{\beta^2 M_P^4}{4M^4} - \sqrt{1 + \frac{(2\beta^2 - \alpha^2)}{4}\left(\frac{M_P}{M}\right)^4 + \frac{\beta^4}{16}\left(\frac{M_P}{M}\right)^8}\right)^{1/2}. \tag{35}$$

This is real for all M providing $\alpha < 2\beta$. The temperature becomes

$$T \approx \frac{\eta\hbar c^3}{2GkM}\left[1 + \left(\frac{\alpha^2 - 4\beta^2}{32}\right)\left(\frac{M_P}{M}\right)^4\right] \tag{36}$$

in the limit $M \gg M_P$ and

$$T \approx \frac{\eta Mc^2}{k\beta}\left[1 + \left(\frac{\alpha^2 - 4\beta^2}{2\beta^4}\right)\left(\frac{M}{M_P}\right)^4\right] \tag{37}$$

in the limit $M \ll M_P$. Since one always has $T < T_P$, there is a sense in which the quantum gravity domain is avoided altogether.

The situation is much simplified if one imposes the BHUP correspondence, as we claim is most natural, because one must then use the same expression for both the GUP and GEH. In this case, using the β-formalism, Eq. (27) is replaced with

$$\left[\left(\frac{\hbar\beta\eta c}{kT}\right)^2 + \left(\frac{2R_P^2 kT}{\hbar\eta c}\right)^2\right]^{1/2} = \left[\left(\frac{\hbar\beta}{Mc}\right)^2 + \left(\frac{2GM}{c^2}\right)^2\right]^{1/2}, \tag{38}$$

which yields the *exact* solutions

$$kT = \frac{\hbar\beta\eta c^3}{2GM}, \quad kT = \eta Mc^2, \tag{39}$$

with no small correction terms for large and small M. The same exact solutions are obtained if one puts $\alpha = 2\beta$ in Eq. (34). The first solution in Eq. (39) is the exact Hawking temperature providing one puts $\eta\beta = 1/(4\pi)$, which might be regarded as supporting the BHUP correspondence. However, one must cross over to the second solution below $M = \sqrt{\beta/2}\,M_P$ in order to avoid the temperature going above T_P. The second solution can be obtained heuristically by putting $\Delta x \approx \beta\hbar/(Mc)$ fin Eq. (27). Since this is less than the Planck temperature, the second equality still applies to a good approximation, as required for consistency.

The different M-dependences for $M \ll M_P$ and $M \gg M_P$ can be understood as arising because there are two different asymptotic spaces in the LBH solution, described by the coordinates r and R, so the quantity Δx needs to be specified more precisely. Since Eq. (25) implies the differential relation

$$\frac{\Delta R}{\Delta r} \approx \begin{cases} 1 & (r \gg r_P) \\ (r/R_P)^{-2} & (r \ll r_P), \end{cases} \tag{40}$$

putting $r = 2GM/c^2$ gives

$$\frac{(\Delta x)_R}{(\Delta x)_r} \approx \begin{cases} 1 & (M \gg M_P) \\ (M/M_P)^{-2} & (M \ll M_P). \end{cases} \tag{41}$$

This explains the different mass dependence of the temperature in the two regimes.

However, one can use another argument which gives a different result in the sub-Planckian regime. Since the temperature is determined by the black hole's surface gravity,[12] Eq. (17) suggests

$$T \propto \frac{GM}{R_S'^2} \propto \begin{cases} M^{-1} & (M \gg M_P) \\ M^3 & (M \ll M_P), \end{cases} \tag{42}$$

so this should scale as M^3 rather than M for $M \ll M_P$. More precisely, the surface gravity at the outer horizon of the LBH solution is

$$\kappa_+ = \frac{4G^3 m^3 c^4 (1 - P^2)}{16 G^4 m^4 + a_o^2 c^8} . \tag{43}$$

The mass dependence of the temperature $\hbar \kappa_+ / (2\pi k c)$ is thus as illustrated in Fig. 4.

Both equations predict that the temperature of a black hole deviates from the Hawking expression when its mass falls below M_P and that it never goes above T_P. But which prediction is correct? The apparent inconsistency between Eq. (39) and Eq. (42) for $M \ll M_P$ may arise because not all the emission from a sub-Planckian black hole can escape through the wormhole throat to reach our asymptotic infinity. Replacing R_S' with R_P in Eq. (42) implies that the temperature associated with the wormhole throat itself is $T \propto M$, which corresponds to the prediction of Eq. (42).

In either case, one has the important implication that there is no value of M for which T becomes zero. However, there are still *effectively* stable relics since the evaporation timescale becomes longer than the age of the universe for sufficiently small M. Indeed, as illustrated in Fig. 4, the temperature falls below the background radiation density – suppressing evaporation altogether – below some critical mass and such relics might even be dark matter candidates.[11]

5. Effects of Higher Dimensions

The black hole boundary in Fig. 2 assumes there are three spatial dimensions but many theories suggest that the dimensionality could increase on sufficiently small scales. Either the extra dimensions are compactified or matter is confined to a brane

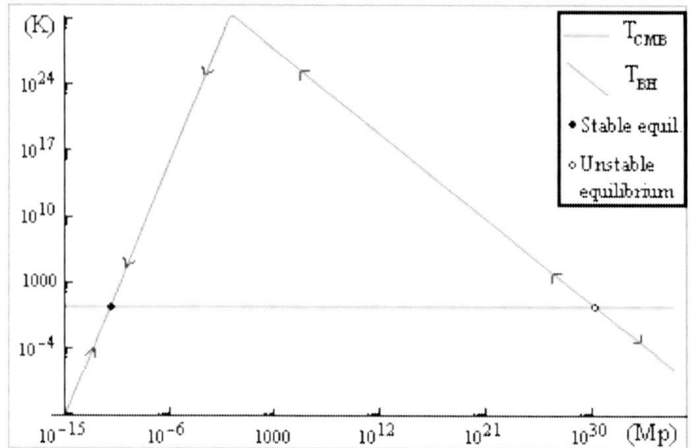

Fig. 4.   Comparing black hole temperature predicted by GUP with CMB temperature.

of finite thickness in the extra dimension due to warping. In both case, the extra dimensions are associated with some scale R_C and only detectable for $R < R_C$. If the number of extra spatial dimensions is n, then the gravitational force between masses m_1 and m_2 is

$$F_{\text{grav}} = \frac{G_D m_1 m_2}{R^{2+n}} \tag{44}$$

where G_D is the higher-dimensional gravitational constant. This becomes

$$F_{\text{grav}} = \frac{G m_1 m_2}{R^2} \quad \text{with} \quad G = \left(\frac{G_D}{R_C^n} \right) \tag{45}$$

for $R > R_C$, so one recovers the inverse-square law there. The gravitational constants at large and small scales are different because of the dilution effect of the extra dimensions. The effective Planck mass (M_P') and Planck length (R_P') in the higher-dimensional space are related to the 4-dimensional Planck scales by

$$M_P' \sim M_P \left(\frac{R_P}{R_C} \right)^{n/(n+2)} , \quad R_P' \sim R_P \left(\frac{R_C}{R_P} \right)^{n/(n+2)} , \tag{46}$$

so $M_P' \sim M_P$ and $R_P' \sim R_P$ for $R_C \sim R_P$ but $M_P' \ll M_P$ and $R_P' \gg R_P$ for $R_C \gg R_P$. For some values of R_C and n, M_P' would be as low as 1 TeV, in which case quantum gravity effects could be detectable in accelerator experiments.

An important implication of extra spatial dimensions is that Eq. (3) no longer applies. If black holes with mass below $M_C = c^2 R_C/(2G)$ are assumed to be spherically symmetric in the higher dimensional space, Eq. (3) must be replaced with

$$R_S = R_C \left(\frac{M}{M_C} \right)^{1/(n+1)} \tag{47}$$

or $M \ll M_P$, so the slope of the black hole boundary in Fig. 2 becomes shallower, as indicated in Fig. 5 for various values of n. The intersect with the Compton boundary

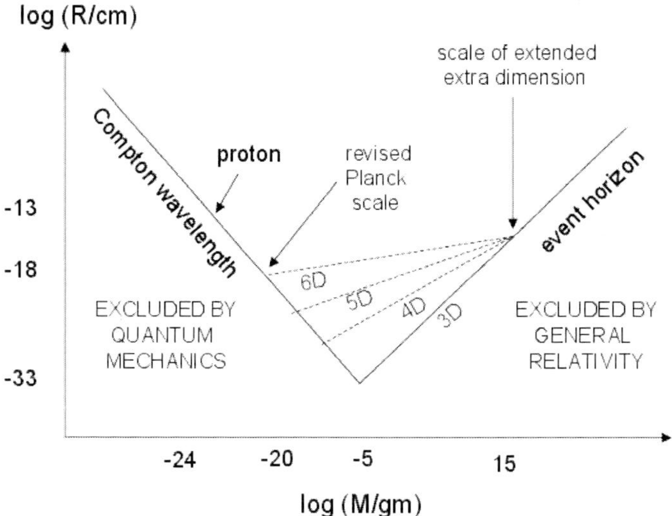

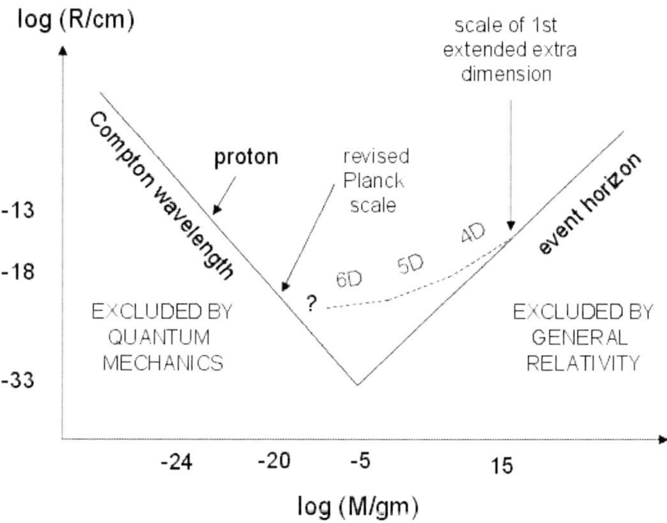

Fig. 5.   Modification to Fig. 2 for extra dimensions with same or hierarchy of scales.

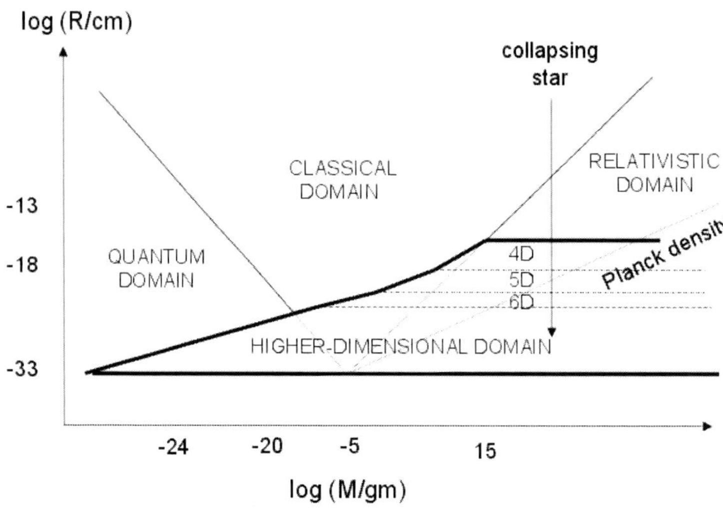

Fig. 6.   Showing classical, quantum, relativistic and higher-dimensional $(M, R$ domains.

just corresponds to the revised Planck scales (46). If the GUP still corresponds to the extension of the Compton line to the Schwarzschild line, the generalized Compton expression in the higher-dimensional case should become

$$R'_C = \frac{\hbar}{Mc} \left[1 + \left(\frac{M}{M'_P} \right)^{(n+2)/(n+1)} \right] \tag{48}$$

rather than Eq. (8). This assumes all the extra dimensions have the same size. Alternatively, one could have a hierarchy of compactification scales, $R_i = \alpha_i R_P$, with decreasing α_i and this situation is also represented in Fig. 5.

If one wants $M'_P \sim 1$ TeV, allowing detectable effects at the LHC, one needs $R_C \sim 10^{(32/n)-17}$cm. The case $n = 1$ gives $R_C \sim 10^{16}$ cm and is clearly excluded but $n = 2$ gives $R_C \sim 0.1$ cm which is marginally allowed. Indeed, it is intriguing that this is roughly the dark energy scale $R_{DE} \sim \sqrt{R_U R_P} \sim 100\,\mu$m. One would have $n = 7$ in M-theory if every extra dimension had the size of a proton. Since this is also the size of a black hole evaporating at the present epoch, this means that such holes are *necessarily* higher-dimensional, so one has the situation indicated in Fig. 6. This means that the standard limits on their number density and the likelihood of their detectability no longer apply.

References

1. R J Adler and D I Santiago, On gravity and the Uncertainty Principle, Mod. Phys. Lett. A14, 1371 (1999) [arXiv:gr-qc/9904026]; R J Adler, P Chen and D I Santiago, The Generalized Uncertainty Principle and black hole remnants, Gen. Rel. Grav. 33, 2101 (2001); P Chen and R J Adler, Black hole remnants and dark matter [arXiv:gr-qc/0205106]; P Chen, Might dark matter be actually black? [arXiv:astro-ph/0303349]; R. J. Adler, Six easy roots to the Planck scale [arXiv:1001.1205 [gr-qc]].

2. B. J. Carr, L Modesto and I Prémont-Schwarz, Generalized Uncertainty Principle and self-dual black holes, arXiv: 1107.0708 [gr-qc] (2011); Black Hole Uncertainty Principle correspondence, preprint (2012); Cosmological implications of quantum particle black hole duality, preprint (2012).

3. W. Heisenberg, Zeitschrift fur Physik 43, 172198 (1927); E.H.Kennard, Zeitschrift fr Physik 44, 326 (1927).

4. J. A. Wheeler, Geons, Phys. Rev. 97, 511-536 (1955)

5. M. Maggiore, Phys. Lett. B 304, 65 (1993); Phys. Lett. B 319, 83 (1993); M. Maggiore, Phys. Rev. D. 49, 5182 (1994).

6. A. Ashtekar, S. Fiarhurst and J. L. Willis, Class. Quant. Grav. 20, 1031 (2003); G.M. Hossain, V. Husain and S.S. Seahra, arXiv:1003.22071 (gr-qc).

7. B. S. Kay, Class. Quant. Grav.15, L89-L98 (1998); B. S. Kay and V. Abyaneh, arXiv:0710.0992 (2007).

8. G. Veneziano, Europhys.Lett. 2, 199 (1986); E.Witten, Phys.Today April 24 (1996); F. Scardigli, Phys.Lett. B452, 39 (1999); D.J. Gross and P.F. Mende, Nuc.Phys.B303, 407 (1988); D. Amati, M. Ciafaloni and G. Veneziano, Phys, Lett. B. 216, 41 (1989); T. Yoneya, Mod.Phys.Lett., A4, 1587 (1989); K.Konishi, G. Paffuti and P. Proverpo, Phys. Lett. B 234, 276 (1990).

9. C. Rovelli, *Quantum Gravity*, Cambridge University Press, Cambridge (2004); A. Ashtekar, Class. Quant. Grav. 21, R53 (2004) [arxiv:gr-qc/0404018]; T Thiemann, [hep-th/0608210]; [gr-qc/0110034]; Lect. Notes Phys. 631, 41-135 (2003) [arxiv: gr-qc/0210094]; A. Ashtekar, Phys. Rev. Lett. **57** (18): 22442247 (1986).

10. M. Bojowald, Living Rev. Rel. 8, 11 (2005) [gr-qc/0601085]; A. Ashtekar, M Bojowald and J Lewandowski, Adv. Theor. Math. Phys. 7, 233-268 (2003) [gr-qc/0304074]; M. Bojowald, Phys. Rev. Lett. 86, 5227-5230 (2001) [gr-qc/0102069].

11. L. Modesto, Space-Time Structure of Loop Quantum Black Hole, Int. J. Theor. Phys 2010 [arXiv:0811.2196 [gr-qc]; L Modesto, I Premont-Schwarz Self-dual Black Holes in LQG: Theory and Phenomenology, Phys. Rev. D 80, 064041 (2009) [arXiv:0905.3170 [hep-th]]; L Modesto, Black hole interior from loop quantum gravity, Adv. High Energy Phys. 2008, 459290 (2008) [gr-qc/0611043]; L Modesto, Loop quantum black hole, Class. Quant. Grav. 23, 5587-5602 (2006) [gr-qc/0509078]; A Ashtekar and M Bojowald, Quantum geometry and Schwarzschild singularity, Class. Quant. Grav. 23, 391-411 (2006) [gr-qc/0509075]; L Modesto, Loop quantum black hole, Class. Quant. Grav. 23, 5587-5602 (2006); [gr-qc/0509078]; L Modesto, Disappearance of black hole singularity in quantum gravity, Phys. Rev. D 70, 124009 (2004) [gr-qc/0407097]; L Modesto, The Kantowski-Sachs Space-Time in Loop Quantum Gravity, Int. J. Theor. Phys. 45, 2235-2246 (2006) [arXiv:gr-qc/0411032]; F Caravelli, L Modesto, Spinning Loop Black Holes [arXiv:1006.0232]; A Model for non-singular black hole collapse and evaporation. Phys. Rev. D81, 044036 (2010) [arXiv:0912.1823 [gr-qc].

12. S W Hawking, Nature 248: 30 (1974); Comm. Math. Phys. 43:199 (1975).

NEW PERSPECTIVE ON SPACE AND TIME FROM LORENTZ VIOLATION

BO-QIANG MA

School of Physics and State Key Laboratory of Nuclear Physics and Technology, Peking University, Beijing 100871, China
Center for High Energy Physics, Peking University, Beijing 100871, China
Center for History and Philosophy of Science, Peking University, Beijing 100871, China
E-mail: mabq@pku.edu.cn

I present a brief review on space and time in different periods of physics, and then talk on the nature of space and time from physical arguments. I discuss the ways to test such a new perspective on space and time through searching for Lorentz violation in some physical processes. I also make an introduce to a newly proposed theory of Lorentz violation from basic considerations.

Keywords: space; time; Lorentz violation.

1. Space and Time in Physics

Space and time have been discussed in human history over thousands of years, and their nature still remains mysterious to human beings. There are many issues concerning the nature of space and time, such as

- Whether space and time are existences or concepts?
- Whether they are objective or subjective?
- Whether they are continuous or discrete?

There are many speculations on these questions from different perspectives, such as from metaphysics, art, philosophy, or science. The speculations are also different in different periods of human history.

In physics, space and time are where all physical events take place. However, there are different understandings concerning the properties of space and time in different periods of physics.

In classical mechanics, space is 3-dimensional and time is 1-dimensional, and they have the following properties

- Space and time are independent from each other;
- Space and time are objective and continuous;
- Time is universal and observer independent;
- Space are observer dependent.

The space and time provide the 3+1=4 dimensions of degrees of freedom where the Newton's laws of motion and gravity act on objects existing in space and time. The space between different inertial frames of reference are connected by the Galileo transformation.

In 1905, Einstein established his theory of special relativity. The special relativity offers a revolution to the concepts of space and time in Newton's mechanics, and provides a theoretical derivation of adopting the Lorentz transformation for the covariance of the equations of electrodynamics, to replace the traditional Galileo transformation in classical physics. Space and time are unified into 4-dimensional space-time. The space and time in special relativity have the following properties

- Space and time are dependent with each other;
- Both space and time are observer dependent;
- Space and time are continuous and flat.

There are two basic principles of special relativity:

- Principle of Relativity: the equations describing the laws of physics have the same form in all inertial frames of reference.
- Principle of constant light speed: the speed of light is the same in all directions in vacuum in all reference frames, regardless whether the source of the light is moving or not.

These two principles lead to the unification of space and time into a 4-dimensional space-time satisfying the Lorentz symmetry.

To unify the relativity with Newton's law of gravity, Einstein developed his theory of general relativity during 1907-1915. Then the curvature of space-time is determined by energy and momentum distribution. The space and time in general relativity have the following properties

- Space and time are dependent with each other;
- Both space and time are observer and also matter-distribution dependent;
- Space and time are continuous and can be curved.

One of the essence of general relativity is the principle of equivalence between gravity and inertial force, and this means that every observer can find a local inertial frame which is free from any gravitational effect. Thus the Lorentz symmetry always holds in such kind of local inertial frames.

Einstein's theories of relativity have been proved to be valid at very high precision and thus have achieved great triumphs. The Lorentz invariance, i.e., that statement that physical laws keep invariant under the Lorentz transformation, becomes a basic theoretical foundation of physics. Then we need to face the question:

- Is there any reason that we seek for Lorentz violation?

The Lorentz symmetry is a symmetry related with space and time, therefore the Lorentz violation should be related to the basic understandings of space and time. From the viewpoint of physics, the origin for the breaking down of conventional concepts of space and time might be traced back to Planck. With three fundamental constants in physics: the Newton gravitational constant G, the light speed in vacuum c, and the Boltzmann constant k_B, Planck introduced a new constant $\hbar$ In 1899, for the purpose to construct a "God-given" unit system.[1] By setting the above four constants as bases, one can construct the Planck unit system with a number of basic quantities, such as the Planck length $l_{\rm P} \equiv \sqrt{G\hbar/c^3} \simeq 1.6 \times 10^{-35}$ m, the Planck time $t_{\rm P} \equiv \sqrt{G\hbar/c^5} \simeq 5.4 \times 10^{-44}$ s, the Planck energy $E_{\rm P} \equiv \sqrt{\hbar c^5/G} \simeq 2.0 \times 10^9$ J, and the Planck temperature $T_{\rm P} \equiv \sqrt{\hbar c^5/Gk_B^2} \simeq 1.4 \times 10^{32}$ K. Therefore one may suspect that conventional understanding of space and time might be breaking down at the Planck scale:[2] i.e., at the Planck length $l_{\rm P}$, or the Planck time $t_{\rm P}$, or the Planck energy $E_{\rm P}$, where new features of existence may emerge. The breaking down of continue space-time was also conjectured.[3,4] The expectation for the existence of a minimal length as the Planck length led also to the establishment of some theories, e.g., the doubly special relativity (DSR).[5,6]

Just recently, Xu and I provided a physical argument for the discreteness of space and time.[7] From two known entropy constraints:

$$S_{\rm matter} \leq 2\pi ER, \quad \text{and} \quad S_{\rm matter} \leq \frac{A}{4}, \tag{1}$$

combined with the black-body entropy,

$$S = \frac{4}{45}\pi^2 T^3 V = \left(\frac{16}{135}\right)\pi^3 R^3 T^3, \tag{2}$$

we arrive at a minimum value of space

$$R \geq \left(\frac{128}{3645\pi}\right)^{\frac{1}{2}} l_{\rm P} \simeq 0.1 l_{\rm P}. \tag{3}$$

Thus we reveal from physical arguments that space-time is discrete rather than continuous. From another point of view, the newly proposed entropic gravity suggests gravity as an emergent force rather than a fundamental one.[8,9] If gravity is emergent, a new fundamental constant should be introduced to replace the Newtonian constant G.[2] It is natural to suggest a fundamental length scale, and such constant can be explained as the smallest length scale of quantum space-time. Its value can be measured through searches of Lorentz violation.[2,7] The existence of an "æther" (or a "vacuum" at rest in a specific frame) can also bring the breaking down of Lorentz invariance.[10,11]

Therefore the research on the Lorentz violation may provide us the chance for new understanding of the nature of basic concepts, such as "space", "time", and "vacuum", through physical ways, rather than from the perspectives of metaphysics or philosophy. It is thus necessary to push forward the studies on Lorentz violation from both theoretical and experimental aspects.

2. A Glimpse on Lorentz Violation Studies

Nowadays, there has been an increasing interest in Lorentz invariance Violation (LV or LIV) both theoretically and experimentally. The possible Lorentz symmetry violation effects have been sought for from various theories, motivated by the unknown underlying theory of quantum gravity together with various phenomenological applications.[12–17] This can happen in many alternative theories, e.g., the doubly special relativity (DSR),[5,6,18] torsion in general relativity,[19–21] and large extra-dimensions[22,23] *et al.* As examples, I list below some phenomenological consequences of the Lorentz violation effects studied by my students and I in the last few years:

- The Lorentz violation could provide an explanation of neutrino oscillation without neutrino mass.[24,25] We carried out Lorentz violation contribution to neutrino oscillation by the effective field theory for Lorentz violation and give out the equations of neutrino oscillation probabilities. In our model, neutrino oscillations do not have drastic oscillation at low energy and oscillations still exist at high energy. It is possible that neutrinos may have small mass and both Lorentz violation and the conventional oscillation mechanisms contribute to neutrino oscillation.

- The modified dispersion relation of the proton could increase the threshold energy of photo-induced meson production of the proton and cause an increase of the GZK cutoff energy. The earlier reports on super-GZK events triggered attention on Lorentz-Violation. The new results of observation of GZK cut-off put strong constraints on Lorentz violation parameters.[24]

- The modified dispersion relation of the photon may cause time lag of photons with different energies when they propagate in space from far-away astro-objects. The Lorentz violation can modify the photon dispersion relation, and consequently the speed of light becomes energy-dependent.[13] This results in a tiny time delay between high energy photons and low energy ones. Very high energy photon emissions from cosmological distance can amplify these tiny LV effects into observable quantities. We analyzed photons from γ-ray bursts from Fermi satellite observations and presented a first robust analysis of these taking the intrinsic time lag caused by sources into account, and gave an estimate to LV energy scale $\sim 2 \times 10^{17}$ GeV for linear energy dependence, and $\sim 5 \times 10^{9}$ GeV for quadratic dependence.[14]

- We also studied recent data on Lorentz violation induced vacuum birefringence from astrophysical consequences.[15] Due to the Lorentz violation, two helicities of a photon have different phase velocities and group velocities, termed as "vacuum birefringence". From recently observed γ-ray polarization from Cygnus X-1, we obtained an upper limit $\sim 8.7 \times 10^{-12}$ for Lorentz-violating parameter χ, which is the most firm constraint from well-known systems.

3. A Newly Proposed Theory of Lorentz Violation from Basic Principles

Among many theoretical investigations of Lorentz violation, it is a powerful framework to discuss various LV effects based on traditional techniques of effective field theory in particle physics. Here we focus our attention on a newly proposed theory of Lorentz violation from basic principles: the Standard Model Supplement (SMS).[26,27]

It is clear that human should not be narrowly focused just on effective theories with some additional terms beyond the conventional theory added by hand. It is a basic requirement that we should find a fundamental theory to derive the Lorentz violation terms from basic consideration. In the standard model supplement (SMS) framework,[26,27] the LV terms are brought about from a basic principle denoted as the physical independence or physical invariance (PI):

- Principle of Physical Invariance: the equations describing the laws of physics have the same form in **all admissible mathematical manifolds**.

The principle leads to the following replacement of the ordinary partial ∂_α and the covariant derivative D_α

$$\partial^\alpha \to M^{\alpha\beta}\partial_\beta, \quad D^\alpha \to M^{\alpha\beta}D_\beta, \tag{4}$$

where $M^{\alpha\beta}$ is a local matrix. The Lorentz violation terms are thus uniquely determined from the standard model Lagrangian without any ambiguity,[26] and their general existence is derived from basic consideration rather than added by hand. The explicit form of the matrices $M^{\alpha\beta}$ demands more basic theories concerning the true nature of space and time, and we suggest to adopt a physical way to explore these matrices through experiments rather than from theory at first. For more generality, we do not make any ad hoc assumption about these matrices. Thus these matrices might be particle dependent corresponding to the standard model particles under consideration, with the elements of these matrices to be measured or constrained from experimental observations.

We separate $M^{\alpha\beta}$ to two matrices like $M^{\alpha\beta} = g^{\alpha\beta} + \Delta^{\alpha\beta}$, where $g^{\alpha\beta}$ is the metric tensor of space-time and $\Delta^{\alpha\beta}$ is a new matrix which is particle-type dependent generally. Since $g^{\alpha\beta}$ is Lorentz invariant, $\Delta^{\alpha\beta}$ contains all the Lorentz violating degrees of freedom from $M^{\alpha\beta}$. Then $\Delta^{\alpha\beta}$ brings new terms violating Lorentz invariance in the standard model and is called Lorentz violation matrix. The theory returns back to the standard model when these Lorentz violation matrices vanish.

More explicitly, the effective Lagrangian $\mathcal{L}_{\mathrm{SM}}$ of the minimal standard model is composed of four parts

$$\mathcal{L}_{\mathrm{SM}} = \mathcal{L}_{\mathrm{G}} + \mathcal{L}_{\mathrm{F}} + \mathcal{L}_{\mathrm{H}} + \mathcal{L}_{\mathrm{HF}}, \tag{5}$$

$$\mathcal{L}_{\mathrm{G}} = -\frac{1}{4}\Gamma^{a\alpha\beta}F^a_{\alpha\beta}, \tag{6}$$

$$\mathcal{L}_{\mathrm{F}} = i\bar{\psi}\gamma^\alpha D_\alpha\psi, \tag{7}$$

$$\mathcal{L}_{\mathrm{H}} = (D^\alpha\phi)^\dagger D_\alpha\phi + V(\phi), \tag{8}$$

where we omit the chiral differences, the summation of chirality and gauge scripts. ψ is the fermion field, ϕ is the Higgs field, and $V(\phi)$ is the Higgs self-interaction. $F^a_{\alpha\beta} = \partial_\alpha A^a_\beta - \partial_\beta A^a_\alpha - gf^{abc}A^b_\alpha A^c_\beta$, $D_\alpha = \partial_\alpha + igA_\alpha$ and $A_\alpha = A^a_\alpha t^a$, with A^a_α being the gauge field. g is the coupling constant, and f^{abc} and t^a are the structure constants and generators of the corresponding gauge group respectively. $\mathcal{L}_{\mathrm{HF}}$ is the Yukawa coupling between the fermions and the Higgs field, and is not related to derivatives, thus it remains unchanged under the replacement (4).

Under (4) and the decomposition $M^{\alpha\beta} = g^{\alpha\beta} + \Delta^{\alpha\beta}$, the Lagrangians in (6)-(8) become

$$\mathcal{L}_{\mathrm{G}} = -\frac{1}{4}(M^{\alpha\mu}\partial_\mu A^{a\beta} - M^{\beta\mu}\partial_\mu A^{a\alpha} - gf^{abc}A^{b\alpha}A^{c\beta})$$
$$\times (M_{\alpha\mu}\partial^\mu A^a_\beta - M_{\beta\mu}\partial^\mu A^a_\alpha - gf^{abc}A^b_\alpha A^c_\beta)$$
$$= -\frac{1}{4}F^{a\alpha\beta}F^a_{\alpha\beta} + \mathcal{L}_{\mathrm{GV}}, \tag{9}$$
$$\mathcal{L}_{\mathrm{F}} = i\bar{\psi}\gamma_\alpha M^{\alpha\beta}D_\beta\psi = i\bar{\psi}\gamma^\alpha D_\alpha\psi + \mathcal{L}_{\mathrm{FV}}, \tag{10}$$
$$\mathcal{L}_{\mathrm{H}} = (M^{\alpha\mu}D_\mu\phi)^\dagger M_{\alpha\nu}D^\nu\phi + V(\phi)$$
$$= (D^\alpha\phi)^\dagger D_\alpha\phi + V(\phi) + \mathcal{L}_{\mathrm{HV}}, \tag{11}$$

with $M^{\alpha\beta}$ being the real matrix to maintain the Lagrangian hermitian. The last three terms $\mathcal{L}_{\mathrm{GV}}$, $\mathcal{L}_{\mathrm{FV}}$ and $\mathcal{L}_{\mathrm{HV}}$ of the equations mentioned above are the supplementary terms for the ordinary Standard Model. The explicit forms of these terms are

$$\mathcal{L}_{\mathrm{GV}} = -\frac{1}{2}\Delta^{\alpha\beta}\Delta^{\mu\nu}(g_{\alpha\mu}\partial_\beta A^{a\rho}\partial_\nu A^a_\rho - \partial_\beta A^a_\mu\partial_\nu A^a_\alpha)$$
$$- F^a_{\mu\nu}\Delta^{\mu\alpha}\partial_\alpha A^{a\nu}, \tag{12}$$
$$\mathcal{L}_{\mathrm{FV}} = i\Delta^{\alpha\beta}\bar{\psi}\gamma_\alpha\partial_\beta\psi - g\Delta^{\alpha\beta}\bar{\psi}\gamma_\alpha A_\beta\psi, \tag{13}$$
$$\mathcal{L}_{\mathrm{HV}} = (g_{\alpha\mu}\Delta^{\alpha\beta}\Delta^{\mu\nu} + \Delta^{\beta\nu} + \Delta^{\nu\beta})(D_\beta\phi)^\dagger D_\nu\phi. \tag{14}$$

Thus we obtain a new effective Lagrangian for the Standard Model with new supplementary terms, denoted by $\mathcal{L}_{\mathrm{SMS}}$

$$\mathcal{L}_{\mathrm{SMS}} = \mathcal{L}_{\mathrm{SM}} + \mathcal{L}_{\mathrm{LV}}, \tag{15}$$
$$\mathcal{L}_{\mathrm{LV}} = \mathcal{L}_{\mathrm{GV}} + \mathcal{L}_{\mathrm{FV}} + \mathcal{L}_{\mathrm{HV}}, \tag{16}$$

where $\mathcal{L}_{\mathrm{SMS}}$ satisfies the Lorentz covariance $(\mathrm{SO}^+(1,3))$, the gauge symmetry invariance of $\mathrm{SU}(3)\times\mathrm{SU}(2)\times\mathrm{U}(1)$ and invariance under the requirement of the principle of physical invariance or independence (PI), under which $\mathcal{L}_{\mathrm{SM}}$ cannot remain unchanged in a general situation.

We can have a better understanding of the LV terms in SMS here. The elements of $M^{\alpha\beta}$ of a particle are mass dimensionless (which is natural for the sign of testifying the Lorentz invariance), and they are not global constants generally. All of the LV terms are expressed in $\mathcal{L}_{\mathrm{LV}}$, and the LV information is measured by the concise matrix $\Delta^{\alpha\beta}$, which is convenient for a systematic study of the LV effects. To determine whether the Lorentz invariance holds exactly, further work is needed to analyze the effective Lagrangian (15) of QED, QCD and EW (ElectroWeak) fields, and more experiments are needed to determine the magnitude of the elements in the matrices $M^{\alpha\beta}$ for different particles. There have been some preliminary progress along this line with the theory of SMS applied to discuss the Lorentz violation effects for protons,[26] photons,[27] and neutrinos.[28] More works are still needed for systematic studies.

Generally, $\Delta^{\alpha\beta}$ might be particle-type and flavor-type dependent. If we use the vacuum expectation values of a $\Delta^{\alpha\beta}$ for the coupling constants in the corresponding effective Lagrangian, not all of the 16 degrees of freedom of $M^{\alpha\beta}$ are physical. For the derivative field $M(\partial_x)\varphi(x)$ of an arbitrary given field, $\varphi(x)$ can be rescaled to absorb one of the 16 degrees of freedom so that only 15 are left. When more fields are involved, there is only one degree of freedom that can be reduced from a rescaling consideration for all fields. Thus for generality, we may keep all 16 degrees of freedom in $M^{\alpha\beta}$ for a specific particle in our study.

The matrix $M^{\alpha\beta}$ in the SMS theory just appears with the derivative terms of Lagrangians, but not with the coordinate terms. Therefore one should not confuse this matrix with the metric of the general relativity. To define a covariant derivative in general relativity, one needs to introduce the concept of connection to reflect the effect due to the curvature of space and time from gravity. Our introduction of the matrix $M^{\alpha\beta}$ can be considered as an alternative choice along the similar philosophy, but with a more general sense by including also possible effects from other interactions other than solely gravity. Therefore our matrix $M^{\alpha\beta}$ can contain the effect due to general relativity or more beyond that, but we do not intend to derive it from theory but to detect it from phenomenological manifestations. This is from the consideration that the space-time structure of nature might be more complicated than just the effect from gravity.

There still exists the question of how to understand and handle the Lorentz violation matrix $\Delta^{\alpha\beta}$. We list here three options for understandings and treatments:[29]

- **Scenario I**: which can be called as fixed scenario in which the Lorentz violation matrices are taken as constant matrices in any inertial frame of reference the observer is working. It means that the Lorentz violation matrices can be taken as approximately the same for any working reference frames such as the earth-rest frame, the sun-rest frame, or the CMB frame. However, there will be the problem of inconsistency for an "absolute physical event" between different reference frames,[30] if one sticks to this scenario. Therefore this scenario can be adopted as a practical approach when one is focused on the Lorentz violation effect within a certain frame and does not care about relationships between different frames.

- **Scenario II**: which can be called as "new æther" scenario in which the Lorentz violation matrices transform as tensors between different inertial frames but keep as constant matrices within the same frame. The Lorentz violation matrices play the roles for the exitance of some kinds of background fields, or the "new æther" (i.e., a "vacuum" at rest in a specific frame), which changes from one frame to another frame by Lorentz transformation. Thus it can be considered as a standard viewpoint to treat the matrix $M^{\alpha\beta}$ as a tensor satisfying Lorentz symmetry.

- **Scenario III**: which can be called as covariant scenario in which the Lorentz violation matrices transform as tensors adhered with the corresponding standard model particles. It means that these Lorentz violation matrices are emergent and covariant with their standard model particles. Such a scenario still needs to be checked for consistency and for applications in future.

Before accepting the SMS as a fundamental theory, one can take the SMS as an effective framework for phenomenological applications by confronting with various experiments to determine and/or constrain the Lorentz violation matrioce $\Delta^{\alpha\beta}$ for various particles. So our idea is to reveal the real structure of Lorentz violation of nature from experiments rather than from theory. We consider this phenomenological way as more appropriate for physical investigations, rather than to derive everything from theory at first. As a comparison, the specific form of the quark mixing matrix is determined from experimental measurements rather than derived from theory.[31] Even after so many years of research and also the elements of the CKM mixing matrix have been measured to very high precision, there is still no a commonly accepted theory to derive the quark mixing matrix from basic principles.

We now provide some remarks concerning the Lorentz violation studies in field theory frameworks. In the effective field theory frameworks, the standard particles transform according to the Lorentz symmetry between different momentum states. The background fields should also transform according to the Lorentz symmetry between different observer working frames from the requirement of consistency. From this sense, there is actually no Lorentz violation for the whole system of the standard model particles together with the background fields. The Lorentz violation exists for the standard model particles within an observer working frame, when these particles have different momenta between each other. From this sense, the Lorentz violation is due to the existence of the background fields, which are treated as fixed parameters in the observer working frame.

The newly proposed theory of SMS not only provides clear relationship between some general LV parameters,[32] but also can be conveniently applied for phenomenological analysis.[33] We would need more experimental investigations to check whether it can meet the criterion of being able to provide a satisfactory description of the physical reality, with simplicity and beauty in formalism, together with the predictive power towards new knowledge for human beings. It is also possible that the nature satisfies the Lorentz symmetry perfectly and we would be unable to find a physical evidence to support the theory. This would imply that the newly introduced Lorentz violation matrix $\Delta^{\alpha\beta}$ would vanish for nature.

4. Conclusion

From the discussions above, we may have some new perspectives on the nature of space-time:

- Space and time are 3+1=4 dimensional or may have extra-dimensions;
- Both space and time are observer and object dependent;
- Space and time might be particle-type dependent and can be curved by the existence of matter.
- Space and time might be discrete and such discreteness can be tested through experiments on Lorentz violation.

Finally we present our conclusion:

- Researches on Lorentz violation have been active for many years;
- There might be some marginal evidences for Lorentz violation yet, but non of them can be considered as confirmed;
- The Lorentz violation study can bring conceptual revolution on the understanding of space-time for human beings;
- Lorentz violation is being an active frontier both theoretically and experimentally.

Acknowledgements

I acknowledge Professor Pisin Chen for his warm invitation and hospitality for attending the 1st LeCosPA Symposium. I am very grateful for the discussions and collaborations with a number of my students: Zhi Xiao, Shi-Min Yang, Lijing Shao, Lingli Zhou, Xinyu Zhang, Yunqi Xu, and Nan Qin, who devoted their wisdoms and enthusiasms bravely on the topic of Lorentz violation in past few years. The work was supported by National Natural Science Foundation of China (Nos. 10975003, 11021092, 11035003 and 11120101004).

References

1. M. Planck, *Sitzber. K. Preuss Aka. Berlin* **5**, 440 (1899).
2. L. Shao, B.-Q. Ma, *Sci. China Phys. Mech. Astro.* **54**, 1771 (2011) [arXiv:1006.3031 [hep-th]].
3. H.S. Snyder, *Phys. Rev.* 71, 38 (1947); 72, 68 (1947).
4. J.A. Wheeler, *Ann. Phys.* 2, 604 (1957).
5. G. Amelino-Camelia, *Int. J. Mod. Phys.* D **11**, 35 (2002) [arXiv:gr-qc/0012051].
6. J. Magueijo and L. Smolin, *Phys. Rev. Lett.* **88**, 190403 (2002) [arXiv:hep-th/0112090].
7. Y. Xu and B.-Q. Ma, *Mod. Phys. Lett.* A **26**, 2101 (2011) [arXiv:1106.1778 [hep-th]].
8. E. P. Verlinde, *JHEP* **1104**, 029 (2011) [arXiv:1001.0785 [hep-th]].
9. X. G. He and B.-Q. Ma, *Chin. Phys. Lett.* **27**, 070402 (2010) [arXiv:1003.1625 [hep-th]].
10. P.A.M. Dirac, *Nature* **168**, 906 (1951).
11. J.D. Bjorken, *Ann. Phys.* **24**, 174 (1963).

12. For a brief review on Lorentz violation effects through very high energy photons of astrophysical sources, see. e.g., L. Shao, B.-Q. Ma, *Mod. Phys. Lett.* A **25**, 3251 (2010) [arXiv:1007.2269], and references therein.

13. Z. Xiao, B.-Q. Ma, *Phys. Rev.* D **80**, 116005 (2009) [arXiv:0909.4927 [hep-ph]].

14. L. Shao, Z. Xiao, B.-Q. Ma, *Astropart. Phys.* **33**, 312 (2010) [arXiv:0911.2276 [hep-ph]].

15. L. Shao and B. -Q. Ma, *Phys. Rev.* D **83**, 127702 (2011) [arXiv:1104.4438 [astro-ph.HE]].

16. Z. Xiao, L. Shao, B.-Q. Ma, *Eur. Phys. J.* C **70**, 1153 (2010) [arXiv:1011.5074 [hep-th]].

17. W. Bietenholz, *Phys. Rept.* **505**, 145 (2011).

18. X. Zhang, L. Shao, B.-Q. Ma, *Astropart. Phys.* **34**, 840 (2011) [arXiv:1102.2613 [hep-th]].

19. W.-T. Ni, *Phys. Rev. Lett.* **35**, 319 (1975).

20. W.-T. Ni, *Rept. Prog. Phys.* **73**, 056901 (2010) [arXiv:0912.5057 [gr-qc]].

21. M.L. Yan, *Commun. Theor. Phys.* **2**, 1281 (1983).

22. V. Ammosov and G. Volkov, hep-ph/0008032.

23. H. Pas, S. Pakvasa, T.J. Weiler, *Phys. Rev.* D **72**, 095017 (2005).

24. Z. Xiao, B.-Q. Ma, *Int. J. Mod. Phys.* A **24**, 1359 (2009) [arXiv:0805.2012].

25. S. Yang, B.-Q. Ma, *Int. J. Mod. Phys.* A **24**, 5861 (2009) [arXiv:0910.0897].

26. Zhou L., B.-Q. Ma, *Mod. Phys. Lett.* A **25**, 2489 (2010) [arXiv:1009.1331].

27. Zhou L., B.-Q. Ma, *Chin. Phys.* C **35**, 987 (2011) [arXiv:1109.6387].

28. Zhou L., B.-Q. Ma, arXiv:1109.6097.

29. B.-Q. Ma, *Mod. Phys. Lett.* A **27**, 1230005 (2012) [arXiv:1111.7050 [hep-ph]].

30. B. Q. Ma, *Int. J. Mod. Phys. Conf. Ser.* (2012) in press [arXiv:1203.0086 [hep-ph]].

31. B.-Q. Ma, *Int. J. Mod. Phys. Conf. Ser.* **1**, 291 (2011) [arXiv:1109.5276 [hep-ph]].

32. Zhou L., B.-Q. Ma, arXiv:1110.1850 [hep-ph].

33. Zhou L., B.-Q. Ma, arXiv:1009.1675.

FOUNDATIONS OF CLASSICAL ELECTRODYNAMICS, EQUIVALENCE PRINCIPLE AND COSMIC INTERACTIONS: A SHORT EXPOSITION AND AN UPDATE[*]

WEI-TOU NI,[1] HSIEN-HAO MEI,[2] SHAN-JYUN WU[3]

Center for Gravitation and Cosmology,
Department of Physics, National Tsing Hua University,
Hsinchu 30010, Taiwan
[1]*E-mail: weitou@gmail.com*
[2]*E-mail: mei@phys.nthu.edu.tw*
[3]*E-mail: lightemit@gmail.com*

We look at the foundations of electromagnetism in this 1st LeCosPA Symposium. For doing this, after some review (constraints on photon mass etc.), we use two approaches. The first one is to formulate a Parametrized Post-Maxwellian (PPM) framework to include QED corrections and a pseudoscalar photon interaction. PPM framework includes lowest corrections to unified electromagnetism-gravity theories based on connection approach. It may also overlap with corrections implemented from generalized uncertainty principle (GUP) when electromagnetism-gravity coupling is considered. We discuss various vacuum birefringence experiments – ongoing and proposed -- to measure these parameters. The second approach -- the χ-g framework is to look at electromagnetism in gravity and various experiments and observations to determine its empirical foundation. The SME (Standard Model Extension) and SMS (Standard Model Supplement) overlap with the χ-g framework in their photon sector. We found that the foundation is solid with the only exception of a potentially possible pseudoscalar-photon interaction. We discussed its experimental constraints and look forward to more future experiments.

Keywords: Classical electrodynamics; Equivalence principle; Cosmic interactions; PPM framework.

1. Introduction

1.1. *Classical electrodynamics*

Classical electrodynamics is based on Maxwell equations and Lorentz force law. It can be derived by a least action with the following Lagrangian density for a system of charged particles in Gaussian units (e.g., Jackson [1]),

$$L_{EMS}=L_{EM}+L_{EM\text{-}P}+L_P=-(1/(16\pi))[(1/2)\eta^{ik}\eta^{jl}-(1/2)\eta^{il}\eta^{kj}]F_{ij}F_{kl}-A_kj^k-\Sigma_I m_I[(ds_I)/(dt)]\delta(\mathbf{x}\text{-}\mathbf{x}_I), \quad (1)$$

---

[*]Plenary talk at First LeCosPA Symposium: Towards Ultimate Understanding of the Universe (LeCosPA2012), National Taiwan University, Taipei, ROC, February 6-9, 2012.

where $F_{ij} \equiv A_{j,i} - A_{i,j}$ is the electromagnetic field strength tensor with A_i the electromagnetic 4-potential and comma denoting partial derivation, η^{ij} is the Minkowskii metric with signature $(+, -, -, -)$, m_I the mass of the Ith charged particle, s_I its 4-line element, and j^k the charge 4-current density. Here, we use Einstein summation convention, i.e., summation over repeated indices. There are three terms in the Lagrangian density L_{EMS} -- (i) L_{EM} for the electromagnetic field, (ii) $L_{EM\text{-}P}$ for the interaction of electromagnetic field and charged particles and (iii) L_P for charged particles.

The electromagnetic field Lagrangian density (1) can be written in terms of the electric field $\mathbf{E}$ $[\equiv (E_1, E_2, E_3) \equiv (F_{01}, F_{02}, F_{03})]$ and magnetic induction $\mathbf{B}$ $[\equiv (B_1, B_2, B_3) \equiv (F_{32}, F_{13}, F_{21})]$ as

$$L_{EM} = (1/8\pi)[\mathbf{E}^2 - \mathbf{B}^2]. \tag{2}$$

1.2. Proca Lagrangian and the photon mass

The classical Lagrangian density (1) is based on the photon having zero mass. To include the effects of nonvanishing photon mass m_{photon}, Proca [2-6] added a mass term L_{Proca},

$$L_{Proca} = (m_{photon}{}^2 c^2/8\pi\hbar^2)(A_k A^k), \tag{3}$$

to the Lagrangian density of classical electrodynamics soon after Yukawa proposed short-range interaction in 1935. We use η^{ij} and its inverse η_{ij} to raise and lower indices. With this term, the Coulomb law is modified to have the electric potential A_0:

$$A_0 = q(e^{-\mu r}/r), \tag{4}$$

where q is the charge of the source particle, r is the distance to the source particle, and μ $(\equiv m_{photon}c/\hbar)$ gives the inverse range of the interaction. The constraints on the mass and range of photons from various experiments are compiled in Table 1. For a comprehensive review, please see Goldhaber and Nieto [11].

Table 1. Constraints on the mass and range of photon.

Experiment/Observation	Mass constraint	Range constraint
Williams, Faller & Hill (1971) [7]: Lab Test	$m_{photon} \leq 10^{-14}$ eV $(= 2 \times 10^{-47}$ g$)$	$\mu^{-1} \geq 2 \times 10^7$ m
Davis, Goldhaber & Nieto (1975) [8]: Jupiter Magnetic field (Pioneer 10 Jupiter flyby)	$m_{photon} \leq 4 \times 10^{-16}$ eV $(= 7 \times 10^{-49}$ g$)$	$\mu^{-1} \geq 5 \times 10^8$ m
Ryutov (2007) [9]: Solar wind magnetic field	$m_{photon} \leq 10^{-18}$ eV $(= 2 \times 10^{-51}$ g$)$	$\mu^{-1} \geq 2 \times 10^{11}$ m
Chibisov (1976) [10]: Galactic sized magnetic field	$m_{photon} \leq 2 \times 10^{-27}$ eV $(= 4 \times 10^{-60}$ g$)$	$\mu^{-1} \geq 10^{20}$ m

As larger scale magnetic field discovered and measured, the constraints on photon mass and on the interaction range may become more stringent. If cosmic scale magnetic field is discovered, the constraint on the interaction range may become bigger or comparable to Hubble distance (of the order of radius of curvature of our observable universe). If this happens, the concept of photon mass may lose significance amid gravity coupling or curvature coupling of photons.

This paper is a short exposition of empirical foundations of electromagnetism with an update to include discussions of relevant recent theories and models. For a longer exposition, please see Ni [12]. The outline is as follows. In section 2, we present the Parametrized Post-Maxwell (PPM) framework for testing the foundations of classical electrodynamics in flat spacetime (including effective quantum corrections, but without gravity coupling), discuss its scope and summarize its usefulness. In section 3, we present the basic equations and discuss wave propagation in the PPM electrodynamics. In section 4, we discuss ultra-high precision laser interferometry experiments to measure the parameters of PPM electrodynamics. In section 5, we discuss empirical tests of electromagnetism in gravity and the χ-g framework, and find pseudoscalar-photon interaction uniquely standing out. In section 6, we discuss the pseudoscalar-photon interaction, its relation to other approaches, and the use of radio galaxy observations and Cosmic Microwave Background (CMB) observations to constrain the cosmic polarization rotation induced by the pseudoscalar-photon interaction. In section 7, we present a summary and an outlook briefly.

2. Parametrized Post-Maxwellian (PPM) Framework

For formulating a phenomenological framework for testing corrections to Maxwell-Lorentz classical electrodynamics, we notice that $(\mathbf{E}^2-\mathbf{B}^2)$ and $(\mathbf{E}\cdot\mathbf{B})$ are the only Lorentz invariants second order in the field strength, and $(\mathbf{E}^2-\mathbf{B}^2)^2$, $(\mathbf{E}\cdot\mathbf{B})^2$ and $(\mathbf{E}^2-\mathbf{B}^2)(\mathbf{E}\cdot\mathbf{B})$ are the only Lorentz invariants fourth order in the field strength. However, $(\mathbf{E}\cdot\mathbf{B})$ is a total divergence and, by itself in the Lagrangian density, does not contribute to the equation of motion (field equation). Multiplying $(\mathbf{E}\cdot\mathbf{B})$ by a pseudoscalar field Φ, the term $\Phi(\mathbf{E}\cdot\mathbf{B})$ is the Lagrangian density for the pseudoscalar-photon (axion-photon) interaction. When this term is included together with the fourth-order invariants, we have the following phenomenological Lagrangian density for our Parametrized Post-Maxwell (PPM) Lagrangian density including various corrections and modifications to be tested by experiments and observations,

$$L_{PPM} = (1/8\pi)\{(\mathbf{E}^2-\mathbf{B}^2)+\xi\Phi(\mathbf{E}\cdot\mathbf{B})+B_c^{-2}[\eta_1(\mathbf{E}^2-\mathbf{B}^2)^2+4\eta_2(\mathbf{E}\cdot\mathbf{B})^2+2\eta_3(\mathbf{E}^2-\mathbf{B}^2)(\mathbf{E}\cdot\mathbf{B})]\}, \qquad (5)$$

where

$$B_c \equiv E_c \equiv m^2c^3/e\hbar = 4.4\text{x}10^{13}\ \text{G=4.4x}10^9\,\text{T=4.4x}10^{13}\ \text{statvolt/cm=1.3x}10^{18}\ \text{V/m}, \qquad (6)$$

with e the absolute value of electron charge and m the electron mass. This PPM Lagrangian density contains 4 parameters ξ, η_1, η_2 & η_3, and is an extension of the two-parameter (η_1 and η_2) post-Maxwellian Lagrangian density of Denisov, Krivchenkov and Kravtsov [13]. If there are absorptions, e.g., due to pair production or conversion to other particles, there would be imaginary part of the Lagrangian density. For example, one could add $L_{PPM}^{(Im)}$ to the Lagrangian density (5):

$$L_{PPM}^{(Im)} = (i/8\pi)\{ B_c^{-2}[\zeta_1(\mathbf{E}^2-\mathbf{B}^2)^2+4\zeta_2(\mathbf{E}\cdot\mathbf{B})^2+2\zeta_3(\mathbf{E}^2-\mathbf{B}^2)(\mathbf{E}\cdot\mathbf{B})]\}. \tag{7}$$

In this exposition, we are mainly concerned ourselves with the real part (5). To test the imaginary part (7), one may look into strong field pair production (e.g., Kim [14, 15]) and astrophysical phenomenon in strong field (e.g., Ruffini, Vereshchagin and Xue [16]). In the Ruffini-Vereshchagin-Xue [16] review of astrophysical phenomenon in strong field, their parameters, $\kappa_{2,0}$ and $\kappa_{2,1}$, corresponds to $\eta_1 = 8\pi B_c^2\kappa_{2,0}$ and $\eta_2 = 2\pi B_c^2\kappa_{0,2}$ in (5) and (7). In passing, we have noticed that in this first LeCosPA Symposium, there are talks related to pair productions and quantum fluctuations on acceleration and temperature (Labun and Rafelski [17]: Unruh [18]; S.Weinfurtner $et\ al.$ [19]) which could be subjected to similar kind of tests.

The manifestly Lorentz covariant form of Eq. (5) is

$$L_{PPM} = (1/(32\pi))\{-2F^{kl}F_{kl} -\xi\Phi F^{*kl}F_{kl}+B_c^{-2}[\eta_1(F^{kl}F_{kl})^2+\eta_2(F^{*kl}F_{kl})^2+\eta_3(F^{kl}F_{kl})(F^{*ij}F_{ij})]\}, \tag{8}$$

where

$$F^{*ij} \equiv (1/2)e^{ijkl} F_{kl}, \tag{9}$$

with e^{ijkl} defined as

$$e^{ijkl} \equiv 1 \text{ if } (ijkl) \text{ is an even permutation of } (0123); -1 \text{ if odd; } 0 \text{ otherwise.} \tag{10}$$

Heisenberg-Euler [20] Lagrangian density including the leading order quantum effects in slowly varying electric and magnetic field

$$L_{Heisenberg-Euler} = [2\alpha^2\hbar^2/(45(4\pi)^2m^4c^6)][(\mathbf{E}^2-\mathbf{B}^2)^2 + 7(\mathbf{E}\cdot\mathbf{B})^2], \tag{11}$$

fits the PPM framework with

$$\eta_1 = \alpha/(45\pi) = 5.1 \times 10^{-5}, \ \eta_2 = 7\alpha/(180\pi) = 9.0 \times 10^{-5}, \ \eta_3 = 0 \text{ and } \xi = 0, \tag{12}$$

where α is the fine structure constant.

Before Heisenberg and Euler [20], Born and Infeld [21, 22] proposed the following (classical) Lagrangian density for the electromagnetic field

130

$$L_{Born-Infeld} = - (b^2/4\pi) \, [1 - (\mathbf{E}^2-\mathbf{B}^2)/b^2 - (\mathbf{E}\cdot\mathbf{B})^2/b^4]^{1/2}, \tag{13}$$

where b is a constant which gives the maximum electric field strength. For field strength small compared with b, (13) can be expanded into

$$L_{Born-Infeld} = (1/8\pi) \, [(\mathbf{E}^2-\mathbf{B}^2) + (\mathbf{E}^2-\mathbf{B}^2)^2/b^2 + (\mathbf{E}\cdot\mathbf{B})^2/b^2 + O(b^{-4})]. \tag{14}$$

The lowest order of Born-Infeld electrodynamics agrees with the classical electrodynamics. The next order corrections fit the PPM framework Eq. (5) with

$$\eta_1 = \eta_2 = B_c^2/b^2, \text{ and } \eta_3 = \xi = 0. \tag{15}$$

In the Born-Infeld electrodynamics, b is the maximum electric field. Electric fields at the edge of heavy nuclei are of the order of 10^{21} V/m. If we take b to be 10^{21} V/m, then, $\eta_1 = \eta_2 = 5.9 \times 10^{-6}$.

The PPM framework is useful in testing various models and theories of both electromagnetism and gravity. A class of unified theories of electromagnetism and gravity with Lagrangian of the BF type (F: Curvature of the connection 1-form A (ω), with the gauge group U(2) (complexified) and with a potential for the B (Σ) field (Lie-algebra valued 2-form)) is proposed by Torres-Gomez, Krasnov and Scarinci [23]. Given a choice of a potential function with parameters α, γ, χ, δ and ξ, the theory is a deformation of (complex) general relativity and electromagnetism. With the reality conditions and using their equations (37), (38), (44), (45), the quadratic order plus quartic order Lagrangian can be put into the following form:

$$L^{(2)}+L^{(4)}=\alpha/(\gamma(\alpha+\gamma))\{ (\mathbf{E}^2-\mathbf{B}^2)+(1/2)[\chi/\alpha(\alpha+\gamma)^3+(2\delta/(\alpha\gamma \, (\alpha+\gamma)) +\xi(\alpha+\gamma)/\alpha\gamma^3)(\mathbf{E}^2-\mathbf{B}^2)^2$$
$$- 2[\chi/\alpha(\alpha+\gamma)^3-2\delta/(\alpha\gamma \, (\alpha+\gamma)) +\xi(\alpha+\gamma)/\alpha\gamma^3] \, (\mathbf{E}\cdot\mathbf{B})^2$$
$$- 8i(\chi/\alpha(\alpha+\gamma)^3-\xi(\alpha+\gamma)/\alpha\gamma^3) \, (\mathbf{E}^2-\mathbf{B}^2)(\mathbf{E}\cdot\mathbf{B})]\}. \tag{16}$$

Comparing with (5) and (7), we have

$$\eta_1 = (1/2)B_c^2[\gamma\chi/\alpha(\alpha+\gamma)^3+2\delta/(\alpha\gamma \, (\alpha+\gamma))+\xi(\alpha+\gamma)/\alpha\gamma^3], \, \eta_3 = \xi = 0,$$

$$\eta_2 =-(1/2)B_c^2[\gamma\chi/\alpha(\alpha+\gamma)^3-2\delta/(\alpha\gamma(\alpha+\gamma))+\xi(\alpha+\gamma)/\alpha\gamma^3], \, \zeta_3=-4B_c^2[\gamma\chi/\alpha(\alpha+\gamma)^3-\xi(\alpha+\gamma)/\alpha\gamma^3].\tag{17}$$

Thus, we see that experiments to measure the PPM parameters will also constrain the parameters of the proposed nonlinear electrodynamics from a class of unified theory of electromagnetism and gravity.

A focus in this Symposium is the Generalized Uncertainty Principle (GUP) as advocated by Bernard Carr [24] and Pisin Chen [25]. GUP affects the black hole entropy and the associated quantum effects in entropic gravity modify the Newton's gravitational law [25]. Although the modification of gravity law is small, when the coupling to electromagnetism is considered/integrated/unified, the quartic corrections in the Lagrangian might not be negligible and, therefore, might be detectable by experiments to measure the PPM parameters.

In section 4, we will discuss how to measure the PPM parameters using birefringence measurements after we give the basic equations and discuss wave propagation in the PPM electrodynamics in section 3 in the following.

3. Basic Equations and Wave Propagation in the PPM Electrodynamics

In analogue with the nonlinear electrodynamics of continuous media, we can define the electric displacement $\mathbf{D}$ and magnetic field $\mathbf{H}$ as follows:

$$\mathbf{D} \equiv 4\pi(\partial L_{PPM}/\partial\mathbf{E}) = [1+2\eta_1(\mathbf{E}^2-\mathbf{B}^2)B_c^{-2}+2\eta_3(\mathbf{E}\cdot\mathbf{B})B_c^{-2}]\mathbf{E}+[\Phi+4\eta_2(\mathbf{E}\cdot\mathbf{B})B_c^{-2}+\eta_3(\mathbf{E}^2-\mathbf{B}^2)B_c^{-2}]\mathbf{B}, \quad (18)$$

$$\mathbf{H} \equiv -4\pi(\partial L_{PPM}/\partial\mathbf{B}) = [1+2\eta_1(\mathbf{E}^2-\mathbf{B}^2)B_c^{-2}+2\eta_3(\mathbf{E}\cdot\mathbf{B})B_c^{-2}]\mathbf{B}-[\Phi+4\eta_2(\mathbf{E}\cdot\mathbf{B})B_c^{-2}+\eta_3(\mathbf{E}^2-\mathbf{B}^2)B_c^{-2}]\mathbf{E}. \quad (19)$$

From $\mathbf{D}$ & $\mathbf{H}$, we can define a second-rank G_{ij} tensor, just like from $\mathbf{E}$ & $\mathbf{B}$ to define F_{ij} tensor. With these definitions and following the standard procedure in electrodynamics [see, e.g., Jackson [1], p. 599], the nonlinear equations of the electromagnetic field are

$$\text{curl } \mathbf{H} = (1/c)\,\partial\mathbf{D}/\partial t + 4\pi\,\mathbf{J}, \tag{20}$$

$$\text{div } \mathbf{D} = 4\pi\,\rho, \tag{21}$$

$$\text{curl } \mathbf{E} = -(1/c)\,\partial\mathbf{B}/\partial t, \tag{22}$$

$$\text{div } \mathbf{B} = 0. \tag{23}$$

We notice that it has the same form as in macroscopic electrodynamics. The Lorentz force law remains the same as in classical electrodynamics:

$$d[(1-\mathbf{v}_I^2/c^2)^{-1/2}m_I\mathbf{v}_I]/dt = q_I[\mathbf{E} + (1/c)\mathbf{v}_I \times \mathbf{B}], \tag{24}$$

for the I-th particle with charge q_I and velocity $\mathbf{v}_I$ in the system. The source of Φ in this system is $(\mathbf{E}\cdot\mathbf{B})$ and the field equation for Φ is

$$\partial^l L_\Phi/\partial(\partial^l\Phi) - \partial L_\Phi/\partial\Phi = \mathbf{E}\cdot\mathbf{B}, \tag{25}$$

where L_Φ is the Lagrangian density of the pseudoscalar field Φ.

Following our previous method [12, 26, 27], i.e., separating the electric field $\mathbf{E}$ and magnetic induction field $\mathbf{B}$ into the wave part $\mathbf{E}^{\text{wave}}$, $\mathbf{B}^{\text{wave}}$ (small compared to external part) and external part $\mathbf{E}^{ext}$, $\mathbf{B}^{ext}$, and linearizing the equations of motion, one can derive the PPM wave propagation equations and obtain the dispersion relations [12]. From the dispersion relations, the principal indices of refraction can be found. The necessary and sufficient conditions of "no birefringence" on the PPM parameters are

$$\eta_1 = \eta_2,\ \eta_3 = 0,\ \text{and no constraint on } \xi. \tag{26}$$

The Born-Infeld electrodynamics satisfies (26) and has no birefringence in the theory.

For $\mathbf{E}^{\text{ext}} = 0$, the (principal) refractive indices in the transverse external magnetic field $\mathbf{B}^{\text{ext}}$ for the linearly polarized lights whose polarizations are parallel and orthogonal to the magnetic field, are as follows:

$$n_{\parallel} = 1 + \{(\eta_1 + \eta_2) + [(\eta_1 - \eta_2)^2 + \eta_3^2]^{1/2}\}\ (\mathbf{B}^{ext})^2 B_c^{-2} \quad (\mathbf{E}^{\text{wave}} \parallel \mathbf{B}^{ext}), \tag{27}$$

$$n_{\perp} = 1 + \{(\eta_1 + \eta_2) - [(\eta_1 - \eta_2)^2 + \eta_3^2]^{1/2}\}\ (\mathbf{B}^{ext})^2 B_c^{-2} \quad (\mathbf{E}^{\text{wave}} \perp \mathbf{B}^{ext}). \tag{28}$$

For $\mathbf{B}^{\text{ext}} = 0$, the (principal) refractive indices in the transverse external electric field $\mathbf{E}^{\text{ext}}$ for the linearly polarized lights whose polarizations are parallel and orthogonal to the magnetic field, are as follows:

$$n_{\parallel} = 1 + \{(\eta_1 + \eta_2) + [(\eta_1 - \eta_2)^2 + \eta_3^2]^{1/2}\}\ (\mathbf{E}^{ext})^2 B_c^{-2} \quad (\mathbf{E}^{\text{wave}} \parallel \mathbf{E}^{ext}), \tag{29}$$

$$n_{\perp} = 1 + \{(\eta_1 + \eta_2) - [(\eta_1 - \eta_2)^2 + \eta_3^2]^{1/2}\}\ (\mathbf{E}^{ext})^2 B_c^{-2} \quad (\mathbf{E}^{\text{wave}} \perp \mathbf{E}^{ext}). \tag{30}$$

The magnetic field near pulsars can reach 10^{12} G, while the magnetic field near magnetars can reach 10^{15} G. The astrophysical processes in these locations need nonlinear electrodynamics to model. In the following section, we turn to experiments to measure the parameters of the PPM electrodynamics.

4. Measuring the Parameters of the PPM Electrodynamics

There are four parameters η_1, η_2, η_3, and ξ in PPM electrodynamics to be measured by experiments. For the QED (Quantum Electrodynamics) corrections to classical electrodynamics, $\eta_1 = \alpha/(45\pi) = 5.1 \times 10^{-5}$, $\eta_2 = 7\alpha/(180\pi) = 9.0 \times 10^{-5}$, $\eta_3 = 0$, and $\xi = 0$. There are three vacuum birefringence experiments on going in the world to measure this QED vacuum birefringence – the BMV experiment [28]), the PVLAS experiment [29] and the Q & A (QED vacuum birefringence and Axion search) experiment [30, 31]. The QED vacuum birefringence Δn in a magnetic field $\mathbf{B}^{\text{ext}}$ is

$$\Delta n = n_{\parallel} - n_{\perp} = 4.0 \times 10^{-24}\ (\mathbf{B}^{\text{ext}}/1\text{T})^2. \tag{31}$$

For 2.3 T field of the Q & A rotating permanent magnet, Δn is 2.1×10^{-23}. This is about the same order of magnitude change in fractional length that ground interferometers for gravitational-wave detection aim at. Quite a lot of techniques developed in the gravitational-wave detection community are readily applicable to vacuum birefringence measurement [26].

The basic principle of these experimental measurements is shown as Figure 1. The laser light goes through a polarizer and becomes polarized. This polarized light goes through a region of magnetic field. Its polarization status is subsequently analyzed by the analyzer-detector subsystem to extract the polarization effect imprinted in the region of the magnetic field. In the actual experiments, one has to multiply the optical pass through the magnetic field by using reflections or Fabry-Perot cavities.

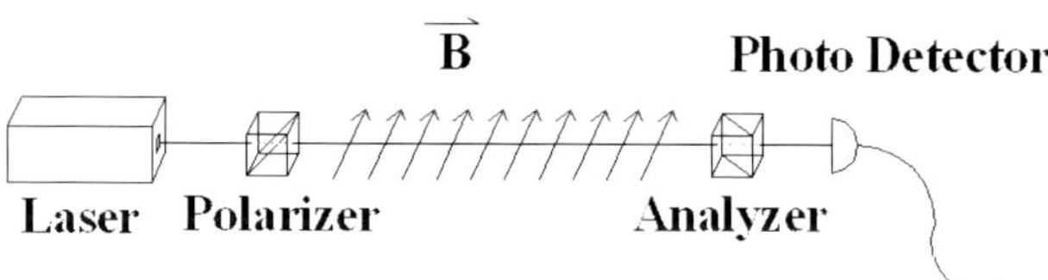

Fig. 1. Principle of vacuum birefringence and dichroism measurement.

For our Q & A experiment, the facility is shown in Figure 2. Photo on the left shows the Q & A apparatus for Phase II experiment [30]; photo in the middle shows the Q & A apparatus for Phase III experiment [31]; the upper right photo shows the mirror suspension; the lower right photo shows the laser injection table. Two vacuum tanks shown on the left photo of Figure 2 house two 5 cm-diameter Fabry-Perot mirrors with their suspensions respectively; the 0.6 m 2.3 T permanent magnet is between two tanks. For Phase III, we double the distance of two Fabry-Perot mirrors to 7 m, and insert another 2.3 T permanent magnet with magnetic field length 1.8 m rotatable up to 13 cycle/s.

All three ongoing experiments – PVLAS, Q & A, and BMV – are measuring the birefringence Δn, and hence, $\eta_1 - \eta_2$ in case η_3 is assumed to be zero. To measure η_1 and η_2 separately, one-arm common path polarization measurement interferometer is not enough. We need a two-arm interferometer with the paths in two arms in magnetic fields with different strengths (or one with no magnetic field).

To measure η_3 in addition, one needs to use both external electric and external magnetic field. One possibility is to let light goes through strong microwave cavity and interferes [12].

134

Fig. 2. Photo on the left shows the Q & A apparatus for Phase II experiment; photo in the middle shows the Q & A apparatus for Phase III experiment; the upper right photo shows a mirror suspension; the lower right photo shows the laser injection table.

As to the term $\xi\Phi$ and parameter ξ, it does not give any change in the index of refraction. However, it gives a polarization rotation and the effect can be measured though observations with astrophysical and cosmological propagation of electromagnetic waves (Section 6).

5.   Empirical Tests of Electromagnetism in Gravity and the χ-g Framework

In section 1, we have discussed the constraints on Proca part of Lagrangian density, i.e., photon mass. In this section, we discuss the empirical foundation of the Maxwell (main) part of electromagnetism. Since gravity is everywhere, for doing this, we use the χ-g framework [32-35] which is summarized in the following interaction Lagrangian density

$$L_I = - (1/(16\pi))\chi^{ijkl} F_{ij} F_{kl} - A_k j^k (-g)^{(1/2)} - \Sigma_I m_I (ds_I)/(dt)\, \delta(\mathbf{x}\text{-}\mathbf{x}_I), \tag{32}$$

with $\chi^{ijkl} = \chi^{klij} = -\chi^{jikl}$ a tensor density of the gravitational fields (e.g., g_{ij}, φ, etc.) or fields to be investigated. The gravitational constitutive tensor density χ^{ijkl} dictates the behaviour of electromagnetism in a gravitational field and has 21 independent components in general. For general relativity or a metric theory (when EEP holds), χ^{ijkl} is determined completely by the metric g_{ij} and equals $(-g)^{1/2}[(1/2)g^{ik}g^{jl}-(1/2)g^{il}g^{jk}]$; when g^{ik} is replaced by η^{ik}, we obtain the special relativistic Lagrangian density (1). The SME (Standard Model Extension [36]) and SMS (Standard Model Supplement [37-39]) overlap the χ-g framework in their photon sector. Hence, our studies are directly relevant to parameter constraints in these models.

In the following, we summarize experimental constraints on the 21 degrees of freedom of χ^{ijkl} to see how close we can reach EEP and metric theory empirically. This procedure also serves to reinforce the empirical foundations of classical electromagnetism as EEP locally is based on special relativity including classical electromagnetism. For a more detailed survey, see [12] and references therein.

Constraints from no birefringence: In the χ-g framework, the theoretical condition for no birefringence (no splitting, no retardation) for electromagnetic wave propagation in all directions is that the constitutive tensor χ^{ijkl} can be written in the following form

$$\chi^{ijkl} = (-H)^{1/2}[(1/2)H^{ik} H^{jl} - (1/2)H^{il} H^{kj}]\psi + \varphi e^{ijkl}, \tag{33}$$

where $H = \det(H_{ij})$ and H_{ij} is a metric which generates the light cone for electromagnetic propagation [32-34, 40]. Polarization measurements of light from pulsars and cosmologically distant astrophysical sources yield stringent constraints agreeing with (33) down to 2×10^{-32} fractionally; for a review, see [35].

In the remaining part of this section, we assume (33) to be valid. Note that (33) has an axion degree of freedom, φe^{ijkl}, and a 'dilaton' degree of freedom, ψ. To fully recover EEP, we still need (i) good constraints on only one physical metric, (ii) good constraints on no ψ ('dilaton'), and (iii) good constraints on no φ (axion) or no pseudoscalar-photon interaction.

Constraints on one physical metric and no 'dilaton' (ψ): Let us now look into the empirical constraints for H^{ij} and ψ. In Eq. (32), ds is the line element determined from the metric g_{ij}. From Eq. (33), the gravitational coupling to electromagnetism is determined by the metric H_{ij} and two (pseudo)scalar fields φ 'axion' and ψ 'dilaton'. If H_{ij} is not proportional to g_{ij}, then the hyperfine levels of the lithium atom, the beryllium atom, the mercury atom and other atoms will have additional shifts. But this is not observed to high accuracy in Hughes-Drever-type experiments. Therefore H_{ij} is proportional to g_{ij} to certain accuracy. Since a change of H^{ik} to λH^{ij} does not affect χ^{ijkl} in Eq. (33), we can define $H_{11} = g_{11}$ to remove this scale freedom [32-34]. For a review, see [35].

Eötvös-Dicke experiments ([41] and references therein) are performed on unpolarized test bodies. In essence, these experiments show that unpolarized electric and magnetic energies follow the same trajectories as other forms of energy to certain accuracy. The constraints on Eq. (33) are

$$|\,1-\psi\,|\,/\,U < 10^{-10}, \tag{34}$$

and

$$|\,H_{00} - g_{00}\,|\,/\,U < 10^{-6}, \tag{35}$$

where $U\,(\sim 10^{-6})$ is the solar gravitational potential at the earth.

In 1976, Vessot *et al.* [42] used an atomic hydrogen maser clock in a space probe to test and confirm the metric gravitational redshift to an accuracy of 1.4×10^{-4}, i.e.,

$$|\,H_{00} - g_{00}\,|\,/\,U \leq 1.4 \times 10^{-4}, \tag{36}$$

where U is the change of earth gravitational field that the maser clock experienced.

With constraints from (i) no birefringence, (ii) no extra physical metric, (iii) no ψ ('dilaton'), we arrive at the theory (32) with χ^{ijkl} given by

$$\chi^{ijkl} = (-g)^{1/2} [(1/2)\, g^{ik}\, g^{jl} - (1/2)\, g^{il}\, g^{kj} + \varphi\, \varepsilon^{ijkl}], \qquad (37)$$

i.e., an axion theory [32, 33, 43]. Here ε^{ijkl} is defined to be $(-g)^{-1/2}\, e^{ijkl}$. The current constraints on φ from astrophysical observations and CMB polarization observations will be discussed in the next section. Thus, from experiments and observations, only one degree of freedom of χ^{ijkl} is not much constrained.

Now let's turn into more formal aspects of equivalence principles. We proved that for a system whose Lagrangian density given by equation (32), the Galileo Equivalence Principle (UFF [Universality of Free Fall; WEP I) holds if and only if equation (37) holds [44, 45].

If $\varphi \neq 0$ in (37), the gravitational coupling to electromagnetism is not minimal and EEP is violated. Hence WEP I does not imply EEP and Schiff's conjecture (which states that WEP I implies EEP) is incorrect [44-46]. However, WEP I does constrain the 21 degrees of freedom of χ to only one degree of freedom (φ), and Schiff's conjecture is largely right in spirit.

The theory with $\varphi \neq 0$ is a pseudoscalar theory with important astrophysical and cosmological consequences (section 6). This is an example that investigations in fundamental physical laws lead to implications in cosmology [45]. Investigations of CP problems in high energy physics leads to a theory with a similar piece of Lagrangian with φ the axion field for QCD [47-49].

In this section, we have shown that the empirical foundation of classical electromagnetism is solid except in the aspect of a possible pseudoscalar photon interaction. This exception has important consequences in cosmology. In the following section, we address this issue.

6. Pseudoscalar-Photon Interaction

In this section, we discuss the modified electromagnetism in gravity with the pseudoscalar-photon interaction which we have reached in the last section, i.e., the theory (32) with the constitutive tensor density (33). Its Lagrangian density is as follows

$$L_I = - (1/(16\pi))(-g)^{1/2}[(1/2)g^{ik}g^{jl}-(1/2)g^{il}g^{kj}+\varphi\, \varepsilon^{ijkl}]F_{ij}F_{kl} - A_k J^k(-g)^{(1/2)} - \Sigma_I m_I (ds_I)/(dt)\delta(\mathbf{x}-\mathbf{x}_I). \qquad (38)$$

In the constitutive tensor density and the Lagrangian density, φ is a scalar or pseudoscalar function of relevant variables. If we assume that the φ-term is local CPT invariant, than φ should be a pseudoscalar (function) since ε^{ijkl} is a pseudotensor. The pseudoscalar(scalar)-photon interaction part (or the nonmetric part) of the Lagrangian density of this theory is

$$L^{(\varphi\gamma\gamma)} = L^{(\text{NM})} = -(1/16\pi)\,\varphi\,e^{ijkl}F_{ij}F_{kl} = -(1/4\pi)\,\varphi_{,i}\,e^{ijkl}A_j A_{k,l} \text{ (mod div)}, \tag{39}$$

where 'mod div' means that the two Lagrangian densities are related by integration by parts in the action integral. This term gives pseudoscalar-photon-photon interaction in the quantum regime and can be denoted by $L^{(\varphi\gamma\gamma)}$. *This term is also the ξ-term in the PPM Lagrangian density L_{PPM} with the $\varphi \equiv (1/4)\xi\Phi$ correspondence.* The modified Maxwell equations [45-46] from Eq. (38) are

$$F^{ik}_{\;;k} + \varepsilon^{ikml}\,F_{km}\varphi_{,l} = -4\pi j^i, \tag{40}$$

where the covariant derivation $_;$ is with respect to the Christoffel connection of the metric. The Lorentz force law is the same as in metric theories of gravity or general relativity. Gauge invariance and charge conservation are guaranteed. The modified Maxwell equations are also conformally invariant.

The rightest term in equation (39) is reminiscent of Chern-Simons (1974 [50]) term $e^{\alpha\beta\gamma}A_\alpha F_{\beta\gamma}$. There are two differences: (i) Chern-Simons term is in 3 dimensional space; (ii) Chern-Simons term as integrand in the integral is a total divergence (Table 2).

Table 2. Various terms in the Lagrangian and their meanings.

Term	Dimension	Reference	Meaning
$e^{\alpha\beta\gamma} A_\alpha F_{\beta\gamma}$	3	Chern-Simons (1974[50])	Integrand for topological invariant
$e^{ijkl}\, \varphi\, F_{ij} F_{kl}$	4	Ni (1973[46], 1974[44], 1977[45])	Pseudoscalar-photon coupling
$e^{ijkl}\, \varphi\, F^{QCD}_{\;ij} F^{QCD}_{\;kl}$	4	Peccei-Quinn (1977[47]), Weinberg (1978[48]) Wilczek (1978[49])	Pseudoscalar-gluon coupling
$e^{ijkl}\, V_i A_j F_{kl}$	4	Carroll-Field-Jackiw (1990[51])	External constant vector coupling

A term similar to the one in equation (39) (axion-gluon interaction) occurs in QCD in an effort to solve the strong CP problem [47-49]. Carroll, Field and Jackiw [51] proposed a modification of electrodynamics with an additional $e^{ijkl}\,V_i A_j F_{kl}$ term with V_i a constant vector (See also [52]). This term is a special case of the term $e^{ijkl}\,\varphi\,F_{ij}F_{kl}$ (mod div) with $\psi_{,i} = -\frac{1}{2}V_i$. Various terms discussed are listed in Table 2.

Polarization rotation is induced in the propagation of linearly polarized electromagnetic wave obeying the modified Maxwell equations (40) in φ-field. This rotation in the long range propagation in cosmos is called cosmic polarization rotation. Empirical tests/constraints of the pseudoscalar-photon interaction come from polarization observations of radio and optical/UV polarization of radio galaxies, and of cosmic microwave background (CMB). The constraints obtained from these observations on the cosmic polarization rotation angle $\Delta\varphi$ are within $\pm$ 30 mrad. Converting to constraints on ξ and $\Delta\Psi$, we have $|\xi\Delta\Psi| = \pm 0.12$. ([12] and references therein).

7. Outlook

We have looked at the foundations of electromagnetism in this short exposition. For doing this, we have used two approaches. The first one is to formulate a Parametrized Post-Maxwellian framework to include QED corrections and a pseudoscalar photon interaction. We discuss various vacuum birefringence experiments -- ongoing and proposed -- to measure these parameters. The second approach is to look at electromagnetism in gravity and various experiments and observations to determine its empirical foundation. We found that the foundation of EEP of the gravity coupling to classical electrodynamics is solid with the only exception of a potentially possible pseudoscalar-photon interaction. This provides the empirical foundation for our first approach to include quantum corrections, possible unification modifications and pseudoscalar-photon interaction. We have discussed various experimental constraints and look forward to more future experiments.

Acknowledgments

We would like to thank Bernard Carr, Pisin Chen, Dah-Wei Chiou, Sang Pyo Kim, Lance Labun, Bo-Qiang Ma, Chiao-Hsuan Wang and She-Sheng Xue for helpful discussions. We would also like to thank the National Science Council (Grants No. NSC100-2119-M-007-008 and No. NSC100-2738-M-007-004) and the National Center for Theoretical Sciences (NCTS) for supporting this work in part. One of us (WTN) would like to thank Leung Center for Cosmology and Particle Astrophysics (LeCosPA Center) for invitation to the First LeCosPA Symposium.

References

1. J. D. Jackson, *Classical Electrodynamics*, John Wiley & Sons, Hoboken (1999).
2. A. Proca, *Comptes Rendus de l'Académie des Sciences* **202**, 1366–1368 (1936).
3. A. Proca, *Journal de Physique et Le Radium* **7**, 347–353 (1936).
4. A. Proca, *Comptes Rendus de l'Académie des Sciences* **203**, 709–711 (1936).
5. A. Proca, *Journal de Physique et Le Radium* **8**, 23–28 (1937).
6. A. Proca, *Journal de Physique et Le Radium.* **9**, 61–66 (1938).
7. E. R. Williams, J. E. Faller and H. A. Hill, *Phys. Rev. Lett.*, **26**, 721–724 (1971).
8. L. Davis, Jr., A. S. Goldhaber and M. M. Nieto, *Phys. Rev. Lett.*, **35**, 1402–1405 (1975).
9. D. D. Ryutov, *Plasma Physics and Controlled Fusion*, **49**, B429–B438 (2007).
10. G. V. Chibisov, *Uspekhi Fizicheskikh Nauk*, **119**, 551–555 (1976) [*Soviet Physics Uspekhi*, **19**, 624–626 (1976)].
11. A. S. Goldhaber and M. M. Nieto, *Rev. Mod. Phys.* **82**, 939-979 (2010).

12. W.-T. Ni, Foundations of Electromagnetism, Equivalence Principles and Cosmic Interactions, Chaper 3 in *Trends in Electromagnetism - From Fundamentals to Applications*, pp. 45-68 (March, 2012), Victor Barsan (Ed.), ISBN: 978-953-51-0267-0, InTech (open access) (2012) [arXiv:1109.5501], Available from: http://www.intechopen.com/books/trends-in-electromagnetism-from-fundamentals-to-applications/foundations-of-electromagnetism-equivalence-principles-and-cosmic-interactions.

13. V. I. Denisov, I. V. Krivchenkov and N. V. Kravtsov, *Phys. Rev.* **D69**, 066008 (2004).

14. S. P. Kim, arXiv:1105.4382v2 [hep-th] (2011).

15. S. P. Kim, *Phys. Rev.* **D84**, 065004 (2011).

16. R. Ruffini, G. Vereshchagin and S.-S. Xue, *Phys. Reports*, **487**, 1-140 (2010).

17. L. Labun and J. Rafelski, arXiv:1203.6148 (2012).

18. W. G. Unruh, *Phys. Rev.* **D14**, 870 (1976).

19. S. Weinfurtner *et al.*, *Phys. Rev. Lett.* **106**, 021302 (2011).

20. W. Heisenberg and E. Euler, *Zeitschrift für Physik*, **98**, 714 (1936).

21. M. Born, *Proc. R. Soc. London,* **A143**, 410 (1934).

22. M. Born and L. Infeld, *Proc. R. Soc. London,* **A144**, 425 (1934).

23. A. Torres-Gomez, K. Krasnov and C. Scarinci, *Phys. Rev.* **D83**, 025023 (2011).

24. B. Carr, L. Modesto and I. Prémont-Schwarz, arXiv:1107.0708 (2011).

25. P. Chen and C.-H. Wang, arXiv:1112.3078 (2011).

26. W.-T. Ni, K. Tsubono, N. Mio, K. Narihara, S.-C. Chen, S.-K. King and S.-s. Pan, *Mod. Phys. Lett. A* **6**, 3671-3678 (1991).

27. W.-T. Ni, *Frontier Test of QED and Physics of the Vacuum*, Eds. E. Zavattini, D. Bakalov, C. Rizzo, 1998, Heron Press, Sofia, pp. 83-97 (1998).

28. R. Battesti *et al.* (BMV Collaboration), *Eur. Phys. J. D* **46**, 323-333 (2008).

29. E. Zavattini *et al.* (PVLAS Collaboration), *Phys. Rev. D* **77**, 032006 (2008).

30. S.-J. Chen, H.-H. Mei and Ni, W.-T. (Q & A Collaboration), *Mod. Phys. Lett. A* **22**, 2815-2831 (2007) [arXiv:hep-ex/0611050].

31. H.-H. Mei *et al.*, *Mod. Phys. Lett.* A 25, 983–993, (2010) [arXiv:1001.4325].

32. W.-T. Ni, Equivalence Principles, Their Empirical Foundations, and the Role of Precision Experiments to Test Them, *Proceedings of the 1983 International School and Symposium on Precision Measurement and Gravity Experiment, Taipei, Republic of China, January 24-February 2, 1983*, W.-T. Ni, (Ed.), (Published by National Tsing Hua University, Hsinchu, Taiwan, Republic of China), pp. 491-517 (1983) [http://astrod.wikispaces.com/].

33. W.-T. Ni, Equivalence Principles and Precision Experiments, *Precision Measurement and Fundamental Constants II*, B. N. Taylor and W. D. Phillips, (Ed.), Natl. Bur. Stand. (U S) Spec. Publ. **617,** pp 647-651 (1984).

34. W.-T. Ni, Timing Observations of the Pulsar Propagations in the Galactic Gravitational Field as Precision Tests of the Einstein Equivalence Principle, *Proceedings of the Second Asian-Pacific Regional Meeting of the International Astronomical Union*, B. Hidayat and M. W. Feast (Ed.), (Published by Tira Pustaka, Jakarta, Indonesia), pp. 441-448 (1984).

35. W.-T. Ni, *Reports on Progress in Physics* **73**, 056901 (2010).

36. V. A. Kostelecky and M. Mewes, *Phys. Rev.* **D66,** 056005 (2002).

37. L. Zhou and B.-Q. Ma, *Mod. Phys. Lett. A* **25,** 2489 (2010) [arXiv:1009.1331].

38. L. Zhou and B.-Q. Ma, *Chin. Phys. Lett. C* **35,** 987 (2011) [arXiv:1109.6387].

39. B.-Q. Ma, arXiv:1203.5852 (2012).

40. C. Lämmerzahl and F. W. Hehl, *Phys. Rev. D* **70**, 105022 (2004).

41. S. Schlamminger *et al.*, *Phys. Rev. Lett.* **100**, 041101 (2008).

42. R. F. C. Vessot *et al.*, *Phys. Rev. Lett.* **45**, 2081-2084 (1980).

43. F. W. Hehl and Yu. N. Obukhov, *Gen. Rel. Grav.*, **40**, 1239-1248 (2008).

44. W.-T. Ni, *Bull. Am. Phys. Soc.* **19**, 655 (1974).

45. W.-T. Ni, *Phys. Rev. Lett.* **38** 301–304 (1977).

46. W.-T. Ni, A Nonmetric Theory of Gravity, preprint, Montana State University (1973) [http://astrod.wikispaces.com/].

47. R. D. Peccei and H. R. Quinn, *Phys. Rev. Lett.* **38**,. 1440-1443 (1977).

48. S. Weinberg, *Phys. Rev. Lett.* **40**, 233 (1978).

49. F. Wilczek, *Phys. Rev. Lett.* **40**, 279 (1978).

50. S.-S. Chern, and J. Simons, *The Annals of Mathematics*, 2nd Ser. **99**, 48 (1974).

51. S. M. Carroll, G. B. Field and R. Jackiw, *Phys. Rev.* **D41**, 1231-1240 (1990).

52. R. Jackiw, Lorentz Violation in a Diffeomorphism-Invariant Theory, *CPT'07 Proceedings* (2007) [arXiv: 0709.2348].

CRITICAL ACCELERATION AND QUANTUM VACUUM

JOHANN RAFELSKI* and LANCE LABUN

*Department of Physics, The University of Arizona,
Tucson, AZ 85721, US*
** E-mail: rafelski@physics.arizona.edu*

Little is known about the physics frontier of strong acceleration; both classical and quantum physics need further development in order to be able to address this newly accessible area of physics. In this lecture we discuss what strong acceleration means and possible experiments using electron-laser collisions and, data available from ultra-relativistic heavy ion collisions. We review the foundations of the current understanding of charged particle dynamics in presence of critical forces and discuss the radiation reaction inconsistency in electromagnetic theory and the apparent relation with quantum physics and strong field particle production phenomena. The role of the quantum vacuum as an inertial reference frame is emphasized, as well as the absence of such a 'Machian' reference frame in the conventional classical limit of quantum field theory.

Keywords: Acceleration; Quantum vacuum; Electron-laser collisions; Mach's principle.

1. Introduction

Special relativity guarantees that all inertial frames of reference are equivalent. However, we do not know which body is inertial and which is accelerated since there is no apparent connection of the classical microscopic laws of physics to a global class of inertial reference frames, which provides the required definition of inertial motion. On the other hand, when formulating the quantum field theory we must introduce in addition to the quantum action also the ground state, the quantum vacuum. This vacuum state provides within the theory the reference to a general inertial frame. In that sense the quantum theory seems to be considerably closer to 'knowing' 'which is the accelerated frame'. However, this information is lost in the present day procedure for passing to the classical limit.

A related challenge in understanding physics at high acceleration is that there is no limit to the strength of force and thus no limit to the acceleration that the inertia of a body can be subject to. To realize acceleration within e.g. electromagnetic (EM) theory, electric and magnetic fields are used. As the strength of the applied electric field increases, so does acceleration imparted to a particle. However, within quantum electrodynamics (QED) when the field strength is too large, a rapid conversion of field energy into pairs ensues, weakening the field and establishing an effective limit to acceleration strength.

In this report we introduce these ideas in greater detail, discuss open questions in the current theoretical framework, and propose methods to explore experimentally new physics emerging. We discuss how the concept of critical acceleration unites different disciplines of physics in which one speaks of critical field strength. Critical acceleration can be today achieved both in ultra-intense laser pulse collisions with relativistic electrons and in ultra relativistic heavy ion collisions at RHIC and at LHC. We survey the classical theory of electromagnetism and discuss the shortcomings related to the problem of radiation reaction and the related problem of electromagnetic mass contained in the field. We discuss in depth the important role of the quantum vacuum as the inertial reference frame.

2. Critical Acceleration

2.1. *Definition*

Critical acceleration of unity in natural units

$$a_c = 1 \equiv \frac{Mc^3}{\hbar} \to 2.331\ 10^{29}\,\frac{\mathrm{m}}{\mathrm{s}^2} \quad \text{for } M = m_e \tag{1}$$

contains implicitly the inertial mass of the particle being accelerated. Therefore one may introduce critical specific acceleration

$$\aleph_c = \frac{a_c}{Mc^2} = \frac{c}{\hbar}. \tag{2}$$

Both a_c and $\aleph_c$ are constructed employing the same fundamental constants we see in defining Planck mass or length. In addition, the gravitational constant G_N, which establishes the relation of mass-energy density and geometry, is required in defining the Planck length and mass. In general relativity, G_N can be expected to connect with acceleration. However, consider a Newtonian force acting at the Planck length

$$\aleph_c^N = \frac{G_N}{L_p^2} = \frac{G_N}{\hbar G_N/c} = \aleph_c \tag{3}$$

The critical specific acceleration $\aleph_c$ arises from a Newtonian force between two Planck masses separated by one Planck length. Notably, the gravitational constant G_N cancels. Even though the value of critical acceleration is gigantic, the absence of G_N opens up the possibility of present day experiments at the 'Planck' acceleration scale.

By virtue of the equivalence principle, we probe particles subject to Planck-scale force in non-gravitational interactions whenever $a_c, \aleph_c$ is achieved. In order to achieve critical accelerations we need strong 'critical' fields made possible by the formation of extended material objects, not present in Einstein's general relativity (GR) theory of dynamics of point particles. It is the quantum theory combined with gauge interactions which creates a resistance to free fall. Free fall would be otherwise the natural state of any GR particle system. Free fall of all particles is evidently the point of view taken by Einstein, e.g. in his study of GR solutions for the case of radial motion of a dust of massive particles.[1]

2.2. *Experimental methods*

2.2.1. *Relativistic heavy ion collisions*

Critical acceleration can be attained in many high energy hadronic and heavy ion collisions. The phenomenon of interest in this context is the rapid stopping of a fraction of matter in the projectile and target hadron in the CM frame of reference. The acceleration required to stop a constituent of colliding hadrons is estimated from the rapidity shift $a \simeq \Delta y / M_i \Delta \tau$ with $M_i \simeq M_N / 3 \simeq 310\,\text{MeV}$. We consider an example collision of ultra-relativistic heavy ions. With $\Delta y = 2.9$ at the SPS or $\Delta y = 5.4$ at RHIC, $\Delta \tau$ must be less than 1.8 fm/c at SPS or 3.4 fm/c in order to have $a > a_c$.

While there is no direct experimental evidence that these limits are satisfied, the global evidence from many related experimental efforts is $\Delta \tau < 1\,\text{fm}$, yielding the preliminary conclusion that critical acceleration phenomena are probed in these interactions. The observation of an excess of soft photons in such experiments[2] may present already a signal of new physics.

2.2.2. *Electrons in strong fields*

Identification of novel physics phenomena in the context of strong interactions is complicated by the many particles created and the different energy scales involved. We can achieve cleaner experimental conditions exploring the behavior of an electron in electromagnetic fields.

For the electron, a_c is achieved subjecting an electron to an electrical field which has the Schwinger critical field strength

$$E_c = \frac{m_e^2 c^3}{e\hbar} = 1.323 \times 10^{18}\,\text{V/m}. \tag{4}$$

This field strength has long been recognized as critical because an electric field of this strength is expected to decay quickly into copious electron-positrons pairs.[3]

2.2.3. *Relativistic laser pulses*

Instead of seeking to create in lab an electric field of magnitude Eq. (4), we can boost the field and hence the acceleration of an electron by setting up a high energy electron-laser collision. The demonstration experiment of this type was undertaken at SLAC in the late 1990s. The electron energy was $E = \gamma m_e c^2 = 46.6$ GeV. A laser of greatest intensity at the time was employed[4,5] with $a_0 = 0.4$, where the dimensionless amplitude is defined by

$$\vec{A} = \mathcal{R}e\left(\vec{A}_0 e^{i(\vec{k}\cdot\vec{r}-\omega t)}\right), \qquad a_0 \equiv |\vec{A}_0|\frac{e}{mc^2}. \tag{5}$$

Together these values of γ, a_0 imply that a peak acceleration of $|\dot{u}^\alpha| = 0.073\,[m_e]$ was achieved. In these conditions the experiment recorded the effects of nonlinear Compton scattering and electron-positron pairs created by the Breit-Wheeler process.

Today high-intensity laser systems are available with $a_0 \gtrsim 50$, that is $(100)^2$ greater intensity. Further, in addition to SLAC, there is CEBAF with a 12 GeV electron beam, radiation-shielded experimental hall and a laser acceleration team already associated with the facility. Renewing this experimental concept provides in our opinion an immediate access to beyond critical acceleration physics.

3. Electromagnetism and Radiation Reaction

3.1. *Independence of fields and particles*

The physics issues arising at high acceleration can be well illustrated considering the dynamics of charged particles interacting with a strong electromagnetic field. This situation is described by the Maxwell-Lorentz action

$$\mathcal{I} = -\frac{1}{4}\int d^4x\, F^2 + \int d^4x \sum_i q_i \int_{\text{path}_i} d\tau\, u_i \cdot A(x)\delta^4(x - s_i(\tau_i)) + \sum_i \frac{m_{\text{I}}^i c}{2} \int_{\text{path}_i} d\tau\, (u_i^2 - 1).$$

$$(6)$$

There are three separate components, but only two at a time are involved in the generation of, respectively, field dynamics (Maxwell equations) and the particle dynamics (Lorentz force).

The Maxwell field equations are obtained by varying the first two components with respect to A^α, where $F^{\beta\alpha} = \partial^\beta A^\alpha - \partial^\alpha A^\beta$,

$$\partial_\beta F^{\beta\alpha} = j^\alpha\,, \qquad \partial_\beta F^{*\,\beta\alpha} = 0 \to \quad F^{\beta\alpha}(x), \tag{7}$$

and the source of the field is due to all charged particles

$$j^\alpha(x) = \sum_i \int d\tau_i\, u_i^\alpha q_i \delta^4(x - s_i(\tau_i)). \tag{8}$$

As indicated in Eq. (7) the solution for a given source is the field $F^{\beta\alpha}$.

The gauge invariance of the first term in the action is assured by the fact that it depends on the fields only. However the middle term which relates particle inertia to the field is not manifestly gauge invariant. Inserting the gauge potential $A \to \partial\Lambda$ we find that this term is a total differential:

$$\int d^4x \sum_i q_i \int_{\text{path}_i} d\tau\, \frac{ds_i}{d\tau} \cdot \frac{\partial\Lambda}{\partial x}\delta^4(x - s_i(\tau_i)) = -\int d^4x \sum_i q_i \int_{\text{path}_i} d\tau\, \frac{d}{d\tau}\delta^4(x - s_i(\tau_i)) \tag{9}$$

For each particle, the initial and final value of τ_i is chosen to correspond to the respective instances that the particle crosses the hypersurface space-time volume integrated over. Then the right hand side is a complicated way to say that the sum of charges entering the integration domain is the same as the sum of charges exiting from it. Therefore the condition that the gauge potential does not contribute to the action is that charge is conserved, that is in differential form, $\partial \cdot j = 0$. Evaluating this by means of Eq. (8) we of course recover the condition Eq. (9).

The Lorentz force is obtained by varying the two components on right in Eq. (6) with respect to the particle world line. However, to preserve the gauge invariance of the result we must not allow a variation at the surface of the domain, so as to guarantee that the result is compatible with both gauge invariance and charge conservation. One finds

$$m_{\mathrm{I}}\frac{du^\alpha}{d\tau} = -qF^{\alpha\beta}u_\beta; \quad \frac{ds^\alpha}{d\tau} \equiv u^\alpha(\tau) \to s^\alpha(\tau), \tag{10}$$

As indicated, for a given field we can obtain the path of each particle.

The above is a summary of book material. Yet there is an obvious challenge not all books discuss. The question is, are the three terms in the Maxwell-Lorentz action Eq. (6) for critical acceleration consistent with each other and leading to consistent dynamics of particles and fields? This question does not have an obvious answer: the action Eq. (6) is relativistically invariant and leads to gauge invariant dynamics by intricate construction. However, otherwise it appears as an ad hoc composition of several terms. In fact we recapitulated here the derivation of particle and field dynamics to emphasize how they arise in a separate and distinct manner from the action and variational principle framework but involve quite distinct objects of variation, fields and paths.

3.2. *Radiation reaction*

A simple example addressing this consistency is the motion of a charged particle in a magnetic field according to Lorentz equation. While the direction of motion changes, the particle energy remains constant. However, the acceleration that is required to change the direction of motion causes radiation, and emission of radiation removes energy from particle motion placing it in the field. The energy loss is a 'small' effect for small fields and accelerations. When the magnitude of the acceleration approaches the critical value Eq. (2), the dynamics of charged particles are decisively influenced by the radiation field. This is called radiation reaction: to describe how a particle moves we must account for its generated radiation field in addition to the applied strong magnetic field.

The dynamical equations Eqs. (7) & (10) can be 'improved' to account for radiation reaction. The idea pursued by Abraham and Lorentz is to solve the Maxwell equations using Green's functions, obtain radiation field, and incorporate the emitted radiation field as an additional force in the Lorentz equation. This program leads to the Lorentz-Abraham-Dirac (LAD) equation written in this form by Dirac[6]

$$\frac{(m_{\mathrm{I}} + m_{\mathrm{EM}})du^\alpha}{d\tau} = u_\beta q(F^{\beta\alpha}_{\mathrm{external}} + F^{\beta\alpha}_{\mathrm{rad}}), \qquad F^{\beta\alpha}_{\mathrm{rad}} = \frac{2q}{3}(\ddot{u}^\beta u^\alpha - \ddot{u}^\alpha u^\beta) \tag{11}$$

Here the $F^{\beta\alpha}_{\mathrm{external}}$ is the field generated by all other charged particles, i.e. a prescribed field in which the particle considered moves. We see two effects, the appearance of $F^{\beta\alpha}_{\mathrm{rad}}$ which is the radiation field generated by the motion of the particle considered, and m_{EM} which is the classical electromagnetic energy content of the field a charged particle generates. Note that at critical acceleration, the radiation reaction correction $F^{\beta\alpha}_{\mathrm{rad}}$ has the same order of magnitude as the Lorentz force, and thus in principle the iterative feed-back used to obtain radiation reaction in LAD from Eq. (11) breaks down.

3.3. *Problems with LAD*

The LAD equation of motion presents two foundational challenges. First, a charged particle mass acquires an electromagnetic component, which arises from the energy content of particle's field. Second, there are solutions of LAD that violate causality. **1)** The electromagnetic mass m_{EM}, appearing next to the inertial mass m_{I}, is divergent in Maxwell's electromagnetism. This divergence can be regulated by establishing a limiting field strength within a modified framework such as in Born-Infeld[7] (BI) or another nonlinear theory of electromagnetism. In practical terms we modify the first term in Eq. (6).

The particular attractiveness of the BI theory is the elegant format of the action, which addresses the possible presence of curved space time:

$$\mathcal{I}_F = -\int d^4x \sqrt{-\det\left(g_{\mu\nu}\right)}\,\frac{F^2}{4} \rightarrow -\int d^4x \left(\sqrt{-\det\left(G_{\mu\nu}\right)} - \sqrt{-\det\left(g_{\mu\nu}\right)} \right) E_{\mathrm{BI}}^2$$

$$G_{\mu\nu} = g_{\mu\nu} + \frac{F_{\mu\nu}}{E_{\mathrm{BI}}}, \quad \det\left(G_{\mu\nu}\right) = -1 + \frac{E^2 - B^2}{E_{\mathrm{BI}}^2} + \frac{(E \cdot B)^2}{E_{\mathrm{BI}}^4} \tag{12}$$

We evaluated the determinant of $G_{\mu\nu}$ in flat space. Like in any theory of nonlinear electromagnetism, the Maxwell equations apply to displacement fields D, H, while the Lorentz force depends on the E, B fields. The nonlinear relation between D, H and E, B within BI theory imposes a limit on the electric and magnetic field strength and thus an upper limit on the force and acceleration.

The problem with the BI approach is that precision tests show that the linear Maxwell theory still applies for very large 'nuclear' field strengths. Therefore even if we were to assume that all of the electron's mass resides in m_{EM}, the predicted electron mass would have to be much larger than observed.[8] That is, the limiting field E_{BI}, required to 'explain' electron mass as being electromagnetic is too small to be consistent with other experimental evidence.

2) The appearance of a third derivative $\ddot{u}^\alpha$, $u^\alpha = \dot{x}^\alpha$ in Eq. (11) requires assumption of an additional boundary condition to arrive at a unique solution describing the motion of a particle. Only a boundary condition in the (infinite) future can eliminate solutions exhibiting self-accelerating motion, that is 'run-away' solutions. Such a constraint is in conflict with the principle of causality.

There is no known natural remedy to the LAD problems in a systematic ab initio process that for example involves choosing a new action for the charged particle system. As noted, LAD itself is not fully consistent as it was derived assuming that the effect of radiation reaction is perturbative. The effort of Born and Infeld to modify the field action in the end did not cure either of the two problems. Modifying the field-particle coupling is very difficult seeing the subtle implementation of gauge invariance, and we know of no modification of the inertial term consistent with the Lorentz invariance. A commonly heard point of view is to negate the existence of a problem arguing that classical dynamics are superseded by quantum physics. However, strong acceleration relates particle dynamics to gravity, which is a classical theory. *The only way that quantum physics could help is to generate a theoretical classical limit that differs from our expectations. We will return to this point below.*

3.4. *Landau-Lifshitz Equation*

The second difficulty with the LAD equation of motion Eq. (11) motivated many ad hoc repair efforts in the intervening years.[9] Among them, the approach of Landau and Lifshitz (LL)[10] has attracted much study because it has no conceptual problems, is semi-analytically soluble,[11–14] and incorporates the Thomson (classical) limit of the Compton scattering process.[15] The LL form of radiation reaction originates in perturbative expansion in the acceleration, with the problematic third derivative replaced according to

$$\ddot{u}^\mu \to \frac{d}{d\tau}\left(-\frac{e}{m}F^{\mu\nu}u_\nu\right). \tag{13}$$

The resulting equation of motion (m is the sum of inertial and electromagnetic mass)

$$m\,\dot{u}^\mu = -\frac{e}{c}F^{\mu\nu}u_\nu - \frac{2e^3}{3m}\left(\partial_\eta F^{\mu\nu}u_\nu u^\eta - \frac{e}{m}F^{\mu\nu}F^\eta_\nu u_\eta\right) + \frac{2e^4}{3m^2}F^{\eta\nu}F_{\eta\delta}u_\nu u^\delta u^\mu \tag{14}$$

is equivalent to LAD only for weak accelerations, a point stated not sufficiently clearly by Landau and Lifshitz.[10] It is important to recognize that the LL equation Eq. (14) implies the field-particle interaction is altered. However, an appropriate fundamental action has not been found. It must therefore be studied at the level of the equation of motion.

We have studied the motion generated by Eq. (14) for a laser-electron collision.[14] In a consistent solution to the coupled LL dynamics, only after radiation loss is accounted for in the electron dynamics can a laser pulse stop an electron in a head-on collision.

3.5. *Caldirola Equation*

Another proposal of considerable elegance to generalize LAD equations of motion was formulated by Caldirola[16,17]

$$\mp\frac{m_e}{\delta t}\left[u^\alpha_\mp + u^\alpha\frac{u\cdot u_\mp}{c^2}\right] = \frac{e}{c}F^{\alpha\beta}u_\beta, \quad u_\mp \equiv u(t \mp \delta t), \quad \delta t \equiv \frac{4}{3}\frac{e^2}{m_e c^3} \tag{15}$$

Here the choice of sign determines if the electron is radiating (-) or absorbing (+) energy from the environment. Thus the LAD difficulty of run-away solutions is resolved by choosing the non-local form with upper sign in Eq. (15). Further, the non-local form also suggests that consistent Maxwell-Lorentz dynamics can be achieved by connecting with a discrete space-time. However, just like with LL modification of LAD, an action for the dynamics described by Eq. (15) has not been discovered.

148

The dynamics generated by each equation of motion Eq. (11), Eq. (14) and Eq. (15) differ already at acceleration well below critical. As a consequence, we can expect the radiation signatures to be distinguishable and the form of dynamical equations to be testable experimentally. However, these proposed resolutions of the inconsistency suffer from their ad hoc format. We are missing a physics principle that would create a compelling format of action and thus particle and field dynamics, and therefore it is to be expected that none of these efforts represents a definitive theoretical description of charged particle dynamics.

3.6. *Experiments in Classical Domain*

Experiments are possible in kinematic domains where particle dynamics are classical rather than quantum and the predictions of Eq. (11), Eq. (14) and Eq. (15) differ. Seeing that the couple of ad hoc modifications noted here are different in this domain, we believe that *any* modified classical dynamics are accessible to experiment.

For the case studied in detail[14] of a laser of amplitude characterized by the strength a_0 (in units of mc^2, see Eq. (5)) colliding with a high energy electron $E = \gamma m_e c^2$, the condition for the relevance of radiation reaction is approximately given by

$$a_0^2 \gamma \gtrsim \frac{3}{2e^2} \frac{m_e}{\omega}, \tag{16}$$

and LAD Eq. (11) and Landau-Lifshitz Eq. (14) dynamics are distinguishable when

$$a_0 \gamma^2 \gtrsim \frac{3}{2e^2} \frac{m_e}{\omega}. \tag{17}$$

On the other hand, the electron experiences critical acceleration when

$$a_0 \gamma = \frac{m_e}{\omega} \tag{18}$$

The different dependencies on electron and laser parameters (γ and a_0, ω, respectively) reveal the domain of interest, seen in Figure 1.

As Figure 1 shows radiation reaction is relevant to classical dynamics, and there is a domain where radiation effects expected differ between forms of dynamical equations. Other observables such as pair production could be even more sensitive probes of radiation reaction classical dynamics. This means that the challenge to create a dynamical theory that consistently incorporates critical acceleration must be already addressed within the classical physics domain. An important reason to retain focus on the classical domain is the evident connection to gravity.

An important outcome of the effort to understand electromagnetism with critical acceleration is control of the abundant radiation produced by accelerated charges: naively one can say that a charge which is strongly 'kicked' loses for some time its electromagnetic mass component hidden in the field and must reestablish it. The purpose of the experiments is to understand how the true inertia and radiation define the electromagnetic part of mass.

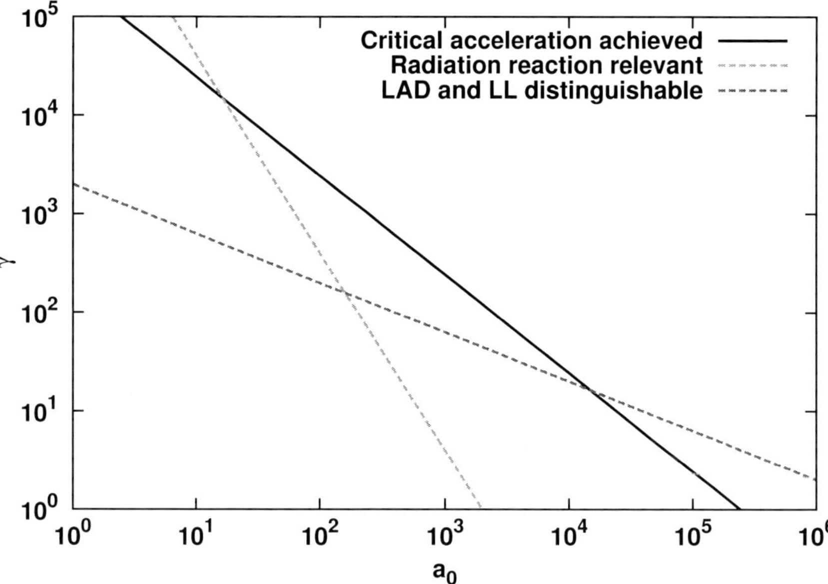

Fig. 1. For the collision of a high energy $E/m_e c^2 = \gamma$ electron with an intense laser pulse with normalized amplitude a_0 and frequency $\hbar\omega = 4.8\,\text{eV}$. Above the steeper dashed (red) line radiation reaction is relevant Eq. (16); and above shallower dashed (blue) line LAD and Landau-Lifshitz equations of motion generate different dynamics. Above the solid line (black), the electron experiences critical acceleration, however, quantum effects are expected to be important.

Nearly 100 years after the difficulty with electromagnetic theory was identified, experimental effort can aid progress today.[18]

4. Mach, Inertia and the Quantum Vacuum

Given that the EM equations of motion are generated from Eq. (6) independently for particle dynamics and for field dynamics, we believe that in order to formulate self-consistent particle-plus-field dynamics a connection of two seemingly separate dynamical elements needs to be found. General relativity accomplishes this goal by removing acceleration altogether; particles are always in free fall, and the geometry of the manifold is the field.

The difference between the nature of classical gravity, described employing general relativity theory, and the electromagnetic forces, provokes the question: *If a charged particle in a "free-fall" orbit bound by the gravitational field to travel around the Earth will not radiate, why should it emit synchrotron radiation if the same orbit is established by other e.g. electromagnetic forces?* This is a paradox of classical field theory arising from inherently different treatment of acceleration under gravitational and electromagnetic forces.

In view of how gravitational theory handles force, one approach to creating a consistent framework is to geometrize the electromagnetic (and in fact all gauge) interactions. This program puts all 'forces' on the same footing as gravity, doing away with acceleration in a higher dimensional Kaluza-Klein space-time. In this formulation, accelerated motion in our space-time may be a consequence of constraints imposed to reveal dynamics by projection onto the 3+1 dimensional hyper-surface. Despite many years of research this program has not been implemented to the point of practical applications, and there is no evidence that it will succeed.

Moreover, geometrization of EM interactions, similarly to gravity, does not resolve the question how to recognize accelerated versus non-accelerated observers in absence of Newton's absolute space. Implicit in the gravitational orbit definition (see question above) is the presence of an (inertial) observer at asymptotically flat spatial infinity. The authors are not aware of an electromagnetic theory that addresses the need to relate acceleration to an inertial Machian reference frame. On the other hand, Mach, Einstein[19] and many others have recognized the need to refer to a class of inertial frames of reference in order to agree on how to measure acceleration. Mach's original proposal introduced the rest frame of the total mass of the universe, and an analogous universal reference frame is the rest frame of the cosmic microwave background.

Moving on from classical EM theory to quantum electrodynamics (QED) we observe a subtle difference even though the QED action is built on the same principles (gauge and Lorentz invariance) as the classical action Eq. (6), and the quantum action is organized to reduce to corresponding classical action. As seen e.g. in canonical quantization, any quantum field theory requires the introduction of a quantum vacuum state. This vacuum state is a natural candidate for the inertial reference frame.

The introduction of quantum theory implicitly embraces reference to an inertial frame, a major conceptual advance compared to the classical theory. Unfortunately, the text-book formulation of the classical limit of the quantum theory loses the information regarding the vacuum state and hence the relationship to an inertial reference frame. We draw attention to the relativistic Wigner function approach to classical limit,[20,21] which addresses this important challenge of retaining the information about the vacuum state in the classical limit. That was the good news, the bad news is that the dynamics of particle motion is connected closely with various possible processes of particle production, for which reason it has been exceedingly hard to make the relativistic Wigner method a practical tool.

It is our belief that the problems addressed here could be partially or even completely resolved by inclusion of the vacuum as an active ingredient in the classical dynamics of charged particles. This agrees well with the vacuum structure defining the nature of the laws of physics; our understanding of the influence of the vacuum has grown considerably in past 75 years. The Casimir force made vacuum fluctuations popular, and spontaneous symmetry breaking defining the structure of electro-weak interactions in the Standard Model showed that interactions yesterday deemed 'fundamental' can be effective. Quark confinement (non-propagation in vacuum of color charge) defines the nature of most of the mass of visible matter in the Universe and arises from non-perturbative properties of the quantum vacuum.

Vacuum properties already play an integral role in electromagnetic theory. Below critical acceleration, the virtual possibility of pair creation leads to well-known vacuum polarization phenomena,[22–24] giving the quantum vacuum dielectric properties. In Feynman diagram language, the real photon is decomposed into a bare photon and a photon turning into a virtual pair. The result is a renormalized electron charge smaller than the bare charge and a slightly stronger Coulomb interaction (0.4% effect at highest accessible experimental scales).

The quantum vacuum state can be probed by experiments at critical acceleration, which in QED corresponds to the critical field strength Eq. (4). An electric field at the critical strength is expected to decay rapidly into particle-anti-particle pairs. The decay process is non-perturbative, involving a large number of quanta ($N\hbar\omega \to \infty$), but has no classical analog or obvious limit. This possibility of electromagnetic field decay gives rise to a temperature parameter analogous to but numerically different from the Unruh temperature detected by an observer accelerated through the vacuum.[25]

5. Conclusions

We have discussed here the new opportunities to study foundational physics involving acceleration and described potential avenues to search for an extension of physics to understand critical acceleration phenomena. Experiments at and much beyond the critical acceleration Eq. (1) can be today performed in electron-laser pulse collisions. This in particular includes the 'stopping' of relativistic charged particles by ultra-intense laser pulses. Such experiments should help resolve the radiation reaction riddle in electromagnetic theory. We can expect a rich field of applications and theoretical insights to follow.

The study of physics phenomena beyond critical acceleration should lead to a better understanding of the relation between General Relativity, Electromagnetism and Quantum Physics, and specifically to understanding of the relation of classical dynamics and inertia to the 'Machian' inertial frame, the quantum vacuum. The (effective) quantum field theory of known interactions, the Standard Model (SM), has incorporated the idea that a theory that describes forces must allow for a universal inertial reference frame. However, this incorporation of an inertial reference frame is lost in the classical particle-motion theory that emerges in the naive classical limit. We have shown that such a classical theory does not provide a consistent framework to address supercritical forces/acceleration.

The inclusion of the quantum vacuum structure into the understanding of laws of physics underpins our interpretation of the SM of all microscopic interactions. Since the quantum vacuum state plays a pivotal role and the SM requires input of tens of parameters, it is the universal belief that the current understanding of fundamental forces is an effective theory. This point of view agrees with our argument that the present framework of classical forces in physics does not adequately describe particle dynamics at critical acceleration as it is not an adequate limit of the effective quantum theory from which supposedly it arises. Both these observations could mean that a more fundamental theory has to be discovered before we can reconcile both SM complexities and classical theory deficiencies with the elegant Universe. The next step leading to this discovery, we suggest, is to study the critical acceleration by the way of relativistic electron-laser pulse collisions.

Acknowledgments

This work was supported by a grant from the U.S. Department of Energy, DE-FG02-04ER41318.

References

1. A. Einstein, Annals of Mathematics **40**, 922 (1939).
2. C. Y. Wong, Phys. Rev. C **81**, 064903 (2010).
3. L. Labun and J. Rafelski, Phys. Rev. D **79** 057901 (2009).
4. D. L. Burke *et al.*, Phys. Rev. Lett. **79**, 1626 (1997).
5. C. Bamber *et al.*, Phys. Rev. D **60**, 092004 (1999).
6. P. A. M. Dirac, Proc. Roy. Soc. Lond. A **167**, 148 (1938).
7. M. Born and L. Infeld, Proc. Roy. Soc. Lond. A **144**, 425 (1934).
8. J. Rafelski, G. Soff and W. Greiner, Phys. Rev. A **7**, 903 (1973).
9. A short catalog of these efforts can be found in Table 1 of
   J. Rafelski, L. Labun and Y. Hadad, "Horizons of Strong Field Physics," AIP Conf. Proc. **1228**, 39 (2010), [arXiv:0911.5556 [physics.gen-ph]].
10. L. D. Landau and E. M. Lifshitz. *The Classical theory of Fields*, 4th ed., (Pergamon, Oxford, 1989).
11. R. Rivera, and D. Villarroel, Phys. Rev. E **66**, 046618 (2002).
12. S. G. Rajeev, Annals Phys. **323**, 2654 (2008).
13. A. Di Piazza, Lett. in Math. Phys. **83**, 305 (2008).
14. Y. Hadad, L. Labun, J. Rafelski, N. Elkina, C. Klier and H. Ruhl, Phys. Rev. D **82**, 096012 (2010).
15. T. Padmanabhan, "Inverse Compton Scattering – Revisited," J. Astrophys. Astr. **18**, 87 (1997).
16. P. Caldirola, Riv. Nuovo Cim. 2N13, 1 (1979).
17. P. Caldirola, G. Casati and A. Prosperetti, Il Nuovo Cim. **43**, 127 (1978).
18. L. Labun and J. Rafelski, Acta Phys. Polon. B **41**, 2763 (2010).
19. A. Einstein, letter to Ernst Mach, in *Gravitation*, C. Misner, K.S. Thorne, and J.A. Wheeler, San Francisco: W. H. Freeman. ISBN 0-7167-0344-0 (1973).
20. I. Bialynicki-Birula, P. Gornicki and J. Rafelski, Phys. Rev. D **44**, 1825 (1991).
21. J. Rafelski and G. R. Shin, Phys. Rev. A **48**, 1869 (1993).
22. E. A. Uehling, Phys.Rev. **48** 55 (1935).
23. H. Euler and B. Kockel, Naturwiss. **23** 246 (1935).
24. W. Heisenberg and H. Euler, Z. Phys. **98**, 714 (1936); translated and placed in arXiv as: [physics/0605038].
25. L. Labun and J. Rafelski, *Temperature of Electron Fluctuations in an Accelerated Vacuum* [arXiv:1204.[ ]] in these proceedings.

TEMPERATURE OF ELECTRON FLUCTUATIONS IN AN ACCELERATED VACUUM

LANCE LABUN* and JOHANN RAFELSKI

Department of Physics, The University of Arizona, Tucson, AZ 85721, USA
** E-mail: lalabun@labunllc.com*

The electron vacuum fluctuations measured by $\langle \bar{\psi}\psi \rangle$ do not vanish in an externally applied electric field $\mathcal{E}$. For an exactly constant field, that is for vacuum fluctuations in presence of a constant accelerating force, we show that $\langle \bar{\psi}\psi \rangle$ has a Boson-like structure with spectral state density $\tanh^{-1}(E/m)$ and temperature $T_\mathrm{M} = e\mathcal{E}/m\pi = a_v/\pi$. Considering the vacuum fluctuations of 'classical' gyromagnetic ratio $g = 1$ particles we find Fermi-like structure with the same spectral state density at a smaller temperature $T_1 = a_v/2\pi$ which corresponds to the Unruh temperature of an accelerated observer.

Keywords: acceleration; quantum vacuum; nonperturbative QED.

1. Introduction

Müller *et al.*[1] showed that in presence of constant external electromagnetic fields the Heisenberg-Euler effective potential of QED can be cast in the form of a thermal background characterized by a spectral density of states $\rho(E)$ and temperature T_M

$$\rho(E) = \frac{m^2}{8\pi^2} \ln(E^2 - m^2 + i\epsilon), \qquad T_\mathrm{M} = \frac{e\mathcal{E}}{m\pi} = \frac{a_v}{\pi}. \tag{1}$$

Since an electric field accelerates all charged particles, real or virtual, this can be understood as a property of the vacuum under constant global acceleration[2] $a_v = e\mathcal{E}/m$. This circumstance has also been discussed by Pauchy Hwang and Kim.[3]

Associated with this result is the quantum statistics: spin-1/2 QED exhibits a thermal distribution as though the fluctuating degrees of freedom are bosons, while spin-0 scalar QED exhibits vacuum fluctuations as though the degrees of freedom are fermionic. The former result could be understood as being due to pairing of electron-positron pairs. However the latter case has no obvious explanation.

The case of 'accelerated vacuum' parallels that of an 'accelerated observer' traveling in a matter- and field-free spacetime. This observer sees a thermal background characterized by the Unruh temperature

$$T_\mathrm{U} = \frac{a_\mathrm{U}}{2\pi} = \frac{T_\mathrm{M}}{2}. \tag{2}$$

The statistics of the thermal distribution are bosonic considering the vacuum of a scalar particle[4,5] and fermionic in the vacuum of a fermi particle.[6] Many readers will expect that there should not be a difference between 'accelerated vacuum' and 'accelerated observer', yet there is a disagreement in relation to particle spin and statistics and the value of the temperatures Eq. (1) & Eq. (2).

We will show that these two results Eq. (1) & Eq. (2) are experimentally distinguishable. This implies the ability to determine which is accelerated, the observer or the vacuum state. The question is whether or not one should be able in-principle to determine the accelerated state within the current formulation of quantum field theory. If the conclusion is that one should *not* be able to determine which is accelerated, then there is additional undiscovered physics content in either the vacuum structure or the Unruh accelerated detector.

It is of considerable interest to find a model achieving agreement in the quasi-thermal properties of the quantum vacuum, in the sense that the accelerated observer registers the same outcome as the 'accelerated' field-filled vacuum. The disagreement in relation to particle spin suggests a closer look at the spin properties of the fluctuations. Altering the magnetic moment of the electron from its Dirac value described by the gyromagnetic ratio $g = 2$ to the 'classical' spinning-particle value $g = 1$ achieves agreement with the Unruh temperature and statistics.[7] We summarize the results obtained and consider the properties of QED vacuum condensate $\langle \bar{\psi}\psi \rangle$ in presence of strong quasi-constant external fields in the thermal framework modeled by the 'classical' $g = 1$ QED.

2. Effective Action

Charge convective current and spin magnetic dipole current are conserved independently and thus in QED the associated integral 'charges' - the electric charge and magnetic dipole moment can be prescribed arbitrarily. The gyromagnetic ratio g combines with particle charge and mass in defining the magnitude of the magnetic dipole moment. The dynamics of a particle ψ with general g is generated by

$$\left[D^2 + m^2 - \frac{g}{2} \frac{e\sigma_{\mu\nu}F^{\mu\nu}}{2} \right] \psi = 0 \tag{3}$$

where $D = i\partial + eA$ is the covariant derivative, $F^{\mu\nu}$ the electromagnetic field strength tensor and $\sigma_{\mu\nu} = (i/2)[\gamma_\mu, \gamma_\nu]$. Only for the specific case $g = 2$ can one choose to write this in the Dirac equation form. Any value of the gyromagnetic ratio g can arise, and as long as the quantization of charge is not understood, it is difficult to claim that the value of g arising in a specific simplified dynamical equation is of greater interest than other values. Like charge, g is the subject of experimental and theoretical effort to determine quantum corrections to an input value, the best known being the QED correction to the Dirac particle $g = 2 + \alpha/\pi + \dots$.

The effect of g on vacuum fluctuations is determined from the effective potential

$$V_{\text{eff}} = -\frac{i}{2}\text{tr} \ln \left[D^2 + m^2 - \frac{g}{2} \frac{e\sigma_{\mu\nu}F^{\mu\nu}}{2} \right] \tag{4}$$

For the Heisenberg-Euler case of a constant electric-only field of strength $\mathcal{E}$, we evaluate the trace using the proper time method[8]

$$V_{\text{eff}} = \frac{\gamma_s}{32\pi^2} \int_0^\infty \frac{du}{u^3} \left(\frac{e\mathcal{E}u \cosh(\frac{g}{2}e\mathcal{E}u)}{\sinh e\mathcal{E}u} - 1 \right) e^{-im^2 u} \tag{5}$$

in which the generalized degeneracy γ_s counts the number and type of degrees of freedom. $\gamma_s = 4$ when $g = 2$ for spin-1/2 Dirac electron, or $\gamma_s = -2$ when $g = 0$ for a spin-0 electron. The -1 inside the parentheses removes the field-independent constant.

Transforming V_{eff} to a statistical format proceeds via meromorphic expansion of the integrand of Eq. (5).[1,7] The finite (regularized and renormalized) effective potential is

$$V_{\text{eff}} = \frac{\gamma_s m^2 T_{\text{M}}^2}{32\pi^2} \int_0^\infty \frac{2u\,du}{u^2 - 1 + i\epsilon} \sum_{n=1}^\infty \frac{e^{-nu\frac{m}{T_{\text{M}}}}}{n^2} \cos\left(n\pi\left(\frac{g}{2} + 1\right) \right) \tag{6}$$

with integration contour defined by the usual assignment

$$m^2 \to m^2 - i\epsilon. \tag{7}$$

This also makes explicit that the effective potential contains an imaginary part, as will be discussed below. Setting $g = 2$ for a spin-1/2 (Dirac) electron, $\cos 2n\pi = 1$ for all n, and setting $g = 0$ for a spin-0 electron, $\cos n\pi = (-1)^n$ producing an alternating sum. For each case, integrating by parts twice and summing the series yields the results of Müller *et al.*[1]

The exponential weights of the terms in the series in Eq. (6) generate a thermal distribution, and the statistics of the distribution are determined by the phase of the terms in the series. Summing with arbitrary g, the effective potential is

$$V_{\text{eff}} = \frac{\gamma_s m^2 T_{\text{M}}}{64\pi^2} \int_0^\infty dE \ln(E^2 - m^2 + i\epsilon) \sum_\pm \ln(1 + e^{\pm i\pi \frac{g}{2}} e^{-E/T_{\text{M}}}) \tag{8}$$

The sum over $\pm$ ensures the distribution is real so that the imaginary part arises only from the branch cut in the first log factor. Restoring $g = 2$ and $\gamma_s = 4$ identifies the spectral density of states Eq. (1) according to $V \equiv T \int_0^\infty \ln(1 - e^{-E/T})\rho(E)dE$.

For $g = 1$ summing over $\pm$ simplifies it to

$$V_{\text{eff}}\Big|_{g=1} = \frac{\gamma_s m^2 T_{\text{U}}}{32\pi^2} \int_0^\infty dE \ln(E^2 - m^2 + i\epsilon) \ln(1 + e^{-E/T_{\text{U}}}) \tag{9}$$

exhibiting in the second log factor a thermal fermionic distribution controlled by the Unruh temperature, T_{U}. The effective potential of a 'classical spinning electron' with $g = 1$ in a constant field thus has the format of a thermodynamic potential with temperature parameter and statistics in agreement with the expectation of an accelerated observer in the (unaccelerated) vacuum of a fermion field.

3. Condensate

The quantum fluctuations induced by the external field are measured by the condensate $\langle \bar{\psi}\psi \rangle$, which is the difference of the Green's functions in the vacuum with the external field and the vacuum with no field present (0 superscript)

$$-\langle \bar{\psi}\psi \rangle = \text{tr}\left[iS_F(x,x) - iS_F^0(x,x) \right], \quad S_F(x,x') = -i\langle \mathcal{T}\psi(x')\bar{\psi}(x) \rangle. \tag{10}$$

$\langle \mathcal{T}... \rangle$ is the vacuum expectation of time-ordered operators, the limit $x' \to x$ is evaluated in the point-splitting procedure to preserve gauge invariance, and the F subscript indicates Feynman boundary conditions. This definition of the condensate displays the implicit definition of the reference, no-field vacuum state.

The condensate is related to the effective potential by

$$-m\langle \bar{\psi}\psi \rangle = m\frac{dV_{\text{eff}}}{dm} \tag{11}$$

Note that the differentiation with respect to m improves the convergence properties of Eq. (5). Evaluating the derivative of Eq. (5), we find the formerly logarithmically divergent contribution is finite. Because $-m\langle \bar{\psi}\psi \rangle$ includes this term quadratic in $\mathcal{E}$ (see discussion in Ref. 9), we use now the meromorphic series

$$1 - \frac{x\cosh(xy)}{\sinh(x)} = -2x^2 \sum_{n=1}^{\infty} \frac{\cos n\pi(y+1)}{x^2 + (n\pi)^2} \tag{12}$$

The resulting expression for $\langle \bar{\psi}\psi \rangle$ for arbitrary g is

$$\langle \bar{\psi}\psi \rangle = \frac{-\gamma_s m^2}{8\pi^2} \int_0^{\infty} dE \frac{\tanh^{-1}(E/m + i\epsilon)(1 + e^{E/T_{\text{M}}}\cos(\frac{g}{2}\pi))}{e^{2E/T_{\text{M}}} + 1 + 2e^{E/T_{\text{M}}}\cos(\frac{g}{2}\pi)} \tag{13}$$

Setting $g = 0$ ($g = 2$) yields a fermionic (bosonic) distribution controlled by T_{M}

$$\langle \bar{\psi}\psi \rangle = -\frac{m^2}{4\pi}\frac{\gamma_s}{2\pi} \int_0^{\infty} \frac{\tanh^{-1}(E/m + i\epsilon)dE}{e^{E/T_{\text{M}}} + (-1)^{g/2}}, \tag{14}$$

Identifying the numerator of the integrand $\tanh^{-1}(z)$ by analogy with statistical physics $\langle N \rangle / V = \int_0^{\infty} \Gamma(E)dE/(e^{E/T} \pm 1)$, the degeneracy of states

$$\Gamma(E) = -\frac{\gamma_s}{4}m^2 \tanh^{-1}(E/m) \tag{15}$$

is the same in each value of g considered.

For $g = 1$, the terms containing cosine in the numerator and denominator of Eq. (13) vanish, leaving the fermi occupancy factor in the denominator with twice the Euler-Heisenberg temperature $2/T_{\text{M}} = 1/T_{\text{U}}$.

$$\langle \bar{\psi}\psi \rangle_{g=1} = \frac{m^2}{4\pi}\frac{\gamma_s}{2\pi} \int_0^{\infty} \frac{\tanh^{-1}(E/m + i\epsilon)dE}{e^{E/T_{\text{U}}} + 1} \tag{16}$$

which displays the fermionic occupancy factor in the denominator with a distribution controlled by the Unruh temperature. The fluctuations of a $g = 1$ 'electron' in an external electric field are thus reconciled with the fluctuations expected by an observer accelerated at $a_v = e\mathcal{E}/\pi m$ through the (unaccelerated) vacuum of a fermion field.

4. Observable Effects from $g = 1$

One observable effect is spontaneous pair production in strong fields discussed by us earlier,[7] and originating in the imaginary part of the effective action Eq. (6). This allows any non-vanishing electric field to spontaneously decay into charged particle pairs. $g = 0, 2$ yield the largest total decay probability and are in this sense the best cases for experiment. For $g = 1$ the reduction in the effective temperature parameter is largest. Due to the exponential dependence, the reduction in the temperature parameter by factor 2 reduces spontaneous pair production below the critical field $\mathcal{E}_c = m^2/e$ by many orders of magnitude.

The real part of the effective potential generates nonlinear field-field interactions. These interactions are exhibited order-by-order in the field by expanding V_{eff} in a (semi-convergent) power series in $(e\mathcal{E})^{2n}$. The power series representation of V_{eff} is obtained by expanding the proper time integrand of Eq. (5) for $e\mathcal{E}u \ll 1$

$$V_{\text{eff}} \simeq \frac{\gamma_s}{32\pi^2} \left\{ \left(7 - \frac{15}{2}g^2 + \frac{45}{48}g^4 \right) \frac{1}{45} \frac{(e\mathcal{E})^4}{m^4} \right. \tag{17}$$
$$\left. + \left(\frac{31}{24} - \frac{49}{32}g^2 + \frac{35}{128}g^4 - \frac{7}{512}g^6 \right) \frac{4}{315} \frac{(e\mathcal{E})^6}{m^8} + \ldots \right.$$

At each order in $(e\mathcal{E})^2$, we have separated the numerical coefficients for $g = 2$ outside the parentheses for ease of comparison to the $g = 1$ result,

$$V_{\text{eff}}\Big|_{g=1} \simeq \frac{\gamma_s}{32\pi^2} \left\{ \frac{7}{5760} \frac{(e\mathcal{E})^4}{m^4} - \frac{31}{161280} \frac{(e\mathcal{E})^6}{m^8} \cdots \right. \tag{18}$$

For example, the ratio of the coefficients of the $(e\mathcal{E})^4$ terms is $V_{\text{eff}}(g = 1)/V_{\text{eff}}(g = 2) \simeq 7/128$. We see that nonlinear field-field interactions generated by this potential are suppressed in the $g = 1$ case relative to the $g = 2$ (or $g = 0$) electron. This outcome is consistent with the suppression of the imaginary part of V_{eff}. Experiments seeking nonlinear field effects[10,11] derived from the Euler-Heisenberg effective potential are thus also sensitive to the effective value of g.

5. Discussion and Conclusions

In summary, we have recalled that in a constant electric field $\mathcal{E}$, the electron fluctuations $\langle \bar{\psi}\psi \rangle$ display a thermal Bose spectrum with temperature $T_{\text{M}} = e\mathcal{E}/m\pi = a_v/\pi$ Eq. (14). This result contrasts with the Fermi spectrum and Unruh temperature $T_{\text{U}} = a_{\text{U}}/2\pi$ expected from viewing the vacuum fluctuations of the electrons as accelerated. We have calculated $\langle \bar{\psi}\psi \rangle$ in an electric field for the gyromagnetic ratio $g = 0, 1, 2$. Setting $g = 1$, as though considering the quantum fluctuations of a 'classical spinning particle', displays the Unruh $T_{\text{U}} = a/2\pi$ and a Fermi spectrum, see Eq. (16).

We highlight the functional dependence of light-light scattering on g because it has implications for future experiments. Any (effective) value of g deviating from the Dirac value $g = 2$ results in a suppression of the rate of light-light scattering.

We note that QED is not yet tested near the critical field strength $\mathcal{E}_c = m^2/e$, and in this strong-field regime, we have still to validate the approach to calculating Eq. (5), which is perturbative in α.[12] Even more to the point there are serious questions about validity of QED in this limit.[13] Therefore, the connection which we established to reconcile the two ways of viewing acceleration could forebear forthcoming theoretical developments.

An observable, physical difference such as now predicted in Eqs. (2) and (1) provides an in-principle test to determine whether the strong field theory is valid, and/or it is the observer or the vacuum that is accelerated. Being able to determine who is accelerated means a fixed reference frame has been selected and defined as inertial. In quantum theory, the quantum vacuum state is a natural candidate for the fixed reference frame,[13] and here we have recalled that the electron condensate contains in its definition Eq. (10) a specific vacuum state as reference. Experimental observation of quantum vacuum phenomena such as light-light scattering offers an important test of our understanding of the vacuum state canonically selected in quantum field theory and may reveal whether or not it is consistent with the vacuum selected in the Unruh accelerated detector situation.

Acknowledgments

L.L. thanks Director Pisin Chen for the opportunity to visit LeCosPA. This work was supported by a grant from the US Department of Energy, DE-FG02-04ER41318.

References

1. B. Muller, W. Greiner and J. Rafelski, Phys. Lett. A **63**, 181 (1977).
2. W. Greiner, B. Muller and J. Rafelski, *Quantum Electrodynamics Of Strong Fields,*, (Springer, Berlin, Germany, 1985). See p.569 ff.
3. W. Y. Pauchy Hwang and S. P. Kim, Phys. Rev. D **80**, 065004 (2009).
4. W. G. Unruh, Phys. Rev. **D14**, 870 (1976).
5. L. C. B. Crispino, A. Higuchi and G. E. A. Matsas, Rev. Mod. Phys. **80**, 787 (2008).
6. P. Candelas and D. Deutsch, Proc. Roy. Soc. Lond. A **362**, 251 (1978).
7. L. Labun and J. Rafelski, "Acceleration and Vacuum Temperature," arXiv:1203.6148 [hep-ph].
8. J. S. Schwinger, Phys. Rev. **82**, 664 (1951).
9. L. Labun and J. Rafelski, Phys. Rev. D **81**, 065026 (2010).
10. G. L. J. A. Rikken and C. Rizzo, Phys. Rev. A **63**, 012107 (2001) and references therein. F. Bielsa, *et al.* "Status of the BMV experiment," arXiv:0911.4567 [physics.optics].
11. S. -J. Chen, H. -H. Mei and W. -T. Ni, Mod. Phys. Lett. A **22**, 2815 (2007). H. -H. Mei, W. -T. Ni, S. -J. Chen and S. -s. Pan, "The Status and prospects of the Q and A experiment with some applications," arXiv:0911.4776 [physics.ins-det].
12. W. Heisenberg and H. Euler, Z. Phys. **98**, 714 (1936) [physics/0605038].
13. J. Rafelski and L. Labun, *Critical Acceleration and Quantum Vacuum,* in these proceedings.

Section III

Cosmology

GENERALIZED G-INFLATION: INFLATION WITH THE MOST GENERAL SECOND-ORDER FIELD EQUATIONS*

TSUTOMU KOBAYASHI

Hakubi Center, Kyoto University, Kyoto 606-8302, Japan
and
Department of Physics, Kyoto University, Kyoto 606-8502, Japan

MASAHIDE YAMAGUCHI

Department of Physics, Tokyo Institute of Technology, Tokyo 152-8551, Japan

JUN'ICHI YOKOYAMA

Research Center for the Early Universe (RESCEU), Graduate School of Science,
The University of Tokyo, Tokyo 113-0033, Japan
and
Institute for the Physics and Mathematics of the Universe (IPMU), The University of Tokyo,
Kashiwa, Chiba, 277-8568, Japan
E-mail: yokoyama@resceu.s.u-tokyo.ac.jp

Generalized G-inflation is the most general single-field inflation model whose gravitational and scalar field equations are of second order. This model contains all the previously known examples such as the conventional potential-driven slow-roll inflation, k-inflation, Higgs inflation, new Higgs inflation, G-inflation, and even $f(R)$ inflation by an appropriate conformal transformation. We present the background and perturbation evolution in this model, calculating the most general quadratic actions for tensor and scalar cosmological perturbations to give the stability criteria and the power spectra of primordial fluctuations.

Keywords: Inflationary Cosmology; Generalized Galileon; G inflation.

1. Introduction

Scalar fields play important roles in cosmology. On one hand, inflation in the early Universe has become a part of the standard cosmology that is driven by a scalar field called the inflaton.[1-3] The conventional inflaton action consists of a canonical kinetic term and a sufficiently flat potential.[3] [See Ref. 4 for the latest review.] Non-canonical kinetic terms[5] also arise naturally in some particle physics models of inflation such as Dirac-Born-Infeld inflation.[6]

---

*Presented by J. Yokoyama.

On the other hand, it is strongly suggested that the present Universe is dominated by mysterious dark energy, and its identity might be a dynamical scalar field.[7] In relation to the present accelerated expansion, modified gravity theories have been studied extensively, and in such theories, an extra gravitational degree of freedom can often be equivalently described by a scalar field coupled non-minimally to gravity or matter. In the decoupling limit of the Dvali-Gabadadze-Porrati brane model,[8] the scalar field has a non-linear derivative self-interaction,[9] which was later generalized to Galileons[10] with a number of applications to various contexts in cosmology.[11–13] Thus, in recent years, there have been growing interests in scalar field theories beyond the canonical one.

The most attractive feature of higher derivative theories possessing the Galilean invariance $\partial_\mu \phi \to \partial_\mu \phi + b_\mu$ is that field equations derived from such a theory contain derivatives only up to second order,[10] so that it can easily avoid ghosts. Unfortunately, however, this desired feature ceases to exist once the background spacetime is curved.[14] To preserve the second-order nature of field equations, the "covariantization" of the Galileon has been proposed by Deffayet *et al.*,[14,15] where the theory is no longer Galilean invariant. In Ref. 16, it is pointed out that the equivalent theory has already been proposed by Horndeski.[17] The equivalence of both theories is explicitly shown in Ref. 18.

The purpose of this presentation is to provide a comprehensive and thorough study of the most general non-canonical and non-minimally coupled single-field inflation models named Generalized G-inflation yielding second-order field equations making use of Ref. 15, which is the most general extension of the Galileons but is no longer based on a symmetry argument.

In this presentation, based on Ref. 18, we clarify the generic behavior of the inflationary background and investigate the nature of primordial tensor and scalar perturbations at linear order. Given a specific model, our formulas are helpful to determine the evolution of cosmological perturbations and its observational consequences.

2. Generalized Higher-order Galileons and Kinetic Gravity Braiding

Recently Galileons[10] and their covariant extension[14] have been further generalized to yield the most general scalar field theories having second-order field equations.[12,13,15]

$$\mathcal{L}_2 = K(\phi, X), \tag{1}$$

$$\mathcal{L}_3 = -G_3(\phi, X)\Box\phi, \tag{2}$$

$$\mathcal{L}_4 = G_4(\phi, X)R + G_{4X}\left[(\Box\phi)^2 - (\nabla_\mu\nabla_\nu\phi)^2\right], \tag{3}$$

$$\mathcal{L}_5 = G_5(\phi, X)G_{\mu\nu}\nabla^\mu\nabla^\nu\phi - \frac{G_{5X}}{6}\left[(\Box\phi)^3 - 3(\Box\phi)(\nabla_\mu\nabla_\nu\phi)^2 + 2(\nabla_\mu\nabla_\nu\phi)^3\right], \tag{4}$$

where $X := -\partial_\mu \phi \partial^\mu \phi / 2$ is the canonical kinetic term, R is the Ricci scalar, $G_{\mu\nu}$ is the Einstein tensor, $(\nabla_\mu \nabla_\nu \phi)^2 = \nabla_\mu \nabla_\nu \phi \nabla^\mu \nabla^\nu \phi$, $(\nabla_\mu \nabla_\nu \phi)^3 = \nabla_\mu \nabla_\nu \phi \nabla^\nu \nabla^\lambda \phi \nabla_\lambda \nabla^\mu \phi$, and $G_{iX} = \partial G_i / \partial X$. Setting $G_3 = X$, $G_4 = X^2$, and $G_5 = X^2$, the above Lagrangians reproduce the covariant Galileons introduced in Ref. 14. The non-minimal couplings to gravity in $\mathcal{L}_4$ and $\mathcal{L}_5$ are necessary to eliminate higher derivatives that would otherwise appear in the field equations. Note that we do not need a separate gravitational Lagrangian other than $\mathcal{L}_4$; for $G_4 = M_{\mathrm{Pl}}^2 / 2$, $\mathcal{L}_4$ reduces to the Einstein-Hilbert term. We obtain a non-minimal coupling of the form $h(\phi)R$ from $\mathcal{L}_4$ by taking $G_4 = h(\phi)$. The non-standard kinetic term $G^{\mu\nu} \partial_\mu \phi \partial_\nu \phi$ as considered in Ref. 19 turns out to be a special case $G_5 \propto \phi$ of $\mathcal{L}_5$ after integration by parts. Equation (24) of Ref. 20, which is obtained from a Kaluza-Klein compactification of higher-dimensional Lovelock gravity, turns out to be equivalent to $\mathcal{L}_5$ with $G_5 = -3X/2$.

We thus consider a gravity + scalar system described by the action

$$S = \sum_{i=2}^{5} \int \mathrm{d}^4 x \sqrt{-g} \mathcal{L}_i, \tag{5}$$

which is the most general single scalar theory resulting in equations of motion containing derivatives up to second order. This action contains only four independent arbitrary functions of ϕ and X. This theory represents a general class of single-field inflation, including models that have not been studied so far, as well as almost all the previously known models such as potential-driven slow-roll inflation,[3] k-inflation,[5] extended inflation,[21] and even new Higgs inflation[19] as special cases.[a]

3. Background Equations

Let us derive the equations of motion describing the background evolution from (5). The easiest way is to substitute $\phi = \phi(t)$ and the metric $\mathrm{d}s^2 = -N^2(t)\mathrm{d}t^2 + a^2(t)\mathrm{d}\mathbf{x}^2$ to the action. Variation with respect to $N(t)$ gives the constraint equation, which can be written as $\sum_{i=2}^{5} \mathcal{E}_i = 0$, where

$$\mathcal{E}_2 = 2X K_X - K, \tag{6}$$
$$\mathcal{E}_3 = 6X\dot{\phi}H G_{3X} - 2X G_{3\phi}, \tag{7}$$
$$\mathcal{E}_4 = 6H^2 G_4 + 24H^2 X(G_{4X} + X G_{4XX}) - 12HX\dot{\phi}G_{4\phi X} - 6H\dot{\phi}G_{4\phi}, \tag{8}$$
$$\mathcal{E}_5 = 2H^3 X\dot{\phi}(5G_{5X} + 2X G_{5XX}) - 6H^2 X(3G_{5\phi} + 2X G_{5\phi X}). \tag{9}$$

---

[a]Even the curvature-square inflation model as well as more general $f(R)$ inflation,[2,22,23] which do not contain any scalar field and result in fourth-order equations of motion, can be recast in the present form by defining a new field as $\phi = \mathrm{d}f/\mathrm{d}R$.

Variation with respect to $a(t)$ yields the evolution equation, $\sum_{i=2}^{5}\mathcal{P}_i = 0$, where

$$\mathcal{P}_2 = K, \tag{10}$$

$$\mathcal{P}_3 = -2X\left(G_{3\phi} + \ddot{\phi}G_{3X}\right), \tag{11}$$

$$\mathcal{P}_4 = 2\left(3H^2 + 2\dot{H}\right)G_4 - 12H^2 X G_{4X} - 4H\dot{X}G_{4X} - 8\dot{H}XG_{4X} - 8HX\dot{X}G_{4XX}$$
$$+ 2\left(\ddot{\phi} + 2H\dot{\phi}\right)G_{4\phi} + 4XG_{4\phi\phi} + 4X\left(\ddot{\phi} - 2H\dot{\phi}\right)G_{4\phi X}, \tag{12}$$

$$\mathcal{P}_5 = -2X\left(2H^3\dot{\phi} + 2H\dot{H}\dot{\phi} + 3H^2\ddot{\phi}\right)G_{5X} - 4H^2 X^2\ddot{\phi}G_{5XX}$$
$$+ 4HX\left(\dot{X} - HX\right)G_{5\phi X} + 2\left[2(HX)^{\boldsymbol{\cdot}} + 3H^2 X\right]G_{5\phi} + 4HX\dot{\phi}G_{5\phi\phi}. \tag{13}$$

The background quantities $\mathcal{E}_i$ and $\mathcal{P}_i$ are defined in an analogous way in which the energy density and the isotropic pressure of a usual scalar field are defined.

Variation with respect to $\phi(t)$ gives the scalar-field equation of motion,

$$\frac{1}{a^3}\frac{\mathrm{d}}{\mathrm{d}t}\left(a^3 J\right) = P_\phi, \tag{14}$$

where

$$J = \dot{\phi}K_X + 6HXG_{3X} - 2\dot{\phi}G_{3\phi} + 6H^2\dot{\phi}\left(G_{4X} + 2XG_{4XX}\right) - 12HXG_{4\phi X}$$
$$+ 2H^3 X\left(3G_{5X} + 2XG_{5XX}\right) - 6H^2\dot{\phi}\left(G_{5\phi} + XG_{5\phi X}\right), \tag{15}$$

$$P_\phi = K_\phi - 2X\left(G_{3\phi\phi} + \ddot{\phi}G_{3\phi X}\right) + 6\left(2H^2 + \dot{H}\right)G_{4\phi} + 6H\left(\dot{X} + 2HX\right)G_{4\phi X}$$
$$- 6H^2 X G_{5\phi\phi} + 2H^3 X\dot{\phi}G_{5\phi X}. \tag{16}$$

4. Quadratic Actions for Tensor and Scalar Perturbations

In this section, our goal is to compute quadratic actions for tensor and scalar cosmological perturbations in Generalized G-inflation. We use the unitary gauge in which $\phi = \phi(t)$ and begin with writing the perturbed metric as

$$\mathrm{d}s^2 = -N^2\mathrm{d}t^2 + \gamma_{ij}\left(\mathrm{d}x^i + N^i\mathrm{d}t\right)\left(\mathrm{d}x^j + N^j\mathrm{d}t\right), \tag{17}$$

where

$$N = 1 + \alpha, \quad N_i = \partial_i\beta, \quad \gamma_{ij} = a^2(t)e^{2\zeta}\left(\delta_{ij} + h_{ij} + \frac{1}{2}h_{ik}h_{kj}\right). \tag{18}$$

Here, α, β, and ζ are scalar perturbations and h_{ij} is a tensor perturbation satisfying $h_{ii} = 0 = h_{ij,j}$.

4.1. *Tensor perturbations*

The quadratic action for the tensor perturbations is found to be

$$S_T^{(2)} = \frac{1}{8}\int \mathrm{d}t\mathrm{d}^3 x\, a^3\left[\mathcal{G}_T\dot{h}_{ij}^2 - \frac{\mathcal{F}_T}{a^2}(\vec{\nabla}h_{ij})^2\right], \tag{19}$$

where

$$\mathcal{F}_T := 2\left[G_4 - X\left(\ddot{\phi}G_{5X} + G_{5\phi}\right)\right], \tag{20}$$

$$\mathcal{G}_T := 2\left[G_4 - 2XG_{4X} - X\left(H\dot{\phi}G_{5X} - G_{5\phi}\right)\right]. \tag{21}$$

The squared sound speed is given by $c_T^2 = \mathcal{F}_T/\mathcal{G}_T$. One sees from the action (19) that ghost and gradient instabilities are avoided provided that $\mathcal{F}_T > 0, \quad \mathcal{G}_T > 0$. Note that c_T^2 is not necessarily unity in general, contrary to the standard cases.

On superhorizon scales, the two independent solutions to the perturbation equation are given by

$$h_{ij} = \text{const} \quad \text{and} \quad \int^t \frac{dt'}{a^3\mathcal{G}_T}. \tag{22}$$

The second solution corresponds to a decaying mode. To evaluate the primordial power spectrum, let us assume that $\epsilon := -\dot{H}/H^2 \simeq \text{const}$,

$$f_T := \frac{\dot{\mathcal{F}}_T}{H\mathcal{F}_T} \simeq \text{const} \quad \text{and} \quad g_T := \frac{\dot{\mathcal{G}}_T}{H\mathcal{G}_T} \simeq \text{const}. \tag{23}$$

We also impose conditions,

$$1 - \epsilon - f_T/2 + g_T/2 > 0, \qquad 3 - \epsilon + g_T > 0. \tag{24}$$

The former guarantees that the time coordinate y_T runs from $-\infty$ to 0 as the Universe expands. The latter implies that the second solution in (22) indeed decays.

The normalized mode solution to the perturbation equation on superhorizon scales is given by

$$k^{3/2}h_{ij} \approx 2^{\nu_T-2}\frac{\Gamma(\nu_T)}{\Gamma(3/2)}\frac{(-y_T)^{1/2-\nu_T}}{z_T}k^{3/2-\nu_T}e_{ij} \tag{25}$$

with

$$\nu_T := \frac{3 - \epsilon + g_T}{2 - 2\epsilon - f_T + g_T}.$$

Thus, we find the power spectrum of the primordial tensor perturbation:

$$\mathcal{P}_T = 8\gamma_T \left.\frac{\mathcal{G}_T^{1/2}}{\mathcal{F}_T^{3/2}}\frac{H^2}{4\pi^2}\right|_{-ky_T=1}, \tag{26}$$

where $\gamma_T = 2^{2\nu_T-3}|\Gamma(\nu_T)/\Gamma(3/2)|^2(1 - \epsilon - f_T/2 + g_T/2)$. The tensor spectral tilt is given by $n_T = 3 - 2\nu_T$. Contrary to the predictions of the conventional inflation models, the blue spectrum $n_T > 0$ can be obtained for $4\epsilon + 3f_T - g_T < 0$. This condition is easily compatible with the conditions (24).

4.2. *Scalar perturbations*

We now focus on scalar fluctuations putting $h_{ij} = 0$. Plugging the perturbed metric into the action and expanding it to second order, we obtain

$$S_S^{(2)} = \int \mathrm{dt d}^3 x a^3 \left[-3\mathcal{G}_T \dot{\zeta}^2 + \frac{\mathcal{F}_T}{a^2} (\vec{\nabla}\zeta)^2 + \Sigma\alpha^2 \right.$$
$$\left. -2\Theta\alpha \frac{\vec{\nabla}^2}{a^2}\beta + 2\mathcal{G}_T \dot{\zeta}\frac{\vec{\nabla}^2}{a^2}\beta + 6\Theta\alpha\dot{\zeta} - 2\mathcal{G}_T\alpha\frac{\vec{\nabla}^2}{a^2}\zeta \right], \quad (27)$$

where

$$\Sigma := XK_X + 2X^2 K_{XX} + 12H\dot{\phi}XG_{3X}$$
$$+6H\dot{\phi}X^2 G_{3XX} - 2XG_{3\phi} - 2X^2 G_{3\phi X} - 6H^2 G_4$$
$$+6\left[H^2 \left(7XG_{4X} + 16X^2 G_{4XX} + 4X^3 G_{4XXX} \right) \right.$$
$$\left. -H\dot{\phi}\left(G_{4\phi} + 5XG_{4\phi X} + 2X^2 G_{4\phi XX} \right) \right]$$
$$+30H^3\dot{\phi}XG_{5X} + 26H^3\dot{\phi}X^2 G_{5XX}$$
$$+4H^3\dot{\phi}X^3 G_{5XXX} - 6H^2 X \left(6G_{5\phi} + 9XG_{5\phi X} + 2X^2 G_{5\phi XX} \right), \quad (28)$$
$$\Theta := -\dot{\phi}XG_{3X} + 2HG_4 - 8HXG_{4X} - 8HX^2 G_{4XX} + \dot{\phi}G_{4\phi} + 2X\dot{\phi}G_{4\phi X}$$
$$-H^2\dot{\phi}\left(5XG_{5X} + 2X^2 G_{5XX} \right) + 2HX \left(3G_{5\phi} + 2XG_{5\phi X} \right). \quad (29)$$

It is interesting to see that even in the most generic case, some of the coefficients are given by $\mathcal{F}_T$ and $\mathcal{G}_T$, i.e., the functions characterizing the tensor perturbation, and only two new functions show up in the scalar quadratic action.

Varying the action (27) with respect to α and β yields the constraint equations,

$$\Sigma\alpha - \Theta\frac{\vec{\nabla}^2}{a^2}\beta + 3\Theta\dot{\zeta} - \mathcal{G}_T\frac{\vec{\nabla}^2}{a^2}\zeta = 0, \qquad \Theta\alpha - \mathcal{G}_T\dot{\zeta} = 0. \quad (30)$$

Using the constraint equations, we eliminate α and β from the action (27) and finally arrive at

$$S_S^{(2)} = \int \mathrm{dt d}^3 x \, a^3 \left[\mathcal{G}_S \dot{\zeta}^2 - \frac{\mathcal{F}_S}{a^2} (\vec{\nabla}\zeta)^2 \right], \quad (31)$$

where

$$\mathcal{F}_S := \frac{1}{a}\frac{\mathrm{d}}{\mathrm{d}t}\left(\frac{a}{\Theta}\mathcal{G}_T^2 \right) - \mathcal{F}_T, \qquad \mathcal{G}_S := \frac{\Sigma}{\Theta^2}\mathcal{G}_T^2 + 3\mathcal{G}_T. \quad (32)$$

The analysis of the curvature perturbation hereafter is completely parallel to that of the tensor perturbation. The squared sound speed is given by $c_S^2 = \mathcal{F}_S/\mathcal{G}_S$, and ghost and gradient instabilities are avoided as long as $\mathcal{F}_S > 0$ $\mathcal{G}_S > 0$.

The two independent solutions on superhorizon scales are

$$\zeta = \mathrm{const} \quad \mathrm{and} \quad \int^t \frac{\mathrm{d}t'}{a^3\mathcal{G}_S}. \quad (33)$$

During inflation, it may be assumed that $\mathcal{G}_S$ is slowly varying. In this case, the second solution decays rapidly.

Closely following the procedure we did in the case of the tensor perturbation, we now evaluate the power spectrum of the primordial curvature perturbation. To do so, we assume that $\epsilon \simeq$ const,

$$f_S := \frac{\dot{\mathcal{F}}_S}{H\mathcal{F}_S} \simeq \text{const}, \quad g_S := \frac{\dot{\mathcal{G}}_S}{H\mathcal{G}_S} \simeq \text{const}, \tag{34}$$

and then define

$$\nu_S := \frac{3 - \epsilon + g_S}{2 - 2\epsilon - f_S + g_S}. \tag{35}$$

The power spectrum is given by

$$\mathcal{P}_\zeta = \frac{\gamma_S}{2} \frac{\mathcal{G}_S^{1/2}}{\mathcal{F}_S^{3/2}} \frac{H^2}{4\pi^2}\bigg|_{-k y_S = 1}, \tag{36}$$

where $\gamma_S = 2^{2\nu_S - 3}|\Gamma(\nu_S)/\Gamma(3/2)|^2(1 - \epsilon - f_S/2 + g_S/2)$. The spectral index is $n_s - 1 = 3 - 2\nu_S$. An exactly scale-invariant spectrum is obtained if $\epsilon + \frac{3}{4}f_S - \frac{1}{4}g_S = 0$. Here again, ϵ, f_S, and g_S are not necessarily very small (as long as $n_s - 1 \simeq 0$).

Taking now the limit $\epsilon, f_T, g_T, f_S, g_S \ll 1$, the tensor-to-scalar ratio is given by

$$r = 16 \left(\frac{\mathcal{F}_S}{\mathcal{F}_T}\right)^{3/2} \left(\frac{\mathcal{G}_S}{\mathcal{G}_T}\right)^{-1/2} = 16\frac{\mathcal{F}_S}{\mathcal{F}_T}\frac{c_S}{c_T}. \tag{37}$$

Note that even in the de Sitter limit where $\epsilon, f_T, g_T, f_S, g_S \to 0$, the scalar perturbation can be produced in general, $r \neq 0$.

5. Summary

In this presentation, we have presented generic inflation models named Generalized G-inflation, driven by a single scalar field. Our gravity + scalar-field system is described by the generalized Galileons, which do not give rise to higher derivatives in the field equations despite the non-minimal coupling. This class of inflation models is the most general ever proposed in the context of single-field inflation.

We have determined the most generic quadratic actions for tensor and scalar cosmological perturbations. Using them, we have presented the stability criteria for both types of perturbations. The primordial power spectra have also been computed. Note that, since the propagation speeds of the two types of fluctuations can be different, we must evaluate the power spectra for the same comoving wavenumber at different epochs, which may have some consequence.[24]

Acknowledgments

I would like to thank Pisin Chen and other organizers of LeCosPA symposium for their hospitality. This work was supported in part by JSPS Grant-in-Aid for Research Activity Start-up No. 22840011 (T.K.), Grant-in-Aid for Scientific Research Nos. 23340058 (J.Y.) and 21740187 (M.Y.), and Grant-in-Aid for Scientific Research on Innovative Areas No. 21111006 (J.Y.).

References

1. A. H. Guth, Phys. Rev. D **23**, 347 (1981); K. Sato, Mon. Not. Roy. Astron. Soc. **195**, 467 (1981).
2. A. A. Starobinsky, Phys. Lett. B **91**, 99 (1980);
3. A. D. Linde, Phys. Lett. B **108**, 389 (1982); A. Albrecht and P. J. Steinhardt, Phys. Rev. Lett. **48**, 1220 (1982); A. D. Linde, Phys. Lett. B **129**, 177 (1983).
4. M. Yamaguchi, Class. Quant. Grav. **28**, 103001 (2011) [arXiv:1101.2488 [astro-ph.CO]].
5. C. Armendariz-Picon, T. Damour and V. F. Mukhanov, Phys. Lett. B **458**, 209 (1999) [arXiv:hep-th/9904075].
6. M. Alishahiha, E. Silverstein and D. Tong, Phys. Rev. D **70**, 123505 (2004) [arXiv:hep-th/0404084].
7. T. Chiba, N. Sugiyama and T. Nakamura, Mon. Not. Roy. Astron. Soc. **289**, L5 (1997) [arXiv:astro-ph/9704199]; R. R. Caldwell, R. Dave, P. J. Steinhardt, Phys. Rev. Lett. **80**, 1582-1585 (1998). [astro-ph/9708069]; T. Chiba, T. Okabe, M. Yamaguchi, Phys. Rev. **D62**, 023511 (2000). [astro-ph/9912463]; C. Armendariz-Picon, V. F. Mukhanov, P. J. Steinhardt, Phys. Rev. Lett. **85**, 4438-4441 (2000). [astro-ph/0004134].
8. G. R. Dvali, G. Gabadadze and M. Porrati, Phys. Lett. B **485**, 208 (2000) [arXiv:hep-th/0005016].
9. M. A. Luty, M. Porrati and R. Rattazzi, JHEP **0309**, 029 (2003) [arXiv:hep-th/0303116]; A. Nicolis and R. Rattazzi, JHEP **0406**, 059 (2004) [arXiv:hep-th/0404159].
10. A. Nicolis, R. Rattazzi, E. Trincherini, Phys. Rev. **D79**, 064036 (2009). [arXiv:0811.2197 [hep-th]].
11. N. Chow and J. Khoury, Phys. Rev. D **80**, 024037 (2009) [arXiv:0905.1325 [hep-th]]; F. P. Silva and K. Koyama, Phys. Rev. D **80**, 121301 (2009) [arXiv:0909.4538 [astro-ph.CO]]; T. Kobayashi, H. Tashiro and D. Suzuki, Phys. Rev. D **81**, 063513 (2010) [arXiv:0912.4641 [astro-ph.CO]]; T. Kobayashi, Phys. Rev. D **81**, 103533 (2010) [arXiv:1003.3281 [astro-ph.CO]]; P. Creminelli, A. Nicolis, E. Trincherini, JCAP **1011**, 021 (2010). [arXiv:1007.0027 [hep-th]]; C. Burrage, C. de Rham, D. Seery, A. J. Tolley, JCAP **1101**, 014 (2011). [arXiv:1009.2497 [hep-th]]; K. Kamada, T. Kobayashi, M. Yamaguchi, J. Yokoyama, Phys. Rev. **D83**, 083515 (2011) [arXiv:1012.4238 [astro-ph.CO]]; T. Kobayashi, M. Yamaguchi and J. Yokoyama, Phys. Rev. D **83**, 103524 (2011) [arXiv:1103.1740 [hep-th]]; O. Pujolas, I. Sawicki and A. Vikman, JHEP **1111**, 156 (2011) [arXiv:1103.5360 [hep-th]]; X. Gao, JCAP **1110**, 021 (2011) [arXiv:1106.0292 [astro-ph.CO]]; X. Gao, T. Kobayashi, M. Yamaguchi and J. Yokoyama, Phys. Rev. Lett. **107**, 211301 (2011) [arXiv:1108.3513 [astro-ph.CO]].
12. C. Deffayet, O. Pujolas, I. Sawicki, A. Vikman, JCAP **1010**, 026 (2010). [arXiv:1008.0048 [hep-th]].
13. T. Kobayashi, M. Yamaguchi, J. Yokoyama, Phys. Rev. Lett. **105**, 231302 (2010). [arXiv:1008.0603 [hep-th]].
14. C. Deffayet, G. Esposito-Farese, A. Vikman, Phys. Rev. **D79**, 084003 (2009). [arXiv:0901.1314 [hep-th]].
15. C. Deffayet, X. Gao, D. A. Steer and G. Zahariade, Phys. Rev. D **84**, 064039 (2011) [arXiv:1103.3260 [hep-th]].
16. C. Charmousis, E. J. Copeland, A. Padilla, P. M. Saffin, [arXiv:1106.2000 [hep-th]].
17. G. W. Horndeski, Int. J. Theor. Phys. 10 (1974) 363-384.
18. T. Kobayashi, M. Yamaguchi and J. Yokoyama, Prog. Theor. Phys. **126**, 511 (2011) [arXiv:1105.5723 [hep-th]].
19. C. Germani, A. Kehagias, Phys. Rev. Lett. **105**, 011302 (2010). [arXiv:1003.2635 [hep-ph]].

20. K. Van Acoleyen, J. Van Doorsselaere, Phys. Rev. **D83**, 084025 (2011). [arXiv:1102.0487 [gr-qc]].

21. D. La and P. J. Steinhardt, Phys. Rev. Lett. **62**, 376 (1989) [Erratum-ibid. **62**, 1066 (1989)].

22. M. B. Mijic, M. S. Morris, W. -M. Suen, Phys. Rev. **D34**, 2934 (1986).

23. J. R. Ellis, N. Kaloper, K. A. Olive and J. Yokoyama, Phys. Rev. D **59**, 103503 (1999) [hep-ph/9807482].

24. L. Lorenz, J. Martin and C. Ringeval, Phys. Rev. D **78**, 083513 (2008) [arXiv:0807.3037 [astro-ph]].

BRANE-WORLD INFLATION: PERTURBATIONS AND COSMOLOGICAL CONSTRAINTS

MARIAM BOUHMADI-LOPEZ[1,2], PISIN CHEN[3−6], and YEN-WEI LIU[3,5*]

[1] *Instituto de Estructura de la Materia, IEM-CSIC, Serrano 121, 28006 Madrid, Spain*
[2] *Centro Multidisciplinar de Astrofísica - CENTRA, Departamento de Física, Instituto Superior Técnico, Av. Rovisco Pais 1,1049-001 Lisboa, Portugal*
[3] *Department of Physics, National Taiwan University, Taipei, Taiwan 10617*
[4] *Graduate Institute of Astrophysics, National Taiwan University, Taipei, Taiwan 10617*
[5] *Leung Center for Cosmology and Particle Astrophysics, National Taiwan University, Taipei, Taiwan 10617*
[6] *Kavli Institute for Particle Astrophysics and Cosmology, SLAC National Accelerator Laboratory, Stanford University, Stanford, CA 94305, U.S.A.*
** E-mail: f97222009@ntu.edu.tw*

A generalization of the Randall-Sundrum single brane-world scenario (RS2) is considered, which is based on adding two curvature corrections: a Gauss-Bonnet (GB) term in the bulk and an induced gravity (IG) term on the brane. Here we are mainly interested in analysing the early inflationary era of the brane, which we model within the extreme slow-roll limit, being the inflaton field confined on the brane. We compute the scalar perturbations on this model and compare our results with those previously obtained for the RS2 scenario with or without a single curvature correction corresponding to a GB or an IG curvature term. Finally, we constrain the model using the latest WMAP7 data.

Keywords: Cosmological perturbations; inflation; brane-worlds; modified theories of gravity.

1. Introduction

Different approaches in cosmology and particle physics imply the possibility that our observable universe is a hypersurface, i.e., a brane, embedded in a higher-dimensional space-time, i.e., a bulk, which is motivated by the superstring/M theory. In this scenario several extra-dimensional models have been proposed (cf. for example Ref. 1). One of the most popular and interesting brane-world scenario is provided by the Randall-Sundrum single brane model (RS2),[2] where our universe corresponds to an observable four-dimensional (4d) single brane embedded in a five-dimensional (5d) anti de-Sitter (AdS_5) bulk.

With regard to the 4d cosmology, there are two important modifications to the RS2 model; the first one is the Gauss-Bonnet (GB) correction to the bulk action, expected at high energy (for example during the inflationary era), which leads to the most general second-order field equation in a 5d bulk.[3] Moreover, this unique combination of the GB term in the bulk action also corresponds to the leading corrections in string theory, and it is a ghost-free combination.[4,5] Besides, it also plays an important role in Chern-Simons gravity,[6–8] which is a gauge theory of gravity. Furthermore, the zero-mode of the 5d gravitons in the GB brane-world is also localized on the brane at low energy as in the RS2 model.[9] The second modification is the induced-gravity (IG) correction to the brane action. This effect is generated due to the quantum loops of matter fields on the brane that couple to the bulk gravitons.[10–13]

Here we investigate the modification of inflation induced by the RS2-type model due to both GB and IG effects, where the spatial curvature and the dark radiation term are rapidly diluted. In addition, in order to compare with the power spectrum of the scalar perturbations in the RS2 model, we consider the normal branch of the model which recovers the standard general relativity at the low energy limit with a vanishing cosmological constant on the brane. Furthermore, this branch reduces to the RS2 model in the absence of GB and IG corrections to the action. When applying the RS2 brane-world to the early universe cosmology, if we consider single-field inflation localized on the brane and in the extreme slow-roll limit, there is no scalar zero-mode contribution from the bulk; moreover, the massive KK scalar mode can be neglected since they are too heavy to be excited during inflation.[14,15] We assume that this remains true on the case under study.

When the GB and IG corrections are both included, we show that the effect from the GB correction in an IG brane-world model is to decrease the amplitude of the scalar perturbations, and a similar result is obtained for the IG effect in a GB brane-world. The same effects have been obtained for the pure RS2 model.[16,17] Finally, we constrain the model by using the latest WMAP7 data. For more details, we refer the reader to Ref. 18.

2. The Setup

We consider a 5d brane-world model where the brane split the bulk into two symmetric pieces. The bulk action contains a GB term in addition to the usual Hilbert-Einstein term, while the brane action is described by an IG term, a brane tension and a Lagrangian for matter. Then the action of the system is given by:[21]

$$S = \frac{1}{2\kappa_5^2} \int_M d^5x \sqrt{-g} \left[R - 2\Lambda_5 + \alpha \left(R^2 - 4R_{\mu\nu}R^{\mu\nu} + R_{\mu\nu\rho\sigma}R^{\mu\nu\rho\sigma} \right) \right]$$
$$+ \int_{\partial M} d^4x \sqrt{-h} \left[\frac{\gamma}{2\kappa_4^2} \hat{R} - \lambda + \mathscr{L}_m \right], \tag{1}$$

where $g_{\mu\nu}$ and $h_{\mu\nu}$ are the bulk metric and the brane metric, respectively, κ_5^2 is the bulk gravitational constant, α the GB parameter which has the dimension of length square, γ is a dimensionless parameter indicating the strength of the IG term, and λ is the brane tension.

We consider the static uncharged black hole solution in 5d GB gravity:[22–25]

$$ds^2 = -f(r)dT^2 + f^{-1}(r)dr^2 + r^2\Omega_{ij}dx^i dx^j, \tag{2}$$

where

$$f = k + \frac{r^2}{4\alpha}\left(1 \pm \sqrt{1 + \frac{4}{3}\alpha\Lambda_5 + 8\alpha\frac{\tilde{\mu}}{r^4}}\right). \tag{3}$$

The parameter $\tilde{\mu}$ is related to the black hole mass. There are two branches for the black hole solutions. However, we disregard the "+" branch as it is unstable. The reason behind this instability is that the graviton degree of freedom is a ghost, in addition, the mass of the black hole is negative in this branch.[22,23] For the bulk black hole solution (3), we impose the junction condition at the brane[26–28] and we obtain the effective 4d Friedmann equation:[29,30]

$$\left[1 + \frac{8}{3}\alpha\left(H^2 + \frac{k}{a^2} + \frac{\phi}{2}\right)\right]^2 \left(H^2 + \frac{k}{a^2} - \phi\right) \tag{4}$$

$$= \frac{\gamma^2}{4}\left(\frac{\kappa_5}{\kappa_4}\right)^4 \left[H^2 + \frac{k}{a^2} - \frac{\kappa_4^2}{3\gamma}(\rho + \lambda)\right]^2, \tag{5}$$

where

$$\phi + 2\alpha\phi^2 = \frac{\Lambda_5}{6} + \frac{C}{a^4}, \tag{6}$$

and $k = 0, \pm 1$. The constant C is related to the mass of the black hole and it measures the strength of the dark radiation on the brane. The condition (6) results in two possible values for ϕ

$$\phi_\pm = \frac{1}{4\alpha}\left[-1 \pm \sqrt{1 + 8\alpha\left(\frac{\Lambda_5}{6} + \frac{C}{a^4}\right)}\right]. \tag{7}$$

From now on, we will restrict our analysis to the solution with ϕ_+, because it corresponds to the stable bulk solution with "−" sign in Eq.(3), moreover, we recover a Hilbert-Einstein action in the bulk if $\alpha \to 0$ and therefore the model we are considering reduces to the RS2 scenario in the absence of any curvature corrections of the GB and IG kind. This is very important because one of our main aims in this paper is to see how the amplitude of the scalar perturbations on RS2 model are modified by including GB and IG terms. For simplicity, from now on we will drop the subscript "+" on ϕ_+.

We are mainly interested in the early inflationary era of the brane, where the spatial curvature and the dark radiation are quickly washed out. Therefore, from now on we will consider a spatially flat brane ($k = 0$) within an AdS$_5$ bulk ($C = 0$). By imposing the RS2 kind of fine-tuning, i.e., the effective cosmological constant on the brane vanishes; in other words in the absence of matter on the brane the Hubble rate vanishes, we obtain

$$\frac{\kappa_5^4}{36}\lambda^2 = -\phi\left(1 + \frac{3}{4}\alpha\phi\right)^2, \tag{8}$$

which implies that ϕ is negative. For later convenience, we introduce a new positive variable $\mu = \sqrt{|\phi|}$, then Eq.(8) can be rewritten as

$$\kappa_5^2\lambda = 2\mu\left(3 - 4\alpha\mu^2\right). \tag{9}$$

Therefore, the parameter μ^2 is bounded as $0 \leq \mu^2 < 1/4\alpha$ (see also Ref. 31), as can then be easily deduced by using Eq.(7) (for $k = 0$ and $C = 0$) and assuming the natural requirement of a positive brane tension.

In order to proceed further, we need first to solve the cubic Friedmann equation (5). This equation was previously analyzed in Refs. 29 and 32. This equation can be solved analytically extending the methodology used in Ref. 33 as we showed in Ref. 18. It is important to solve this equation analytically as it allow us to pick up the right branch for analyzing the brane scalar perturbations in the extreme slow-roll limit, i.e., the branch that reduce to RS2 model in the absence of curvature corrections of the sort IG and GB.

3. The Scalar Perturbations

In this section we consider single field inflation on the brane and study the lowest order of the scalar perturbations in the extreme slow-roll limit. Within this approximation, we assume there is no additional scalar zero-mode contribution from the bulk while the massive scalar mode from the bulk are too heavy to be excited during the inflationary era which is true in RS2 brane-world models.[14] Therefore, the massive KK scalar mode can be neglected in the extreme slow-roll inflation. With these assumptions one can follow the standard procedure to calculate the power spectrum for the scalar perturbations,[34] and we will follow the procedures for brane inflation used in 15–17.

Before calculating the power spectrum of the scalar perturbations, we need first to find out 4d effective gravitational constant on the brane, which is an essential ingredient when calculating the scalar perturbations. Therefore, we focus on the generalized RS2 solutions modified by GB and IG effects and impose the generalized fine-tuning condition Eq.(9). In the low energy region when $\rho \to 0$, we find the relation between the effective gravitational constant on the brane κ_4^2 and bulk gravitational constant κ_5^2:

$$\kappa_4^2 = \left(\frac{1-\gamma}{1+4\alpha\mu^2} \right) \mu\kappa_5^2. \tag{10}$$

The relation between κ_4^2 and κ_5^2 implies: (i) γ is bounded; $0 \leq \gamma < 1$, (ii) μ must be strictly positive, i.e., $\mu > 0$.

Now, we proceed to calculate the power spectrum of the scalar perturbations. The normalized amplitude of the scalar perturbations for a given mode that reenter the horizon after inflation is given by[34]

$$A_S^2 = \frac{4}{25}\langle\zeta^2\rangle = \frac{H^4}{25\pi^2\dot{\phi}^2}. \tag{11}$$

In addition, in the extreme slow-roll inflation, $\dot{\phi} \simeq -V'(\phi)/3H$; we can therefore substitute this approximation into the amplitude of the scalar perturbations

Eq.(11), resulting in

$$A_S^2 = \frac{9}{25\pi^2}\frac{H^6}{V'^2},\tag{12}$$

which is independent of the gravitational field equation.[35] In order to compare conveniently with the standard 4d general relativity results, we rewrite Eq.(12) as:

$$A_S^2 = \frac{\kappa_4^6}{75\pi^2}\left(\frac{V^3}{V'^2}\right)G_{\alpha,\gamma}^2 = [A_S^2]_{4D}G_{\alpha,\gamma}^2,\tag{13}$$

where $[A_S^2]_{4D}$ is the standard 4d result and the correction term $G_{\alpha,\gamma}^2$ is obtained in the extreme slow-roll limit (see also Fig. 1):

$$G_{\alpha,\gamma}^2(x) = \left[\frac{3(1+\beta)x^2}{2(1-\gamma)(3-\beta+2\beta x^2)\sqrt{1+x^2}+3\gamma x^2(1+\beta)+2(1-\gamma)(\beta-3)}\right]^3,\tag{14}$$

where $x = H/\mu$ and $\beta = 4\alpha\mu^2$. This result is consistent with Ref. 16 when $\beta \to 0$ and consistent with Ref. 17 when $\gamma \to 0$.

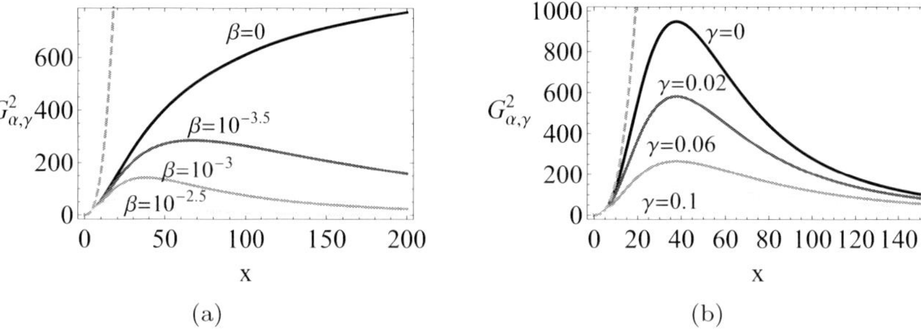

Fig. 1. The amplitude of the scalar perturbations normalized to the standard 4d results in the extreme slow-roll limit (see Eq.(13) and Eq.(14)) against the scale of inflation on the brane normalized to the square root of the absolute value of the effective cosmological constant in the bulk. In figure (a), we have fixed the IG parameter, more precisely, $\gamma = 0.1$, and changed the GB parameter β as shown on the plot. In figure (b), we have fixed the GB parameter, more precisely, $\beta = 4\alpha\mu^2 = 10^{-3}$, and changed the IG parameter γ as shown on the plot. We can see that the effect from the GB correction in an IG brane-world model is to decrease the amplitude of the scalar perturbations, and a similar result is obtained for the IG effect in a GB brane-world.

In Fig. 1 the dashed-grey line corresponds to the amplitude of the RS2 model without GB and IG corrections, which is monotonically increasing with respect to the dimensionless energy scale H/μ. If the GB and IG corrections are both included, we see that the effect from the GB correction in an IG brane-world model is to decrease the amplitude of the scalar perturbations, and a similar result is obtained for the IG effect in a GB brane-world. In the very low energy limit, i.e., the Hubble parameter $H \ll \mu$ or $x \to 0$, the correction to the standard 4d result corresponds to $G_{\alpha,\gamma}^2 \sim 1$. Therefore, the amplitude of the scalar perturbations recovers the 4d standard result. During the intermediate energy scale, the amplitude of the scalar perturbations is enhanced with respect to the energy scale; while in the very high energy regime, i.e., the Hubble parameter $\mu \ll H$ or $x \to \infty$, we obtain the following approximation:

$$G^2_{\alpha,\gamma} \sim \frac{27}{64} \left[\frac{1+\beta}{\beta(1-\gamma)} \right]^3 \frac{1}{x^3}, \qquad (15)$$

which means that in the high energy limit the perturbation will be highly suppressed by the GB effect.

Finally, we impose observational constraints by using the latest WMAP7 data,[36] i.e., for the power spectrum of the scalar perturbations: P_s (normalized amplitude of the scalar perturbation $A_s^2 \equiv 4/25\,P_s$). More precisely, we impose $P_s = 2.45 \times 10^{-9}$ at the pivot scale $k_0 = 0.002$ Mpc^{-1} (Fig. 2 shows a constraint of the potential). Notice that at this large scale we expect the extreme slow-roll approximation to be valid. Our constraints can be extended by considering the measure of the spectral index n_s, and we check that our results are in full agreement with the extreme slow-roll approximation for a wide rage of the parameters β, γ, and x.

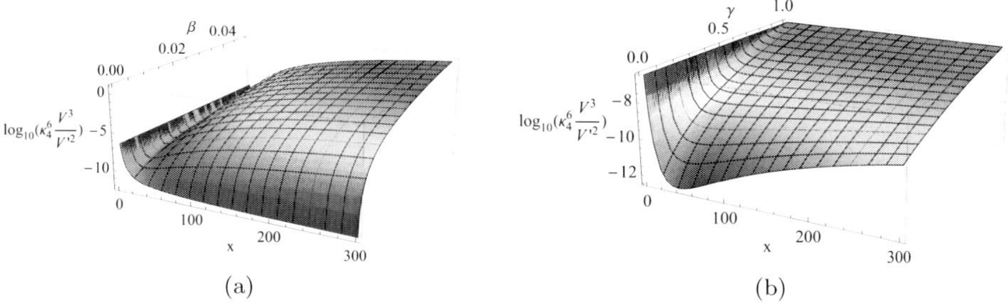

(a)   (b)

Fig. 2. We constrain the inflaton potential by means of the amplitude of the scalar perturbations as measured by WMAP7, i.e., $P_s = 2.45 \times 10^{-9}$ at the pivot scale $k_0 = 0.002$ Mpc^{-1}. More precisely, in plot (a) we show $\log_{10}(\kappa_4^6 V^3/V'^2)$ versus the scale of inflation, $x = H/\mu$, and the dimensionless GB parameter, β, for a fixed value of the dimensionless IG parameter γ such that $\gamma = 0.1$; in plot (b) we show $\log_{10}(\kappa_4^6 V^3/V'^2)$ versus the scale of inflation, $x = H/\mu$, and the dimensionless IG parameter, γ, for a fixed value of the dimensionless GB parameter β such that $\beta = 10^{-3}$.

4. Summary

Brane inflation provides an interesting scenario for the early universe allowing us to explore the properties of the higher-dimensional cosmology. At such high energy scales, the non-conventional brane-world effects become dominant. Therefore, by investigating the scalar perturbations during inflation, we can see the modification from the brane effects relative to the standard general relativistic result. Here, we focus on the RS2-type brane-world modified by a GB correction term in the bulk as well as the strength of the IG effect on the brane. In order to compare with the RS2 model, we choose the normal branch for the cosmic evolution to compute the amplitude of the scalar perturbations, which reduces to the RS2 model in the absence of GB and IG corrections, in the slow-roll limit.

In such brane-world inflationary model and in the extreme slow-roll limit, we assume that there is no scalar zero-mode contribution from the bulk, and moreover the massive scalar modes are too heavy to be excited during inflation.[14,16,17] We can therefore safely disregard any extra scalar degree of freedom from the bulk and we can simply take into account the standard GR result for the amplitude of single field scalar perturbations. Along this line of thought, in this paper we have calculated the corrections to the standard GR results for the scalar perturbations in a brane-world model with the curvature effects mentioned earlier. The amplitude of the scalar perturbations in the RS2 model is monotonically increasing with respect to the energy scale. However, unlike the RS2 case the effect from the GB correction in an IG brane-world model is to decrease the amplitude of the scalar perturbations, and a similar result is obtained for the IG effect in a GB brane-world. Furthermore, in the high energy limit the perturbation will be highly suppressed by the GB effect. Finally, we constrain the model by using WMAP7 data. Here we show a constraint of the potential (see Fig. 2), and we check that our results are in full agreement with the extreme slow-roll approximation we used.

Acknowledgments

M.B.L. is supported by the Spanish Agency "Consejo Superior de Investigaciones Científicas" through JAEDOC064. She also wishes to acknowledge the hospitality of LeCosPA Center at the National Taiwan University during the completion of part of this work and the support of the Portuguese Agency "Fundção para a Ciência e Tecnologia" through PTDC/FIS/111032/2009. P.C. and Y.W.L. are supported by Taiwan National Science Council under Project No. NSC 97-2112-M-002-026-MY3 and by Taiwan's National Center for Theoretical Sciences (NCTS). P.C. is in addition supported by US Department of Energy under Contract No. DE-AC03-76SF00515. This work has been supported by a Spanish-Taiwanese Interchange Program with reference 2011TW0010 (Spain) and NSC 101-2923-M-002-006-MY3 (Taiwan).

References

1. R. Maartens and K. Koyama, Living Rev. Rel. **13**, 5 (2010) [arXiv:1004.3962 [hep-th]].
2. L. Randall and R. Sundrum, Phys. Rev. Lett. **83**, 4690 (1999) [hep-th/9906064].
3. D. Lovelock, J. Math. Phys. **12**, 498 (1971).
4. B. Zwiebach, Phys. Lett. B **156**, 315 (1985).
5. B. Zumino, Phys. Rept. **137**, 109 (1986).
6. E. Witten, Nucl. Phys. **B311**, 46 (1988).
7. A. H. Chamseddine, Phys. Lett. **B233**, 291 (1989).
8. J. Zanelli, [hep-th/0502193].
9. N. Deruelle and M. Sasaki, Prog. Theor. Phys. **110**, 441 (2003) [gr-qc/0306032].
10. V. I. Zakharov, JETP Lett. **12**, 312 (1970).

11. H. Collins, B. Holdom, Phys. Rev. **D62**, 105009 (2000). [hep-ph/0003173].

12. G. R. Dvali, G. Gabadadze, M. Porrati, Phys. Lett. **B485**, 208-214 (2000). [hep-th/0005016].

13. G. R. Dvali, G. Gabadadze, Phys. Rev. **D63**, 065007 (2001). [hep-th/0008054].

14. D. Langlois, R. Maartens and D. Wands, Phys. Lett. B **489**, 259 (2000) [arXiv:hep-th/0006007].

15. R. Maartens, D. Wands, B. A. Bassett and I. Heard, Phys. Rev. D **62**, 041301 (2000) [hep-ph/9912464].

16. M. Bouhmadi-López, R. Maartens, D. Wands, Phys. Rev. **D70**, 123519 (2004). [hep-th/0407162].

17. J. -F. Dufaux, J. E. Lidsey, R. Maartens, M. Sami, Phys. Rev. **D70**, 083525 (2004). [hep-th/0404161].

18. M. Bouhmadi-López, P. Chen, Y.-W. Liu, submitted for publication.

19. J. W. York, Jr., "Role of conformal three geometry in the dynamics of gravitation," Phys. Rev. Lett. **28**, 1082-1085 (1972).

20. G. W. Gibbons, S. W. Hawking, "Action Integrals and Partition Functions in Quantum Gravity," Phys. Rev. **D15**, 2752-2756 (1977).

21. For simplicity, we have omitted the York-Gibbons-Hawking surface term.[19,20] However, we would like to stress that it is quite important to get the equation of motions and the junction conditions at the brane.

22. D. G. Boulware and S. Deser, Phys. Rev. Lett. **55**, 2656 (1985).

23. R. G. Cai, Phys. Rev. D **65**, 084014 (2002) [arXiv:hep-th/0109133].

24. M. Bañados, C. Teitelboim, J. Zanelli, Phys. Rev. **D49**, 975-986 (1994). [gr-qc/9307033].

25. J. Crisostomo, R. Troncoso, J. Zanelli, Phys. Rev. **D62**, 084013 (2000). [hep-th/0003271].

26. S. C. Davis, Phys. Rev. D **67**, 024030 (2003) [arXiv:hep-th/0208205].

27. K. i. Maeda and T. Torii, Phys. Rev. D **69**, 024002 (2004) [arXiv:hep-th/0309152].

28. C. Charmousis and J. F. Dufaux, Class. Quant. Grav. **19**, 4671 (2002) [arXiv:hep-th/0202107].

29. G. Kofinas, R. Maartens, E. Papantonopoulos, JHEP **0310**, 066 (2003). [hep-th/0307138].

30. Please note that we have left the induced gravity parameter γ arbitrary and not set to one as was done in Ref. 29.

31. Notice that we are not considering the limiting case corresponding to the Chern-Simons gravity because on that case a homogenous and isotropic brane cannot be embedded in the bulk.[28]

32. R. A. Brown, Gen. Rel. Grav. **39**, 477-500 (2007). [gr-qc/0602050].

33. M. Bouhmadi-López, P. V. Moniz, Phys. Rev. **D78**, 084019 (2008). [arXiv:0804.4484 [gr-qc]]; M. Bouhmadi-López, Y. Tavakoli and P. V. Moniz, JCAP **1004**, 016 (2010) [arXiv:0911.1428 [gr-qc]]; M. Bouhmadi-López, A. Errahmani and T. Ouali, Phys. Rev. D **84**, 083508 (2011) [arXiv:1104.1181 [astro-ph.CO]]; M. -H. Belkacemi, M. Bouhmadi-López, A. Errahmani and T. Ouali, Phys. Rev. D **85**, 083503 (2012) [arXiv:1112.5836 [gr-qc]].

34. J. E. Lidsey, A. R. Liddle, E. W. Kolb, E. J. Copeland, T. Barreiro, M. Abney, Rev. Mod. Phys. **69**, 373-410 (1997). [astro-ph/9508078].

35. D. Wands, K. A. Malik, D. H. Lyth and A. R. Liddle, Phys. Rev. D **62**, 043527 (2000) [astro-ph/0003278].

36. E. Komatsu *et al.* [WMAP Collaboration], Astrophys. J. Suppl. **192**, 18 (2011) [arXiv:1001.4538 [astro-ph.CO]].

TESTING THE ORIGIN OF PRIMORDIAL PERTURBATION – USE OF BISPECTRUM AND TRISPECTRUM

TERUAKI SUYAMA

Research Center for the Early Universe, Graduate School of Science,
The University of Tokyo, Tokyo 113-0033, Japan
E-mail: suyama@resceu.s.u-tokyo.ac.jp

It has become a standard paradigm that the primordial curvature fluctuations originate from quantum fluctuations of scalar fields that are generated during inflation. Thanks to the development of cosmic observations, next decade will be an era for clarifying conversion mechanism from scalar field perturbations to the curvature perturbation and also the number of such scalar fields. My talk is about the effectiveness of using the bispectrum and trispectrum of the curvature perturbations for elucidating those issues. In particular, I will put much focus on the consistency relations that hold for the parameters of the bi and trispectrum and for each inflationary model and show how one can discriminate models by using such relations.

Keywords: Primordial perturbation; non-Gaussianity.

1. Introduction

It is now widely believed that in the very early Universe, there was a phase of accelerated expansion, called inflation. A standard paradigm for achieving inflation is to use potential energy of scalar fields (not necessarily a single field) that mimics the cosmological constant when the scalar field is changing very slowly (slow-roll). On the background of the accelerating expansion, it is known that any scalar field whose mass is smaller than the Hubble parameter during inflation acquires long wavelength classical fluctuations of order $\frac{H}{2\pi}$. Here by long wavelength, it means the length is comparable to the current cosmological scales, thus much larger than the Hubble size during inflation. It is thought that this scalar field fluctuations (not necesarily a single field) somehow converted to the curvature perturbation, which is observable, resulting in the temperature anisotropy of the Cosmic Microwave Background and inhomogeneous distribution of matters. However, we still do not know what field caused inflation and what field generated curvature perturbation and how it occurred. Therefore, it is important and fundamental to elucidate the nature of inflation and origin of the primordial perturbation. Thanks to the development of cosmic observations which enables us to measure cosmological inhomogeneities to a very good accuracy than ever, it is expected that much progress toward understanding what had happened in the early Universe will be made in the forthcoming decade.

With the improvement of observational accuracy, non-Gaussianity of the primordial fluctuations has become important observable to probe the nature of fluctuations. Since statistics of zero-mean Gaussian fluctuations can be characterized by a variance which corresponds to two-point function. the non-Gaussianity of fluctuations can be linked to the higher-order correlation functions. It is known that primordial fluctuations are almost Gaussian for the simplest model where a single scalar field both causes inflation and generate the primordial fluctuations. However, in more contrived models, the fluctuations can be non-Gaussian to level that is enough distinguishable from Gaussianity.

The degrees of non-Gaussianity are often represented by a so-called non-linearity parameter f_{NL}, which characterizes the size of bispectrum of the curvature perturbation. Depending on the momentum distribution of the bispectrum or the shape of three point function, three types of f_{NL} have been discussed in the literature:[1,2] local, equilateral and orthogonal types. The limits on these f_{NL}s have been obtained as:[1] $-10 < f_{\mathrm{NL}}^{\mathrm{local}} < 74$ for the local type, $-214 < f_{\mathrm{NL}}^{\mathrm{equil}} < 266$ for the equilateral type and $-410 < f_{\mathrm{NL}}^{\mathrm{orthog}} < 6$ for the orthogonal type (95 % C.L.). Detectin of non-zero f_{NL} surely excludes at least as a dominant mechanism of the generation of density fluctuations. However, many other mechanisms have also been discussed in the literature and quite a few of them can generate large non-Gaussianity.

Thus the question we should ask next is "how can we differentiate these models?" In this paper, we discuss this issue by using bispectrum and trispectrum of the curvature perturbation. Although, as mentioned above, we can divide models into some categories depending on the shape of the three point functions (local, equilateral and orthogonal types), the shape is not enough to differentiate models since there remain many models for each type of the shape. Furthermore, f_{NL} predicted in those models can fall onto almost the same value by tuning some model parameters. Thus, obviously, the determination of f_{NL} is not enough to pin down the models of large non-Gaussianity even if f_{NL} is found to be large in the future. The purpose of this paper is to pursue the strategy of how one can differentiate models of large non-Gaussianity. To this end, we consider higher order statistics such as the trispectrum in addition to the bispectrum. The size of the trispectrum can be parametrized by other non-linearity parameters τ_{NL} and g_{NL}[a] and the importance of the trispectrum has been emphasized in some literature.[3–7,9] However, here we give a systematic study of the bispectrum and the trispectrum of models with large non-Gaussianity and make some classifications by using the "consistency relations" between the non-linearity parameters f_{NL}, τ_{NL} and g_{NL}.

Among three types mentioned above (local, equilateral and orthogonal types), we

---

[a]Current observational limits for $\tau_{\mathrm{NL}}^{\mathrm{local}}$ and $g_{\mathrm{NL}}^{\mathrm{local}}$ are given by $-0.6 \times 10^4 < \tau_{\mathrm{NL}}^{\mathrm{local}} < 3.3 \times 10^4$ (95% C.L.) and $-7.4 \times 10^5 < g_{\mathrm{NL}}^{\mathrm{local}} < 8.2 \times 10^5$ (95% C.L.) from cosmic microwave background observations.[6] By using planck data, the bound on τ_{NL} is expected to be improved as $|\tau_{\mathrm{NL}}| < 560$.[10]

focus on local-type models in this paper. As will be shown, by using the "consistency relations," we may be able to discriminate models of large non-Gaussianity.

This talk is mostly based on a paper(Suyama et al.[8]). If a reader wants to know status of this field in more detail, I recommend the reader to have a look at it.[8]

2. Non-Linearity Parameters

In the so-called δN formalism,[11–14] the curvature perturbation is given, up to the third order in scalar field fluctuations, as

$$\zeta(t_f) \simeq N_a \delta\varphi_*^a + \frac{1}{2} N_{ab} \delta\varphi_*^a \delta\varphi_*^b + \frac{1}{6} N_{abc} \delta\varphi_*^a \delta\varphi_*^b \delta\varphi_*^c , \tag{1}$$

where a subscript a, b and c labels a scalar field, which is assumed to be Gaussian fluctuations $\delta\varphi^a$ at $t = t_*$ in the following discussion, and $N_a = dN/d\varphi_*^a$ and so on. The summation is implied for repeated indices.

Then B_ζ and T_ζ can be written as

$$B_\zeta(k_1, k_2, k_3) = \frac{6}{5} f_{\mathrm{NL}}^{\mathrm{local}} \left(P_\zeta(k_1) P_\zeta(k_2) + P_\zeta(k_2) P_\zeta(k_3) + P_\zeta(k_3) P_\zeta(k_1) \right), \tag{2}$$

$$T_\zeta(k_1, k_2, k_3, k_4) = \tau_{\mathrm{NL}}^{\mathrm{local}} \left(P_\zeta(k_{13}) P_\zeta(k_3) P_\zeta(k_4) + 11 \text{ perms.} \right)$$
$$+ \frac{54}{25} g_{\mathrm{NL}}^{\mathrm{local}} \left(P_\zeta(k_2) P_\zeta(k_3) P_\zeta(k_4) + 3 \text{ perms.} \right), \tag{3}$$

with $k_{13} = |\vec{k}_1 + \vec{k}_3|$. Here $f_{\mathrm{NL}}^{\mathrm{local}}, \tau_{\mathrm{NL}}^{\mathrm{local}}$ and $g_{\mathrm{NL}}^{\mathrm{local}}$ are non-linearity parameters of the local type. From Eq. (1), the power spectrum of the curvature perturbation is given by

$$P_\zeta(k) = N_a N^a P_\delta(k) , \tag{4}$$

where $P_\delta(k)$ is the power spectrum for fluctuations of a scalar field:

$$\langle \delta\varphi_{*\vec{k}_1}^a \, \delta\varphi_{*\vec{k}_2}^b \rangle \equiv (2\pi)^3 \delta^{ab} \delta\left(\vec{k}_1 + \vec{k}_2\right) P_\delta(k_1) = (2\pi)^3 \delta^{ab} \delta\left(\vec{k}_1 + \vec{k}_2\right) \frac{2\pi^2}{k_1^3} \mathcal{P}_\delta(k_1) \tag{5}$$

with $\mathcal{P}_\delta(k) = (H_*/2\pi)^2$ and H_* being the Hubble parameter at $t = t_*$.

In terms of the expansion coefficients $N_a, \cdots$, the non-linearity parameters are given by[3,15,16]

$$\frac{6}{5} f_{\mathrm{NL}}^{\mathrm{local}} = \frac{N_a N_b N^{ab}}{(N_c N^c)^2}, \tag{6}$$

$$\tau_{\mathrm{NL}}^{\mathrm{local}} = \frac{N_a N_b N^{ac} N_c^{\ b}}{(N_d N^d)^3}, \qquad \frac{54}{25} g_{\mathrm{NL}}^{\mathrm{local}} = \frac{N_{abc} N^a N^b N^c}{(N_d N^d)^3}. \tag{7}$$

In the following, since we concentrate on non-Gaussianity of the local-type models, we omit a superscript "local" unless some confusions arise.

3. Classification and Consistency Relations among Non-linearity Parameters

Even if we limit ourselves to models generating large local-type non-Gaussianity, there still remain many possibilities. To discuss how we discriminate those models, here we classify the local-type models into some categories. For this purpose, we start from providing the following general inequality that holds between $f_{\rm NL}$–$\tau_{\rm NL}$ plane:

$$\tau_{\rm NL} \geq \left(\frac{6}{5} f_{\rm NL}\right)^2, \tag{8}$$

which can be obtained by using Cauchy-Schwarz inequality.[17] Since all models generating large local-type non-Gaussianity known to date satisfy the above assumptions, the inequality (8) is very important to test the local-type models. Thus we call the inequality (8) "local-type inequality" in the following.

Let's have a look at this inequality in more detail. If the curvature perturbation ζ is sourced by a single field fluctuation $\delta\sigma$, then it corresponds to the border of the local type inequality;

$$\tau_{\rm NL} = \left(\frac{6}{5} f_{\rm NL}\right)^2. \tag{9}$$

Once $f_{\rm NL}$ is fixed, $\tau_{\rm NL}$ is uniquely determined from this equation. In other words, if observation shows $\tau_{\rm NL}$ which deviates from Eq. (9), that means the curvature perturbation is sourced by multiple fields. Therefore, a ratio $\frac{25}{36}\frac{\tau_{\rm NL}}{f_{\rm NL}^2}$ is a very powerful quantity to determine the number of fields contributing to the curvature perturbation. To see in which case this ratio deviates from unity, let us consider a case in which ζ is sourced by two uncorrelated fields;

$$\zeta = N_\phi \delta\phi + N_\sigma \delta\sigma + \frac{1}{2} N_{\sigma\sigma} \delta\sigma^2. \tag{10}$$

Just for simplicity, we have neglected second order term in $\delta\phi$. This type of fluctuation can be, for example, realized if inflaton ϕ and curvaton σ contribute to ζ. In this case, the ratio becomes

$$\frac{25}{36}\frac{\tau_{\rm NL}}{f_{\rm NL}^2} = 1 + \left(\frac{N_\phi}{N_\sigma}\right)^2. \tag{11}$$

The second term on the right hand side represents contribution of inflaton fluctuation relative to that of curvaton one. We see that the ratio is enhanced if the Gaussian field fluctuation contributes more to the curvature perturbation than the non-Gaussian one [b]. This simple example suggests that searching for non-Gaussian signals should include the use of trispectrum. The local type inequality allows a

---

[b]Of course, if we increase the Gaussian field contribution, both $f_{\rm NL}$ and $\tau_{\rm NL}$ decrease. Therefore, there is a maximum limit of N_ϕ above which non-Gaussian signal becomes too weak to be detected.

possibility that the bispectrum is small but trispectrum is large enough so that the first detection of non-Gaussianity comes from trispectrum.

For convenience, we call models in which the local-type inequality becomes equality as "single-source model." Notice that $\tau_{\rm NL}$ is determined once $f_{\rm NL}$ is given in models of this class. We call models in which the $\tau_{\rm NL}$ is larger than $\frac{25}{36}\frac{\tau_{\rm NL}}{f_{\rm NL}^2}$ "multi-source model". As we saw earlier, an example of this type includes mixed fluctuation models, where fluctuations from both of the inflaton and another scalar field such as the curvaton can be responsible for density fluctuations. Since the inflaton gives almost Gaussian fluctuations, non-Gaussianity mostly originates from fluctuations of the other source in such a case.

Now we have categorized models of local-type into two classes by using the key quantity $\tau_{\rm NL}/(6f_{\rm NL}/5)^2$. However, since each category still includes some (or many) possible models, we may need another quantity to discriminate them. For this purpose, we can further utilize the relation between $f_{\rm NL}$ and $g_{\rm NL}$. Although the $f_{\rm NL}$–$g_{\rm NL}$ relation can change depending on the model parameters, we can roughly divide models into three types further by looking at their relative size. As we will argue in the following sections, the relation $|g_{\rm NL}| \sim |f_{\rm NL}|$ holds in some models, then we call such models "linear $g_{\rm NL}$ type." In other models, $g_{\rm NL}$ could be suppressed compared to $f_{\rm NL}$, i.e. $g_{\rm NL} \sim$ (suppression factor)$\times f_{\rm NL}$, which we denote this type of models as "suppressed $g_{\rm NL}$ type." The other type is "enhanced $g_{\rm NL}$ type" in which the relation between $f_{\rm NL}$ and $g_{\rm NL}$ can be given as $g_{\rm NL} \sim f_{\rm NL}^n$ with $n > 1$ (but in most models discussed in this paper, $n = 2$). By using the $f_{\rm NL}$–$\tau_{\rm NL}$ and $f_{\rm NL}$–$g_{\rm NL}$ relations, we may be able to discriminate models well. In Table 1, we provide consistency relations among non-linearity parameters for representative models proposed so far. We see that different models predict different consistency relations. These relations would be useful to test those models and to clarify the origin of primordial perturbations.

4. Discussion and Summary

In this paper, we made a classification of models generating large local-type non-Gaussianity by using some consistency relations between the non-linearity parameters $f_{\rm NL}, \tau_{\rm NL}$ and $g_{\rm NL}$. The first key relation is the ratio of $\tau_{\rm NL}/(6f_{\rm NL}/5)^2$, by which we classify local-type models into two categories:

- single-source model $(\tau_{\rm NL}/(6f_{\rm NL}/5)^2 = 1)$
- multi-source model $(\tau_{\rm NL}/(6f_{\rm NL}/5)^2 > 1)$

To our knowledge, since all models generating local-type large non-Gaussianity known today should satisfy the "local-type inequality," if future observations confirm that this inequality does not hold, local-type models would be practically ruled out as a mechanism of generating large non-Gaussian primordial fluctuations.

On the other hand, if future observations find large $f_{\rm NL}$ of local type and probe the relation between $\tau_{\rm NL}$ and $f_{\rm NL}$ with some accuracy, satisfying the local-type inequality, we can see what category of models would be favored. However, even if

Table 1. Summary of the categories and their examples.

Category	f_{NL}–τ_{NL} relation	Examples and f_{NL}–g_{NL} relation
Single-source	$\tau_{\mathrm{NL}} = (6f_{\mathrm{NL}}/5)^2$	(pure) curvaton (w/o self-interaction) $\left[g_{\mathrm{NL}} = -(10/3)f_{\mathrm{NL}} - (575/108)\right]^{(a)}$
		(pure) curvaton (w/ self-interaction) $\left[g_{\mathrm{NL}} = A_{\mathrm{NQ}}f_{\mathrm{NL}}^2 + B_{\mathrm{NQ}}f_{\mathrm{NL}} + C_{\mathrm{NQ}}\right]^{(b)}$
		(pure) modulated reheating $\left[g_{\mathrm{NL}} = 10f_{\mathrm{NL}} - (50/3)\right]^{(c)}$.
		modulated-curvaton scenario $\left[g_{\mathrm{NL}} = 3r_{\mathrm{dec}}^{1/2}f_{\mathrm{NL}}^{3/2}\right]^{(d)}$
		Inhomogeneous end of hybrid inflation $\left[g_{\mathrm{NL}} = (10/3)\eta_{\mathrm{cr}}f_{\mathrm{NL}}\right]$
		Inhomogeneous end of thermal inflation $\left[g_{\mathrm{NL}} = -(10/3)f_{\mathrm{NL}} - (50/27)\right]^{(e)}$
		Modulated trapping $\left[g_{\mathrm{NL}} = (2/9)f_{\mathrm{NL}}^2\right]^{(f)}$
Multi-source	$\tau_{\mathrm{NL}} > (6f_{\mathrm{NL}}/5)^2$	mixed curvaton and inflaton $\left[g_{\mathrm{NL}} = -(10/3)(R/(1+R))f_{\mathrm{NL}} - (575/108)(R/(1+R))^3\right]^{(g)}$
		mixed modulated and inflaton $\left[g_{\mathrm{NL}} = 10(R/(1+R))f_{\mathrm{NL}} - (50/3)(R/(1+R))^3\right]^{(h)}$
		mixed modulated trapping and inflaton $\left[g_{\mathrm{NL}} = (2/9)((1+R)/R)f_{\mathrm{NL}}^2 = (25/162)\tau_{\mathrm{NL}}\right]^{(i)}$
		multi-curvaton $\left[g_{\mathrm{NL}} = C_{\mathrm{mc}}f_{\mathrm{NL}}, \quad g_{\mathrm{NL}} = (4/15)f_{\mathrm{NL}}^2\right]^{(j)}$
		Multi-brid inflation (quadratic potential) $\left[g_{\mathrm{NL}} = -(10/3)\eta f_{\mathrm{NL}}, \quad g_{\mathrm{NL}} = 2f_{\mathrm{NL}}^2\right]^{(k)}$
		Multi-brid inflation (linear potential) $\left[g_{\mathrm{NL}} = 2f_{\mathrm{NL}}^2\right]^{(l)}$
Constrained multi-source	$\tau_{\mathrm{NL}} = Cf_{\mathrm{NL}}^n$	ungaussiton ($C \simeq 10^3$, $n = 4/3$)

Note:

[a] For the case with $r_{\mathrm{dec}} \ll 1$.

[b] A_{NQ}, B_{NQ} and C_{NQ} are given in[8] and this expression is for $r_{\mathrm{dec}} \ll 1$.

[c] $\Gamma_{\sigma\sigma\sigma} = 0$ is assumed.

[d] This relation holds in the Region 2. For other cases, see text.

[e] $g''' = 0$ is assumed.

[f] $\lambda = \sigma/M$ and $m = g\sigma$ are assumed.

[g] A quadratic potential and $r_{\mathrm{dec}} \ll 1$ are assumed for the curvaton sector. $R \equiv P_\zeta^{(\sigma)}/P_\zeta^{(\phi)}$ is the ratio of the power spectra. This relation can also be written as
$g_{\mathrm{NL}} \sim -(24/5)(f_{\mathrm{NL}}^3/\tau_{\mathrm{NL}}) - (9936/625)(f_{\mathrm{NL}}^6/\tau_{\mathrm{NL}}^3)$.

[h] $\Gamma_{\sigma\sigma\sigma} = 0$ is assumed for the modulated reheating sector. This relation can also be written as
$g_{\mathrm{NL}} \simeq (72/5)(f_{\mathrm{NL}}^3/\tau_{\mathrm{NL}}) - (31104/625)(f_{\mathrm{NL}}^6/\tau_{\mathrm{NL}}^3)$.

[i] $\lambda = \sigma/M$ and $m = g\sigma$ are assumed for the modulaton sector.

[j] The former and the latter relations are for the cases where both curvatons are subdominant and dominant at their decay, respectively. C_{mc} is $\mathcal{O}(1)$ coefficient and always negative.

[k] The former and the latter relations are for the equal mass and the large mass ratio cases, respectively.

[l] For the equal mass case with $g_1 = g_2$.

we can pick up the one of these categories, as we have discussed, there still remain many possibilities for each. Thus we need a further classification to pin down the model of large non-Gaussianity. For this purpose, we can make use of the relation between f_{NL} and g_{NL}. We have shown that models can be further divided into three types according to the relative size of g_{NL} compared to that of f_{NL} as follows:

- Suppressed g_{NL} type ($g_{\mathrm{NL}} \sim$ [suppression factor] $\times f_{\mathrm{NL}}$)
- Linear g_{NL} type ($g_{\mathrm{NL}} \sim f_{\mathrm{NL}}$)
- Enhanced g_{NL} type ($g_{\mathrm{NL}} \sim f_{\mathrm{NL}}^n$ with $n > 1$ or $n = 2$ for many models)

Thus if we further probe the relation f_{NL} and g_{NL} in future observations, we may find that only a few models survive by using the above categorizations. Then we can figure out what type of models are favored as a mechanism of the generation of primordial fluctuations.

We have also worked out the above mentioned relations for various concrete models in this paper. Although models can be categorized rather rigorously by using the ratio $\tau_{\mathrm{NL}}/(6 f_{\mathrm{NL}} 5)^2$, the relation between f_{NL} and g_{NL} can significantly differ depending on some model parameters, in particular, in multi-source models. In other words, the relation between f_{NL} and g_{NL} should be carefully investigated to discriminate a model because a model can predict quite different relations depending on its model parameters. However, it also means that the relation would be useful to explore the parameters of a model.

If three non-linearity parameters $f_{\mathrm{NL}}, \tau_{\mathrm{NL}}$ and g_{NL} are well determined in future observations, we may be able to pin down the model of large non-Gaussianity and pick up a right model of generating primordial fluctuations. The classification by using the consistency relation among the above three parameters would be very useful to pursue the origin of the structure of the Universe and give a deep understanding of the early Universe.

Acknowledgments

T. S. is supported by a Grant-in-Aid for JSPS Fellows No. 1008477.

References

1. E. Komatsu *et al.*, arXiv:1001.4538 [astro-ph.CO].
2. L. Senatore, K. M. Smith and M. Zaldarriaga, JCAP **1001**, 028 (2010) [arXiv:0905.3746 [astro-ph.CO]].
3. C. T. Byrnes, M. Sasaki and D. Wands, Phys. Rev. D **74**, 123519 (2006) [arXiv:astro-ph/0611075].
4. K. Enqvist and T. Takahashi, JCAP **0809**, 012 (2008) [arXiv:0807.3069 [astro-ph]].
5. K. Enqvist and T. Takahashi, JCAP **0912**, 001 (2009) [arXiv:0909.5362 [astro-ph.CO]].
6. J. Smidt, A. Amblard, A. Cooray, A. Heavens, D. Munshi and P. Serra, arXiv:1001.5026 [astro-ph.CO].
7. V. Desjacques and U. Seljak, Phys. Rev. D **81**, 023006 (2010) [arXiv:0907.2257 [astro-ph.CO]].

8. T. Suyama, T. Takahashi, M. Yamaguchi and S. Yokoyama, JCAP **1012**, 030 (2010) [arXiv:1009.1979 [astro-ph.CO]].

9. J. Smidt, A. Amblard, C. T. Byrnes, A. Cooray and D. Munshi, arXiv:1004.1409 [astro-ph.CO].

10. N. Kogo and E. Komatsu, Phys. Rev. D **73**, 083007 (2006) [astro-ph/0602099].

11. A. A. Starobinsky, JETP Lett. **42** (1985) 152 [Pisma Zh. Eksp. Teor. Fiz. **42** (1985) 124];

12. M. Sasaki and E. D. Stewart, Prog. Theor. Phys. **95**, 71 (1996). [arXiv:astro-ph/9507001];

13. M. Sasaki and T. Tanaka, Prog. Theor. Phys. **99**, 763 (1998). [arXiv:gr-qc/9801017].

14. D. H. Lyth, K. A. Malik and M. Sasaki, JCAP **0505**, 004 (2005) [arXiv:astro-ph/0411220].

15. D. H. Lyth and Y. Rodriguez, Phys. Rev. Lett. **95**, 121302 (2005) [arXiv:astro-ph/0504045].

16. L. Alabidi and D. H. Lyth, JCAP **0605**, 016 (2006) [arXiv:astro-ph/0510441].

17. T. Suyama and M. Yamaguchi, Phys. Rev. D **77**, 023505 (2008) [arXiv:0709.2545 [astro-ph]];

SELF-ACCELERATING UNIVERSE IN NONLINEAR MASSIVE GRAVITY*

A. E. GUMRUKCUOGLU*, C. LIN[†] and S. MUKOHYAMA[‡]

*IPMU, The University of Tokyo,
Kashiwa, Chiba 277-8582, Japan*
E-mail: emir.gumrukcuoglu@ipmu.jp
[†] *E-mail: chunshan.lin@ipmu.jp*
[‡] *E-mail: shinji.mukohyama@ipmu.jp*

We discuss the self-accelerating universe solutions in the framework of the potentially ghost-free, nonlinear massive gravity theory recently proposed by de Rham–Gabadadze–Tolley. The theory allows general Friedmann–Robertson–Walker solutions with negative curvature. The contribution of the mass terms at the background level is an effective cosmological constant term depending on the parameters of the theory. We also discuss the cosmological perturbations in these backgrounds, as well as similar solutions of the extended versions of the theory; the actions of the scalar and vector degrees do not undergo a modification with respect to general relativity, while the two polarizations of gravity waves acquire a time-dependent effective mass term. This may lead to a modification of the stochastic gravitational wave spectrum.

Keywords: Modified gravity; Cosmological perturbation theory; Dark energy theory.

1. Introduction

The construction of a massive gravity theory which reduces smoothly to general relativity (GR) in the massless limit has been a challenge of classical field theory for more than seventy years. The first model of massive gravity was introduced by Fierz and Pauli,[1] where the linearized Einstein-Hilbert action was extended by a linear mass term. Although it has the correct properties of a massive spin-2 theory (e.g. has five degrees of freedom, as required by Poincaré representations), its predictions fail to recover those of GR[2,3] in the massless limit. The resolution of this issue is to consider nonlinear mass terms,[4] but with the cost of introducing an additional degree of freedom, the Boulware–Deser (BD) ghost.[5]

The nature of the discontinuity and the extra degree became evident in the gauge-invariant construction of massive gravity with Stückelberg formalism and the introduction of the decoupling limit in.[6] In this effective field theory perspective, it is possible tune the coefficients to remove the ghost degree,[7] analogous to the

---

*Talk given by A. Emir GUMRUKCUOGLU at the First LeCosPA Symposium: Towards Ultimate Understanding of the Universe, Taipei, February 2012.

cancellation in the Fierz–Pauli theory. Equipped with this strategy, a nonlinear massive gravity theory was recently developed.[8,9] The theory is constructed by removing the extra degree at each order in the decoupling limit. The theory was later shown to be free of the BD ghost at any order, away from the decoupling limit.[10–12]

The main goal of this presentation is to show that the nonlinear massive gravity theory admits Friedmann-Robertson-Walker (FRW) solutions. These solutions contain an effective cosmological constant term, which may potentially act as a source for dark energy. We further discuss the dynamics of cosmological perturbations in these backgrounds. In Section 2, we review the theory and derive the open universe solutions. Next, we generalize these to an extended version of the theory. In Section 3, we introduce perturbations, and analyze the action in a gauge invariant language. We conclude with a discussion of our results in Section 4. This presentation is based on Refs. 13 and 14.

2. Nonlinear Massive Gravity and Cosmological Solutions

In this section, we consider the nonlinear massive gravity[9] described by the 4-dimensional metric $g_{\mu\nu}$ and scalar fields ϕ^a ($a = 0, \cdots, 3$), coupled to arbitrary matter source. The role of the scalar fields ϕ^a is to maintain the general covariance.[6] By construction, the matter action S_m is independent of the ϕ^a fields. The total action is

$$S = S_g + S_m,$$

$$S_g = M_{Pl}^2 \int d^4x \sqrt{-g} \left[\frac{R}{2} + m_g^2(\mathcal{L}_2 + \alpha_3\mathcal{L}_3 + \alpha_4\mathcal{L}_4) \right], \tag{1}$$

where

$$\mathcal{L}_2 = \frac{1}{2} \left([\mathcal{K}]^2 - [\mathcal{K}^2] \right),$$

$$\mathcal{L}_3 = \frac{1}{6} \left([\mathcal{K}]^3 - 3[\mathcal{K}][\mathcal{K}^2] + 2[\mathcal{K}^3] \right),$$

$$\mathcal{L}_4 = \frac{1}{24} \left([\mathcal{K}]^4 - 6[\mathcal{K}]^2[\mathcal{K}^2] + 3[\mathcal{K}^2]^2 + 8[\mathcal{K}][\mathcal{K}^3] - 6[\mathcal{K}^4] \right), \tag{2}$$

and,

$$\mathcal{K}^\mu_\nu = \delta^\mu_\nu - \sqrt{g^{\mu\rho}\eta_{ab}\partial_\rho\phi^a\partial_\nu\phi^b}. \tag{3}$$

In the above, the squared brackets denote the trace operation, and for now, $\eta_{ab} = \text{diag}(-1, 1, 1, 1)$.

For the physical metric $g_{\mu\nu}$, we consider an open ($K < 0$) FRW universe

$$g_{\mu\nu}dx^\mu dx^\nu = -N(t)^2dt^2 + a(t)^2\Omega_{ij}dx^i dx^j,$$

$$\Omega_{ij}dx^i dx^j = dx^2 + dy^2 + dz^2 - \frac{|K|(xdx + ydy + zdz)^2}{1 + |K|(x^2 + y^2 + z^2)}, \tag{4}$$

where $x^0 = t$, $x^1 = x$, $x^2 = y$, $x^3 = z$; $\mu, \nu = 0, \cdots, 3$; and $i, j = 1, 2, 3$. As for the scalar fields ϕ^a ($a = 0, \cdots, 3$), we adopt the following ansatz, motivated by the coordinate transformation from the Minkowski coordinates to the open FRW chart of the Minkowski spacetime:

$$
\begin{aligned}
\phi^0 &= f(t)\sqrt{1 + |K|(x^2 + y^2 + z^2)}, \\
\phi^1 &= \sqrt{|K|}f(t)x, \\
\phi^2 &= \sqrt{|K|}f(t)y, \\
\phi^3 &= \sqrt{|K|}f(t)z.
\end{aligned}
\tag{5}
$$

This leads to the following diagonal form for $\eta_{ab}\partial_\mu\phi^a\partial_\nu\phi^b$.

$$
\eta_{ab}\partial_\mu\phi^a\partial_\nu\phi^b = -(\dot{f}(t))^2\delta^0_\mu\delta^0_\nu + |K|f(t)^2\Omega_{ij}\delta^i_\mu\delta^j_\nu,
\tag{6}
$$

where a dot represents differentiation with respect to t.

The equation of motion for the Stückelberg fields yields the constraint[13]

$$
(\dot{a} - \sqrt{|K|}N)\left[\left(3 - \frac{2\sqrt{|K|}f}{a}\right) + \alpha_3\left(3 - \frac{\sqrt{|K|}f}{a}\right)\left(1 - \frac{\sqrt{|K|}f}{a}\right)\right.
$$
$$
\left. + \alpha_4\left(1 - \frac{\sqrt{|K|}f}{a}\right)^2\right] = 0.
\tag{7}
$$

This equation has three solutions. The first solution, $\dot{a} = \sqrt{|K|}N$, implies that the physical metric $g_{\mu\nu}$ is Minkowski spacetime in the open FRW chart; it is therefore not a realistic representation of our universe. Reducing the above equation to remove this solution, we obtain

$$
\left(3 - \frac{2\sqrt{|K|}f}{a}\right) + \alpha_3\left(3 - \frac{\sqrt{|K|}f}{a}\right)\left(1 - \frac{\sqrt{|K|}f}{a}\right) + \alpha_4\left(1 - \frac{\sqrt{|K|}f}{a}\right)^2 = 0,
\tag{8}
$$

which is solved by

$$
f = \frac{a}{\sqrt{|K|}}X_\pm, \quad X_\pm \equiv \frac{1 + 2\alpha_3 + \alpha_4 \pm \sqrt{1 + \alpha_3 + \alpha_3^2 - \alpha_4}}{\alpha_3 + \alpha_4}.
\tag{9}
$$

Note that these two solutions do not exist if $K = 0$ is set. This is consistent with the fact that there is no nontrivial flat FRW solution.[15] On the other hand, for $K < 0$, these solutions are well-defined.

The equations of motion for the physical metric yields the dynamics of the solution

$$
3H^2 + \frac{3K}{a^2} = \Lambda_\pm + \frac{1}{M_{Pl}^2}\rho, \quad -\frac{2\dot{H}}{N} + \frac{2K}{a^2} = \frac{1}{M_{Pl}^2}(\rho + P),
\tag{10}
$$

where $H \equiv \dot{a}/aN$ is the expansion rate defined using the physical time parameter, ρ and P are the energy density and pressure of the matter content. The effect of the mass terms are contained in the effective cosmological constant, which is related to the parameters of the theory through

$$\Lambda_\pm \equiv -\frac{m_g^2}{(\alpha_3 + \alpha_4)^2} \left[(1 + \alpha_3)\left(2 + \alpha_3 + 2\,\alpha_3^2 - 3\,\alpha_4\right) \pm 2\left(1 + \alpha_3 + \alpha_3^2 - \alpha_4\right)^{3/2} \right].$$

(11)

Depending on the parameters, $\Lambda_\pm$ can be either positive or negative, for both cosmological solutions.

As pointed out above, the construction with a Minkowski fiducial metric only allows for an open universe solution. The solutions (9) above are then viable representations of the cosmology, provided that the curvature term is sufficiently small. The lack of a flat solution is related to the fact that a closed chart of Minkowski does not exist. As a result, a discontinuity occurs at zero curvature, leading only to an open universe in this setup.

On the other hand, one can extend the theory to accommodate more general fiducial metrics. In fact, such a setup has recently been shown to be free of the BD ghost.[16] For a more general $f_{\mu\nu}$, the no-go result for flat and closed universe solutions does not necessarily hold. For instance, for a de Sitter type fiducial metric, one can choose the Stückelberg fields in a way to accommodate any curvature. For this reason, we now discuss a more general fiducial metric. The only restriction we impose is the FRW symmetry,

$$f_{\mu\nu} = -n^2(\varphi^0)\partial_\mu\varphi^0\partial_\nu\varphi^0 + \alpha^2(\varphi^0)\Omega_{ij}(\varphi^k)\partial_\mu\varphi^i\partial_\nu\varphi^j,$$

(12)

where the Stückelberg fields in unitary gauge are $\varphi^a = \delta^a_\mu x^\mu$. For the Minkowski fiducial, the fields (5) can be written in this language as

$$\phi^0 = f(\varphi^0)\sqrt{1 - K\delta_{ij}\varphi^i\varphi^j}, \qquad \phi^i = \sqrt{-K}\,f(\varphi^0)\,\varphi^i,$$

(13)

and,

$$n(\phi^0) = \dot{f}(\varphi^0), \qquad \alpha(\phi^0) = \sqrt{-K}\,f(\varphi^0).$$

(14)

For a general $f_{\mu\nu}$ with FRW symmetry, the equations of motion of the Stückelberg fields yield the same branch of solutions as in the Minkowski fiducial; there is again a trivial branch, where the evolution of the physical metric is determined solely by the fiducial metric, rather than the matter content. For the two cosmological branches, we have

$$\alpha(t) = X_\pm\, a(t),$$

(15)

where $X_\pm$ is the constant term defined in (9). This gives rise to an effective cosmological constant which is exactly the same as in Minkowski fiducial case, i.e. (11). In other words, the general FRW fiducial metric of the form (12) leads exactly to the same background dynamics as in the Minkowski case, with the exception of the curvature term, which can now be zero, negative or positive.

3. Cosmological Perturbations

Although the theory is free of the BD-ghost, this does not guarantee that its solutions are stable. For instance, in Fierz-Pauli theory on de Sitter backgrounds, the helicity–0 graviton is known to become a ghost if $2\,H^2 > m_g^2$.[17] Moreover, since there are more degrees of freedom with respect to the massless case, these may have additional couplings, resulting in a modification in the strength of gravity, and leading to possible conflicts with classical tests. If these effects are under control, then one needs a handle to distinguish the massive gravity theory from other large scale modifications of GR, or models of dark energy.

To address some of these questions, we consider linear perturbations around the cosmological solutions discussed in the previous section. We introduce metric perturbations

$$g_{00} = -N^2(t)\left[1 + 2\phi\right] , \quad g_{0i} = N(t)a(t)\beta_i , \quad g_{ij} = a^2(t)\left[\Omega_{ij}(x^k) + h_{ij}\right] \tag{16}$$

and perturbations in the Stückelberg fields

$$\varphi^a = x^a + \pi^a + \frac{1}{2}\pi^b\partial_b\pi^a + O(\epsilon^3) . \tag{17}$$

We also consider an arbitrary set of independent degrees of freedom $\{\sigma_I\}$ which represent the matter content

$$\sigma_I = \sigma_I^{(0)} + \delta\sigma_I . \tag{18}$$

Notice that, although the action (1) has general coordinate invariance, we choose not to fix the gauge. Instead, we work with gauge invariant variables to be able to keep track of the source of each perturbation, as we show later in this section.

Since the fiducial metric $f_{\mu\nu}$ (12) does not depend on the physical metric, the FRW symmetry is preserved even in the presence of Stückelberg field perturbations. This allows us to decompose all perturbations based on the rotational invariance:

$$\beta_i = D_i\beta + S_i , \qquad \pi_i = D_i\pi + \pi_i^T ,$$

$$h_{ij} = 2\psi\Omega_{ij} + \left(D_iD_j - \frac{1}{3}\Omega_{ij}\triangle\right)E + \frac{1}{2}(D_iF_j + D_jF_i) + \gamma_{ij} , \tag{19}$$

where D_i is the covariant derivative associate with the spatial metric Ω_{ij} and $\triangle \equiv D_iD^i$. The vectors in the above decomposition are transverse, i.e. $D^iS_i = D^i\pi_i^T = D^iF_i = 0$, while the tensor is transverse and trace-free, $D^i\gamma_{ij} = \gamma_i^i = 0$.

We now construct gauge invariant variables. First, there is a set of variables we can define without referring to Stückelberg fields, as

$$Q_I \equiv \delta\sigma_I - \mathcal{L}_Z\sigma_I^{(0)} , \qquad \Phi \equiv \phi - \frac{1}{N}\partial_t(NZ^0) ,$$

$$\Psi \equiv \psi - \frac{\dot{a}}{a}Z^0 - \frac{1}{6}\triangle E , \qquad B_i \equiv S_i - \frac{a}{2N}\dot{F}_i , \tag{20}$$

where Q_I are constructed out of the matter fields σ_I and are analogues of the Sasaki-Mukhanov variable. Since we did not specify the nature of these fields, we define Q_I with a Lie derivative with respect to the vector Z^μ given by

$$Z^0 \equiv -\frac{a}{N}\beta + \frac{a^2}{2N^2}\dot{E}\,, \qquad Z^i \equiv \frac{1}{2}\Omega^{ij}(D_j E + F_j)\,, \tag{21}$$

and which transforms like coordinates at linear order.

Since the set (20) do not span all independent physical degrees, we need to define a second set of gauge invariant variables, now built out of Stückelberg perturbations

$$\psi^\pi \equiv \psi - \frac{1}{3}\triangle\pi - \frac{\dot{a}}{a}\pi^0\,, \qquad E^\pi \equiv E - 2\,\pi\,, \qquad F_i^\pi \equiv F_i - 2\,\pi_i^T\,. \tag{22}$$

The two sets (20) and (22) combined, exhaust all physical degrees of freedom in the system.

Using the definitions above, then using the background constraints, the action quadratic in perturbations can formally be written as[14]

$$S^{(2)} = S^{(2)}_{\mathrm{EH}} + S^{(2)}_{\mathrm{matter}} + S^{(2)}_{\Lambda_\pm} + \tilde{S}^{(2)}_{\mathrm{mass}}\,, \tag{23}$$

where $S^{(2)}_{\mathrm{EH}}$ is the Einstein-Hilbert action, $S^{(2)}_{\mathrm{matter}}$ is the matter action and $S^{(2)}_{\Lambda_\pm}$ corresponds to the action of the effective cosmological constant $\Lambda_\pm$ which we extracted out of the mass term as:

$$S^{(2)}_{\mathrm{mass}} = \tilde{S}^{(2)}_{\mathrm{mass}} + S^{(2)}_{\Lambda_\pm}\,. \tag{24}$$

We note that the first three terms in (23) depend only on the set of gauge invariant variables (20), which are built out of physical metric perturbations and matter field perturbations. The additional term has the form:

$$\tilde{S}^{(2)}_{\mathrm{mass}} = M_p^2 \int d^4x\, N\, a^3 \sqrt{\Omega}\, M_{GW}^2$$
$$\times \left[3(\psi^\pi)^2 - \frac{1}{12}E^\pi\triangle(\triangle+3K)E^\pi + \frac{1}{16}F_\pi^i(\triangle+2K)F_i^\pi - \frac{1}{8}\gamma^{ij}\gamma_{ij}\right]\,, \tag{25}$$

where M_{GW} is a time dependent function defined as

$$M_{GW}^2 \equiv \pm(r-1)m_g^2 X_\pm^2 \sqrt{1+\alpha_3+\alpha_3^2-\alpha_4}\,, \qquad r \equiv \frac{na}{N\alpha} = \frac{1}{X_\pm}\frac{H}{H_f}\,, \tag{26}$$

while $H_f \equiv \dot{\alpha}/\alpha n$ is the analogue of the Hubble rate in the fiducial metric. We note that the only common variable between the term $\tilde{S}^{(2)}_{\mathrm{mass}}$ and the rest of the action is the tensor perturbation γ_{ij}. Furthermore, the two scalar degrees ψ^π and E^π, as well as the two vector degrees F_i^π do not have kinetic terms. For now, we take the quadratic theory at face value and assume that these are infinitely heavy modes with no dynamics. With this assumption, we can integrate them out and obtain

$$\tilde{S}^{(2)}_{\mathrm{mass}} = -\frac{M_p^2}{8}\int d^4x\, N\, a^3 \sqrt{\Omega}\, M_{GW}^2\, \gamma^{ij}\gamma_{ij}\,. \tag{27}$$

With the cancellation of the extra degrees, we find that the scalar and vector sectors do not undergo any modification; their action is identical to the standard

GR, with a cosmological constant $\Lambda_\pm$ and same matter content. On the other hand, the tensor sector acquires an additional mass term. Assuming that the matter sector does not have any tensor contribution, the tensor action can be written as

$$S_{tensor}^{(2)} = \frac{M_{Pl}^2}{8} \int d^4x \, N \, a^3 \sqrt{\Omega} \left[\frac{1}{N^2}\dot{\gamma}^{ij}\dot{\gamma}_{ij} + \frac{1}{a^2}\gamma^{ij}(\triangle - 2K)\gamma_{ij} - M_{GW}^2 \gamma^{ij}\gamma_{ij} \right].$$

$$(28)$$

The stability of these modes are defined by the sign of $(r-1)m_g^2$ in (26). Assuming that $M_{GW}^2 > 0$, the stochastic gravitational waves will undergo a suppression compared to what we would expect from a GR signal at large scales. If M_{GW} is of the order of Hubble rate today, the signal will be in the observable horizon. The analysis of potential observational signal is in progress.[18]

4. Conclusion

We have shown that the recent construction of potentially ghost free nonlinear massive gravity theory of Ref. 9 admits open universe cosmological solutions, which self-accelerate with an effective cosmological constant determined by the parameters of theory. We extended these solutions to a theory with arbitrary fiducial metric and found that in general, the spatial metric can be also closed or flat. We then introduced perturbations around these solutions and found that, at the level of the quadratic action, the scalar and vector sectors to are identical to the case of GR with the same effective cosmological constant and same matter fields. The only modification arises in the tensor sector; these modes acquire a time dependent mass term, whose evolution depends on the fiducial metric and the physical metric.

On the other hand, since the theory describes a massive spin-2 field, we expected to have 5 degrees of freedom in the gravity sector. Instead, due to the cancellation of some kinetic terms, we have exactly the same degrees in GR, i.e. the two gravity wave polarizations. The physical nature of this cancellation is not yet well understood. Since the additional degrees exist in the trivial (noncosmological) branch, the lack of kinetic terms seems to be related to some symmetry in the cosmological branch of solutions. The vanishing kinetic terms may be an indication of an infinitely strong coupling in the scalar and vector sectors. If this is the case, these degrees cannot be described without a knowledge of UV completion. On the other hand, these modes also have nonvanishing mass terms; it may be possible that these are just infinitely heavy modes without low-energy dynamics, and can be safely integrated out, as done in the previous section. In order to determine whether these degrees are strongly coupled or nondynamical, the linear perturbation theory approach adopted here is not sufficient, and nonlinear methods are needed.

Hence, at the level of the linear theory, the only possible signature of the cosmological massive gravity solution is in the gravity waves. A positive but large contribution in the time dependent mass term may give rise to a suppression in stochastic gravity wave spectrum. The deviation from GR signal may allow the signal to be potentially observable in the upcoming space-based observatories.

It is important to note that the present analysis is purely classical and special care is needed when discussing the evolution of cosmological perturbations which start off in quantum mechanical vacuum. To address issues such as radiative stability, an analogue of the decoupling limit for the cosmological branches (9) is needed. This is certainly one of the most important directions in the future research.

References

1. M. Fierz and W. Pauli, *Proc.Roy.Soc.Lond.* **A173**, 211 (1939).
2. H. van Dam and M. Veltman, *Nucl.Phys.* **B22**, 397 (1970).
3. V. Zakharov, *JETP Lett.* **12**, p. 312 (1970).
4. A. Vainshtein, *Phys.Lett.* **B39**, 393 (1972).
5. D. Boulware and S. Deser, *Phys.Rev.* **D6**, 3368 (1972).
6. N. Arkani-Hamed, H. Georgi and M. D. Schwartz, *Annals Phys.* **305**, 96 (2003).
7. P. Creminelli, A. Nicolis, M. Papucci and E. Trincherini, *JHEP* **0509**, p. 003 (2005).
8. C. de Rham and G. Gabadadze, *Phys.Rev.* **D82**, p. 044020 (2010).
9. C. de Rham, G. Gabadadze and A. J. Tolley, *Phys.Rev.Lett.* **106**, p. 231101 (2011).
10. S. Hassan and R. A. Rosen, *Phys.Rev.Lett.* **108**, p. 041101 (2012), v2: 4 pages, results extended, comments added.
11. C. de Rham, G. Gabadadze and A. Tolley (2011).
12. C. de Rham, G. Gabadadze and A. J. Tolley, *JHEP* **1111**, p. 093 (2011).
13. A. Gumrukcuoglu, C. Lin and S. Mukohyama, *JCAP* **1111**, p. 030 (2011).
14. A. Gumrukcuoglu, C. Lin and S. Mukohyama, *JCAP* **1203**, p. 006 (2012).
15. G. D'Amico, C. de Rham, S. Dubovsky, G. Gabadadze, D. Pirtskhalava and A. J. Tolley, *Phys.Rev.* **D84**, p. 124046 (2011), 21 pages.
16. S. Hassan, R. A. Rosen and A. Schmidt-May, *JHEP* **1202**, p. 026 (2012).
17. A. Higuchi, *Nucl.Phys.* **B282**, p. 397 (1987).
18. A. E. Gumrukcuoglu, S. Kuroyanagi, S. Mukohyama, C. Lin and N. Tanahashi, work in progress, (2012).

INHOMOGENEOUS BAROTROPIC FRW COSMOLOGIES WITH CONSTANT-SHIFTED CONFORMAL HUBBLE PARAMETERS

HARET C. ROSU*

*IPICyT, Instituto Potosino de Investigacion Científica y Tecnológica,
Apdo Postal 3-74 Tangamanga, 78231 San Luis Potosí, S.L.P., Mexico*
**E-mail: hcr@ipicyt.edu.mx*

KIRA V. KHMELNYTSKAYA

*Faculty of Engineering, Autonomous University of Queretaro,
Apdo Postal 1-798, Arteaga # 5, Col. Centro, Queretaro, Qro. 76001, Mexico*
E-mail: khmel@uaq.mx

It is known that the barotropic FRW system of differential equations can be reduced to simple harmonic oscillator (HO) differential equations in the conformal time variable. This is due to the fact that the Hubble rate parameter in conformal time is the solution of a simple Riccati equation of constant coefficients. In previous works, we have used this mathematical result to set the barotropic HO equations in the nonrelativistic supersymmetric approach by factorizing them. If a constant additive parameter, denoted by S, is added to the common Riccati solution of these supersymmetric partner cosmologies one obtains inhomogeneous barotropic cosmologies with periodic singularities in their spatial curvature indices that are counterparts of the non-shifted supersymmetric partners. The zero-mode solutions of these cyclic singular cosmologies are reviewed here as a function of real and imaginary shift parameter. We also notice the modulated zero modes obtained by using the general Riccati solution and comment on their cosmological application.

Keywords: barotropic FRW cosmologies; cosmological zero-modes; shifted Riccati procedure; factorization.

1. Introduction

Riccati nonlinear equations are together with Bernoulli equations the oldest and the simplest nonlinear differential equations with many applications in the realm of physics. In recent years, their solutions, under the name of superpotentials, have played an important role in supersymmetric quantum mechanics.[1,2] In quantum cosmology, the supersymmetric methods have been employed in the last 40 years and have been reviewed in the book of Moniz.[3] Besides, there are frequent occurrences of Riccati equations spread over the many areas of cosmology and astrophysics.[4] Riccati equations as simple as

$$R' + cR^2 + f = 0 \, , \tag{1}$$

where c is a real constant and f is a function of the independent variable play a central role in barotropic FRW cosmologies, which will be reviewed in the following. We only need to recall that particular solutions of the Riccati equation enter the factorization brackets of second order linear differential equations (usually with $c = 1$)

$$\left(\frac{d}{dt} + cR \right) \left(\frac{d}{dt} - cR \right) u = 0 \equiv u'' - c(cR^2 + R')u = 0 \equiv u'' + cfu = 0 \ . \tag{2}$$

The connections $R = \frac{1}{c}\frac{u'}{u}$ or $u = e^{c\int^t R}$ between the particular solutions of the two equations are also basic results of the factorization method together with the construction of the so-called supersymmetric partner equation of equation (2) obtained by reverting the order of the factorization brackets:

$$\left(\frac{d}{dt} - cR \right) \left(\frac{d}{dt} + cR \right) v = 0 \equiv v'' - c(cR^2 - R')v = 0 \equiv v'' + c(f + 2R')v = 0 \ . \tag{3}$$

2. Cosmological Riccati Equation of FRW Barotropic Cosmologies

As first shown by Faraoni,[5] the well-known comoving Einstein-Friedmann dynamical equations of barotropic FRW cosmologies can be turned into a single Riccati equation for the Hubble parameter in conformal time $\mathcal{H}(\eta)$ that is the simpler case of (1) with $c = \tilde{\gamma}$ and $f = \kappa\tilde{\gamma}$

$$\mathcal{H}' + \tilde{\gamma}\mathcal{H}^2 + \kappa\tilde{\gamma} = 0 \ , \tag{4}$$

where henceforth the prime and also $\frac{d}{d\eta}$ stand for the derivative with respect to η, $\tilde{\gamma} = \frac{3}{2}\gamma - 1$ is related to the adiabatic index γ of the cosmological fluid and $\kappa = \pm 1$ is the curvature parameter for the closed and open universe, respectively. In the following, we will consider only the $\kappa = 1$ case since it allows us to focus on the periodic features of the problem. This conformal-time Riccati equation is valid for any barotropic fluid except for $\gamma = \frac{2}{3}$ which leads to the simple linear equation $\mathcal{H}' = 0$. In addition, equation (4) is just the Riccati equation of the classical harmonic oscillator:

$$\dot{R} + \omega_0 R^2 + \omega_0 = 0 \ , \tag{5}$$

if one sets $\tilde{\gamma} \equiv \omega_0$ for the closed universe case. For the open case the analogy is with the up-side down harmonic oscillator. However, the fact that the independent variable is the conformal time and not the usual Newtonian time makes a substantial difference from the physical point of view. Since the conformal time is related to the comoving time through $t(\eta) = \int^\eta a(\eta)d\eta$, one can see that this type of time depends on the spatial scaling parameter. Therefore, the second order differential equations with which the conformal Riccati equation is connected are not really laws of force as in classical mechanics but can be treated as Schroedinger equations in quantum mechanics. We thus define the conformal Hubble parameter directly as the logarithmic derivative $\mathcal{H}_u(\eta) = \frac{1}{\tilde{\gamma}}\frac{u'}{u}$ because we know that by substituting $\mathcal{H}_u$ in equation (4) we get the very simple harmonic oscillator equation

$$u'' + \tilde{\gamma}_u^2 u = 0 , \qquad \tilde{\gamma}_u^2 = \tilde{\gamma}^2 = \text{const} . \tag{6}$$

Due to the similitude with supersymmetric quantum mechanics, one can call the u modes as bosonic zero modes.[6] They are $\tilde{\gamma}$ powers of the scale factor parameters $a(\eta)$, i.e., $u = a^{\tilde{\gamma}}(\eta)$. Using the particular solution

$$u_1 \sim \cos\tilde{\gamma}\eta \qquad \rightarrow \qquad a_{1u}(\eta) \sim u_1^{1/\tilde{\gamma}}$$

in the definition of $\mathcal{H}_u$ one gets $\mathcal{H}_{u_1} = -\tan\tilde{\gamma}\eta$.

A supersymmetric partner equation (3) of equation (6) leads immediately to a class of cosmologies with inverse scale factors with respect to the standard barotropic ones but with a conformal-time-dependent curvature index,[6]

$$v'' + \kappa_v(\eta)\tilde{\gamma}^2 v = 0 , \tag{7}$$

where

$$\kappa_v(\eta) = -(1 + 2\tan^2\tilde{\gamma}\eta) \tag{8}$$

denotes the conformal time dependent supersymmetric partner curvature index of fermionic type associated through the mathematical scheme to the constant bosonic curvature index.

A particular fermionic solution v is of the following type

$$v_1 = \frac{\tilde{\gamma}}{\cos\tilde{\gamma}\eta} \qquad \rightarrow \qquad a_{1v}(\eta) \sim v_1^{1/\tilde{\gamma}} .$$

We can see that the u and v barotropic cosmologies are dual to each other from the standpoint of these particular solutions, in the sense that $u_1 v_1 = \tilde{\gamma}$ and therefore the geometric mean of their scale parameters

$$a_g = (a_{1u}a_{1v})^{1/2} = (\tilde{\gamma})^{1/2\tilde{\gamma}}$$

is constant. Thus, a joint evolution of a u cosmology of constant curvature index and a v cosmology of the time-dependent curvature index (8) is stationary in conformal time from the standpoint of their geometric mean scale parameter a_g. The fermionic metric is of the form:

$$ds^2 = a_{1v}^2(\eta)\left[-d\eta^2 + \frac{dr^2}{1 - \kappa_v(\eta)r^2} + r^2 d\Omega^2\right] .$$

This metric should be thought of as an averaged metric in an inhomogeneous cosmology and so it does not even have to satisfy the Einstein field equations.[7] The only thing it has in common with the standard FRW model is the Riccati solution. Such metrics, with other time-dependent curvature indices have been used to mimic the backreaction of small scale density perturbations on the large scale spacetime geometry.[7,8]

3. Barotropic Cosmologies with Conformal Hubble Parameters Having Non-Zero Initial Conditions

Since the Riccati equation is a first-order differential equation, its solution is entirely determined by one initial condition. In the case of the cosmological Riccati equation (4), one cannot be sure that the initial condition is $\mathcal{H}(0) = 0$.

We thus reexamine the consequences of a constant shift $S \neq 0$ of the conformal Hubble parameter that provides a non-zero initial condition,[9]

$$\mathcal{H}_S(\eta) = \mathcal{H}_{u_1}(\eta) + S \qquad\qquad \mathcal{H}_S(0) = S \ .$$

It is easy to obtain the corresponding second-order differential equations in this case by the factorization method

$$\left(\frac{d}{d\eta} + \tilde{\gamma}\mathcal{H}_S \right) \left(\frac{d}{d\eta} - \tilde{\gamma}\mathcal{H}_S \right) \mathcal{U} = 0 \equiv \mathcal{U}'' + \kappa_{S,u}(\eta)\tilde{\gamma}^2\mathcal{U} = 0 \ , \qquad (9)$$

where

$$\kappa_{S,u}(\eta) = \left[1 - S^2 + 2S \tan \tilde{\gamma}\eta \right] \qquad (10)$$

and

$$\left(\frac{d}{d\eta} - \tilde{\gamma}\mathcal{H}_S \right) \left(\frac{d}{d\eta} + \tilde{\gamma}\mathcal{H}_S \right) \mathcal{V} = 0 \equiv \mathcal{V}'' + \kappa_{S,v}(\eta)\tilde{\gamma}^2\mathcal{V} = 0 \ , \qquad (11)$$

where

$$\kappa_{S,v}(\eta) = - \left[1 + S^2 - 2S \tan \tilde{\gamma}\eta + 2 \tan^2 \tilde{\gamma}\eta \right] \ . \qquad (12)$$

One can see that when $S = 0$, the initial pair of unshifted partner equations are obtained and the corresponding spatial curvature indices are recovered

$$\kappa_{0,u}(\eta) = 1 \ , \qquad\qquad \kappa_{0,v}(\eta) = - \left[1 + 2 \tan^2 \tilde{\gamma}\eta \right] \ . \qquad (13)$$

While in classical mechanics, these equations define parametric oscillators,[10] in cosmology they describe two new classes of barotropic-like cosmological universes that we call $\mathcal{U}$ and $\mathcal{V}$ cosmologies, respectively. One can also write averaged conformal-like metrics with the corresponding variable curvature indices. These two modified cosmologies are periodic, of period $T = \frac{\pi}{\tilde{\gamma}}$, and have the same conformal Hubble parameter given by $\mathcal{H}_S(\eta)$. Both cosmologies have periodic singularities in their time-dependent curvature indices.

The linear independent solutions $\mathcal{U}_1$ and $\mathcal{U}_2$ have the following form

$$\mathcal{U}_1(\eta) = e^{-i\Omega_S\eta} \, {}_2F_1 \left(1, -iS; 2 - iS; -e^{-2i\tilde{\gamma}\eta} \right) \qquad (14)$$

and

$$\mathcal{U}_2(\eta) = e^{i\Omega_S\eta}(2 \cos^2 \tilde{\gamma}\eta - i \sin 2\tilde{\gamma}\eta) \qquad (15)$$

On the other hand, the linear independent solutions $\mathcal{V}_1$ and $\mathcal{V}_2$ are given by:

$$\mathcal{V}_1(\eta) = \frac{e^{-i\Omega_S\eta}}{(2 \cos^2 \tilde{\gamma}\eta - i \sin 2\tilde{\gamma}\eta)} \qquad (16)$$

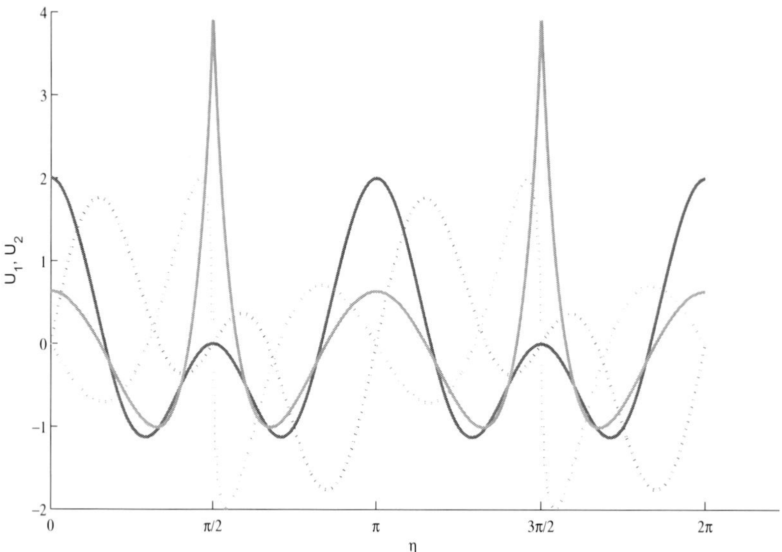

Fig. 1. The real (solid lines) and imaginary (dotted lines) parts of the periodic zero modes $\mathcal{U}_1(\eta)$ (in red) and $\mathcal{U}_2(\eta)$ (in blue) for the shift parameter $S = 3i$.

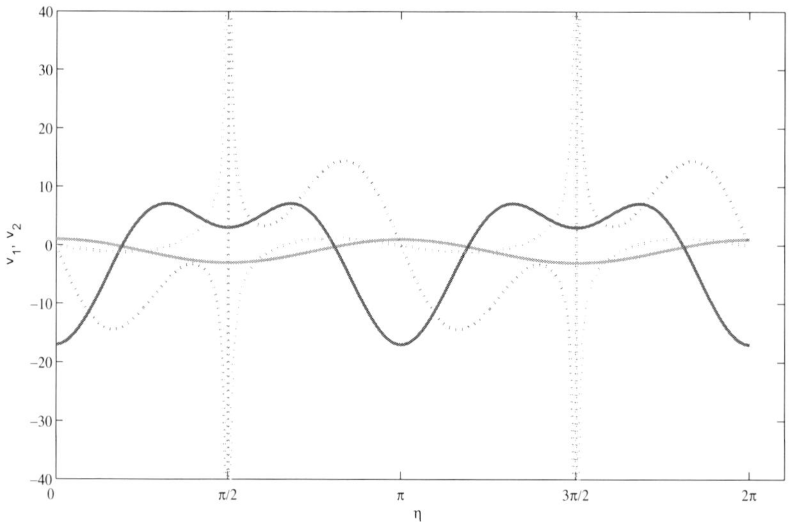

Fig. 2. The real (solid line) and imaginary (dotted lines) parts of the periodic zero modes $\mathcal{V}_1(\eta)$ (in red) and $\mathcal{V}_2(\eta)$ (in blue) for the shift parameter $S = 3i$.

and

$$\mathcal{V}_2(\eta) = \tilde{\gamma}^2 e^{i\Omega_S \eta} \left[1 - i \tan \tilde{\gamma}\eta + (S^2 + iS)(2\cos^2 \tilde{\gamma}\eta - i\sin 2\tilde{\gamma}\eta) - 2iS \right] . \qquad (17)$$

In these zero mode solutions $\Omega_S = \tilde{\gamma}(1 - iS)$ and the simple convergence condition $\Re(a + b - c) < 0$ for the hypergeometric series in (14) is fulfilled for all real values of η. These solutions can also be written in the simpler form:

$$\mathcal{U}_1(\eta) = \frac{e^{-S\tilde{\gamma}\eta}}{2\cos\tilde{\gamma}\eta} \, {}_2F_1\left(1, 2; 2 - iS; \frac{1}{1 + e^{2i\tilde{\gamma}\eta}}\right) \,, \qquad \mathcal{U}_2(\eta) = e^{S\tilde{\gamma}\eta}\cos\tilde{\gamma}\eta$$

and

$$\mathcal{V}_1(\eta) = \frac{e^{-S\tilde{\gamma}\eta}}{2\cos\tilde{\gamma}\eta} \,, \qquad \mathcal{V}_2(\eta) = 2\tilde{\gamma}^2 e^{S\tilde{\gamma}\eta}\left[\frac{1}{2\cos\tilde{\gamma}\eta} + S(S\cos\tilde{\gamma}\eta + \sin\tilde{\gamma}\eta)\right] \,.$$

However, the more complicated form of these zero-mode solutions is directly their Floquet-Bloch form that leads to the following considerations. The parameter S affects only the period of the phases $e^{\pm i\Omega_S\eta}$ of the solutions but not that of their periodic part. The solutions (14),(15) and the Bloch factors $e^{\pm i\Omega_S\eta}$ in (16),(17) are bounded if and only if the "quasifrequency" Ω_S has a real value, or equivalently

$$\tilde{\gamma}(1 - iS) \in \Re \,. \tag{18}$$

Taking into account that $\tilde{\gamma} \in \Re$, the last condition reads as $S = is$, $s \in \Re$. Thus, equation(9) has bounded solutions $\forall s \in \Re$. It is also worth to notice that for a purely imaginary shift parameter, the curvature indices $\kappa_{S,u}$ and $\kappa_{S,v}$ are related through

$$\kappa_{S,u}(-\eta) = \kappa^*_{S,u}(\eta) \,, \qquad \kappa_{S,v}(-\eta) = \kappa^*_{S,v}(\eta) \,,$$

where $*$ denotes the complex conjugation operation. This means that we can have in this cosmological context the parity-conformal time (PT) symmetry.[11] Additionally, by inspecting the solutions (14)and (15) and (16) and (17) we note that they are periodic for $s_p = (2m - 1)$, $m = 0, 1, 2, \dots$ and antiperiodic for $s_a = 2m$, $m = \pm 1, \pm 2, \dots$. For $\tilde{\gamma} = 1$ (radiation-filled universe), the solutions in the periodic case $m = 2$, i.e.,

$$\mathcal{U}_1(\eta) = \frac{e^{-i3\eta}}{2\cos\eta} \, {}_2F_1\left(1, 2; 5; \frac{1}{1 + e^{2i\eta}}\right) \,, \qquad \mathcal{U}_2(\eta) = e^{i3\tilde{\gamma}\eta}\cos\eta \tag{19}$$

and

$$\mathcal{V}_1(\eta) = \frac{e^{-i3\eta}}{2\cos\eta} \,, \qquad \mathcal{V}_2(\eta) = \left[\frac{e^{i3\eta}}{2\cos\eta} - 12e^{i3\eta}\cos\eta - 6e^{i2\eta}\right] \tag{20}$$

are displayed in Figs. 1 and 2, respectively, whereas the zero-mode solutions in the antiperiodic case $m = 2$

$$\mathcal{U}_1(\eta) = \frac{e^{-i4\eta}}{2\cos\eta} \, {}_2F_1\left(1, 2; 6; \frac{1}{1 + e^{2i\eta}}\right) \,, \qquad \mathcal{U}_2(\eta) = e^{i4\eta}\cos\eta \tag{21}$$

and

$$\mathcal{V}_1(\eta) = \frac{e^{-i4\eta}}{2\cos\eta} \,, \qquad \mathcal{V}_2(\eta) = \left[\frac{e^{i4\eta}}{2\cos\eta} - 24e^{i4\eta}\cos\eta - 8e^{i3\eta}\right] \tag{22}$$

can be found in Figs. 3 and 4, respectively. In general, the $\mathcal{U}$ modes are regular indicating that these shifted cosmologies are not sensitive to the singularities of their curvature indices at the level of their zero modes. On the other hand, the $\mathcal{V}$ cosmologies have periodic singularities in their imaginary parts but not in their real parts. In addition, the duality property is maintained for the pair of zero-modes $\mathcal{U}_2\mathcal{V}_1 = \text{const}$.

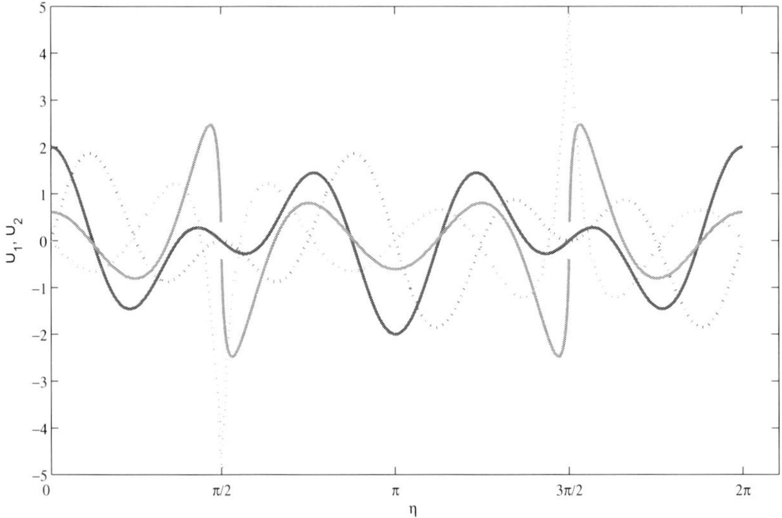

Fig. 3.   The real (solid lines) and imaginary (dotted lines) parts of the antiperiodic zero modes $\mathcal{U}_1(\eta)$ (in red) and $\mathcal{U}_2(\eta)$ (in blue) for the shift parameter $S = 4i$.

The dualities introduced by the supersymmetric approach have certain similarities with the superstring dualities,[12] and the phantom duality.[13] In the first case, there is an invariance of the action with respect to an inversion of the cosmological scale factor and special shifts of the value of the dilaton field ($a \rightarrow a^{-1}$ and $\Phi \rightarrow \Phi - 6\ln a$, respectively). In our case, the scale factor duality is the same but the shift is done in the conformal Hubble parameter. Moreover, the string action with such properties corresponds to cosmologies that are spatially flat and homogeneous. Thus, our scale factor duality connecting homogeneous and inhomogeneous nonflat cosmologies look more general. On the other hand, Dąbrowski et al.,[13] discussed the symmetry $\gamma \rightarrow -\gamma$ of a nonlinear oscillator equation while in our case, the $\tilde{\gamma}$ parameter occurs mostly trigonometrically. For the unshifted bosonic and fermionic cosmologies, we see readily that $\tilde{\gamma} \rightarrow -\tilde{\gamma}$ is a symmetry preserving their curvature indices, although we have now $u_1v_1 = -\tilde{\gamma}$ and $a_g = (-\tilde{\gamma})^{-1/2\tilde{\gamma}}$, a different constant. In the case of the shifted cosmologies, the sign changes $\tilde{\gamma} \rightarrow -\tilde{\gamma}$, $\kappa_{S,u}(\eta) \rightarrow \kappa_{-S,u}(\eta)$, $\kappa_{S,v}(\eta) \rightarrow \kappa_{-S,v}(\eta)$ leave the equations unchanged. We also draw the attention of

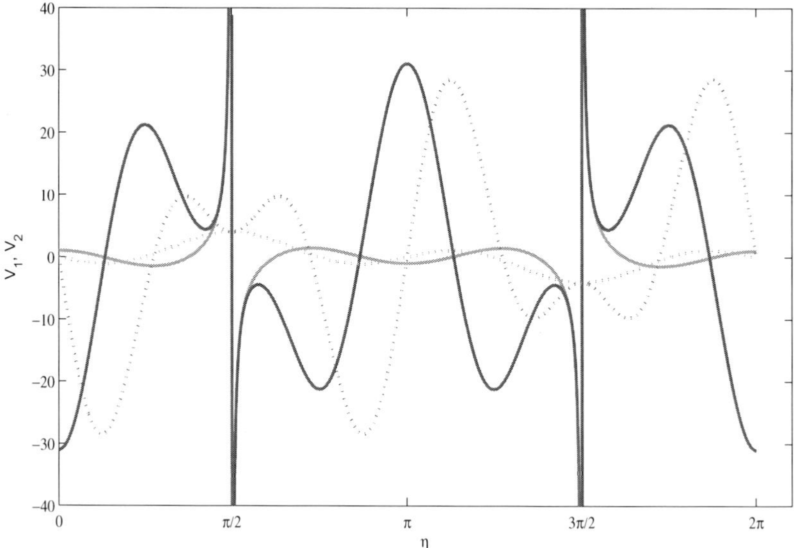

Fig. 4. The real (solid lines) and imaginary (dotted lines) parts of the antiperiodic zero modes $\mathcal{V}_1(\eta)$ (in red) and $\mathcal{V}_2(\eta)$ (in blue) for the shift parameter $S = 4i$.

the interested reader to a recent paper of Faraoni,[14] for a discussion of scale factor dualities of the spatially flat Friedmann equations with barotropic fluids, where also many references on such dualities are provided.

If we move now to real S values, we see that the $\mathcal{U}_2(\eta)$ zero mode is bound and nonsingular for negative real values of S. So, we can obtain damped cyclic behavior of the universe mimicking viscous effects directly by tuning the S parameter in the underdamped regime. Plots of such underdamped zero modes are given in Fig. 5.

In addition, damped cyclic behavior can be also obtained using the general Riccati solution of the shifted cosmologies, which introduces one-parameter zero modes of the form,[2,15]

$$\mathcal{U}_{2;\lambda}(\eta) = \frac{\mathcal{U}_2(\eta)}{\lambda + \int^\eta \mathcal{U}_2^2(\eta)} \; . \tag{23}$$

In this case, λ plays the role of a damping lag in conformal time after which the weighting effect of the integral starts to dominate, see Fig. 6. The advantage of this deformed zero mode is that one can give a physical interpretation to the parameter λ, which is related to the introduction of finite-interval boundaries on the conformal time axis. This is very well described in Section II A of a paper by Monthus $et\ al.$,[16] in the case of quantum mechanics where it is clearly shown how the introduction of boundary conditions at certain points on the axis generates a modulation of the particular solution as given by (23) and the λ parameter can be fixed through the boundary conditions. This can be used to simulate the effects of voids directly on the cosmological zero modes.[17]

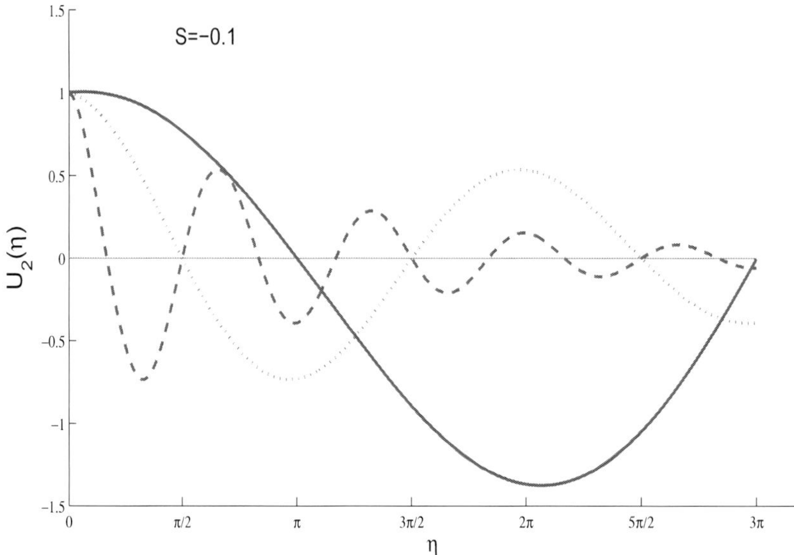

Fig. 5. The $\mathcal{U}_2(\eta)$ zero mode for $S = -0.1$ in the cases $\tilde{\gamma} = -1/2$ (solid line) corresponding to the vacuum, $\tilde{\gamma} = 1$ (dotted line) corresponding to the radiation case, and $\tilde{\gamma} = 3$ (dashed line) corresponding to a stiff fluid.

4. Summary and Final Remarks

We have reviewed here the procedures that led us to introduce inhomogeneous supersymmetric-partner classes of barotropic cosmologies of variable spatial curvature index, by considering non zero initial conditions of the conformal Hubble parameter of FRW barotropic cosmologies. These results have been obtained in a supersymmetric context similar to supersymmetric quantum mechanics and can be applied to any type of cosmological fluid except for $\gamma = \frac{2}{3}$. In other words, these classes of inhomogeneous cosmologies together with the unshifted supersymmetric partner cosmology can be associated to any barotropic cosmological fluid with the only one exception of the coasting (non-accelerating and non-decelerating) universe. Interestingly, we have found that even purely imaginary initial conditions can be considered. It is known that such 'unnatural' initial conditions are required to explain why a thermodynamic arrow of time exists. Cyclic behavior in conformal time of the cosmological zero modes can be obtained in addition to that of their curvature indices that in the pure imaginary case are also related through the parity-time (PT) property. Both inhomogeneity,[18] and cyclicity,[19] are hot issues since the accelerated expansion of the universe got experimental evidence at the end of the past century.

On the other hand, if we forget about the topology of the universe, the same results can be interpreted as due to the time dependence of the adiabatic indices of the cosmological fluids.[20] Along this line, Dąbrowski and Denkievicz,[21] have provided a

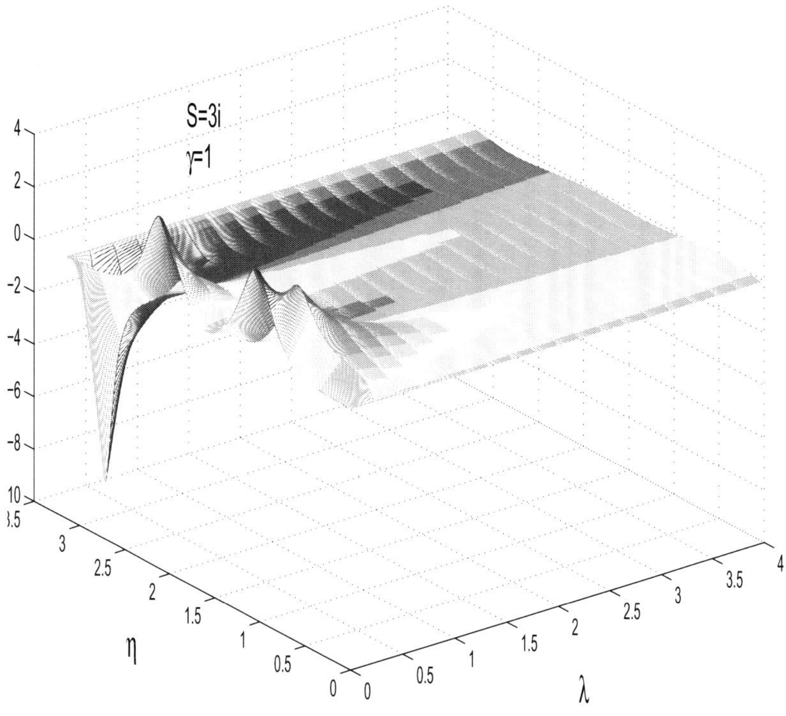

Fig. 6. The imaginary part of the $\mathcal{U}_{2;\lambda}(\eta)$ zero mode for $S = 3i$ and $\tilde{\gamma} = 1$ showing damped oscillations at small values of λ.

discussion of an explicit barotropic model in which the cosmological singularity occurs only in the singular time-dependent barotropic index. They assert that physical examples of such singularities appear in $f(R)$, scalar field, and brane cosmologies.[22] We think that the barotropic inhomogeneous cosmologies of supersymmetric type with periodic singularities in the curvature index could be related to the same physical examples. Moreover, the periodic singularities are not an impediment to build appropriate cosmological zero modes along the whole conformal time axis,[23] which can be used to define novel classes of scale factors corresponding to these generalized barotropic universes within any chosen period of the curvature index.

Acknowledgements

One of the authors (HCR) would like to thank Professor Pisin Chen and the Organizing Committee for the kind invitation and administration of the first symposium of LeCosPA center at NTU as well as for their warm hospitality.

References

1. Cooper F., Khare A., Sukhatme U., Supersymmetry in quantum mechanics, World Scientific, Singapore, 2001.
2. Rosu H.C., Short survey of Darboux transformations, in "Symmetries in Quantum Mechanics and Quantum Optics", Burgos, Spain, Sept. 21-24, 1998, Eds. F.J. Herranz, A. Ballesteros, L.M. Nieto, J. Negro, C.M. Perena, Serv. de Publ. Univ. Burgos, Burgos, Spain, 1999, arXiv:quant-ph/9809056.
3. Moniz P.V., Quantum cosmology - The supersymmetric perspective, vol. 1: Fundamentals, Lecture Notes in Phys. **803**, Springer Verlag, Berlin, 2010.
4. Mak M.K., Harko T., Brans-Dicke cosmology with a scalar field potential, *Europhys. Lett.* **60** (2002) 155-161.
   Saha B., Nonlinear spinor field in cosmology, *Phys. Rev. D* **69** (2004) 124006.
   Nowakowski M., Rosu H.C., Newton's laws of motion in the form of a Riccati equation, *Phys. Rev. E* **65** (2002) 047602.
   Bender C.M., Sarkar S., Asymptotic analysis of the Boltzmann equation for dark matter relics, arXiv:1203.1822.
5. Faraoni V., Solving for the dynamics of the universe, *Am. J. Phys.* **67** (1999), 732-734.
   Lima J.A.S., Note on solving for the dynamics of the universe, *Am. J. Phys.* **69** (2001), 1245-1247.
   Holanda R.F.L., Note on solving for the dynamics of the universe, arXiv:0707.3387.
   Man'ko V.I., Marmo G., Stornaiolo C., Cosmological dynamics in tomographic probability representation, *Gen. Rel. Grav.* **37** (2005), 2003-2014.
   Khandai S., A method to solve Friedmann equations for time dependent equation of state, *Prayas* **3** (2008), 78-88.
6. Rosu H. C., Cornejo-Pérez O., López-Sandoval R., Classical harmonic oscillator with Dirac-like parameters and possible applications, *J. Phys. A* **37** (2004), 11699-11710.
   Rosu H. C., López-Sandoval R., Barotropic FRW cosmologies with a Dirac-like parameter, *Mod. Phys. Lett. A* **19** (2004), 1529-1535.
   Rosu H. C., Ojeda-May P., Supersymmetry of FRW barotropic cosmologies, *Int. J. Theor. Phys.* **45** (2006), 1191-1196.
7. Rosenthal E., Flanagan É.É., Cosmological backreaction and spatially averaged spatial curvature, arXiv: 0809.2107.
8. Larena J., Alimi J.-M., Buchert T., Kunz M., Corasaniti P.-S., Testing backreaction effects with observations, *Phys. Rev. D* **79** (2009), 083011.
9. Rosu H.C., Khmelnytskaya K.V., Shifted Riccati procedure: application to conformal barotropic FRW cosmologies, *SIGMA* **7** (2011), 013.
10. Rosu H.C., Khmelnytskaya K.V., Parametric oscillators from factorizations employing a constant-shifted Riccati solution of the classical harmonic oscillator, *Phys. Lett. A* **375** (2011), 3491-3495.
11. Bender C.M., Introduction to PT-symmetric quantum mechanics, *Contemp. Phys.* **46** (2005), 277-292.
12. Gasperini M., Veneziano G., The pre-big bang scenario in string cosmology, *Phys. Rep.* **373** (2003), 1-212.
13. Dąbrowski M. P., Stachowiak T., Szydłowski M., Phantom cosmologies, *Phys. Rev. D* **68** (2003), 103519.
14. Faraoni V., A symmetry of the spatially flat Friedmann equations with barotropic fluids, *Phys. Lett. B* **703** (2011), 228-231.
15. Rosu H.C., Socorro J., One-parameter family of closed, radiation-filled FRW quantum universes, *Phys. Lett. A* **223** (1996), 28-30.
16. Monthus C., Oshanin G., Comtet A., Burlatsky S. F., Sample-size dependence of the

ground-state energy in a one-dimensional localization problem, *Phys. Rev. E* **54** (1996) 231-242.

17. de Lavallaz A., Fairbain M., Effects of voids on the reconstruction of the equation of state of dark energy, *Phys. Rev. D* **84** 083005; Di Dio E., Vonlanthen M., Durrer R., Back reaction from walls, *JCAP* **02** (2012) 036.

18. Bolejko K., Célérier M.-N., Krasiński A., Inhomogeneous cosmological models: exact solutions and their applications, *Class. Quant. Grav.* **28** (2011) 164002.

19. Sahni V., Toporensky A., Cosmological hysteresis and the cyclic universe, arXiv:1203.0395; Jamil M., Myrzakulov N.A., Yerzhanov K.K., Momeni D., Myrzakulov R., Some models of cyclic and knot universes, arXiv:1201.4360.

20. Rosu H.C., Darboux class of cosmological fluids with time-dependent adiabatic indices, *Mod. Phys. Lett. A* **15** (2000), 979-990.

21. Dąbrowski M. P., Denkiewicz T., Barotropic index w-singularities in cosmology, *Phys. Rev. D* **79** (2009), 063521.
Dąbrowski M.P., Dark energy from temporal and spatial singularities of pressure, *Ann. Phys. (Berlin)* **19** (2010), 299-303.
Fernández-Jambrina L., w-cosmological singularities, *Phys. Rev. D* **82** (2010), 124004.
Nojiri S., Odintsov S.D., Tsujikawa S., Properties of singularities in (phantom) dark energy universe, *Phys. Rev. D* **71** (2005), 063004.

22. Nojiri S., Odintsov S.D., Unified cosmic history in modified gravity: from $f(R)$ theory to Lorentz non-invariant models, *Phys. Rep.* **505** (2011) 59-144.

23. Khmelnytskaya K. V., Rosu H. C., González A., Periodic Sturm-Liouville problems related to two Riccati equations of constant coefficients, *Ann. Phys.* **325** (2010), 596-606.

MYSTERIOUS ANTI-GRAVITY AND DARK-ESSENCE

JE-AN GU

Leung Center for Cosmology and Particle Astrophysics, National Taiwan University,
Taipei 10617, Taiwan (R.O.C.)
E-mail: jagu@ntu.edu.tw

The need of anti-gravity and dark-essence in cosmology is the greatest scientific mystery in the 21st century. This paper presents a personal view of several relevant issues, including the long-standing cosmological constant problem, the newly emerging dark radiation issue, and the basic stability issue of the general-relativity limit in modified gravity.

Keywords: Dark Energy; Modified Gravity; Dark Radiation; Cosmological Constant Problem.

1. Introduction

Cosmology is a science of the evolution, the structures and the compositions of the universe. It has recently become an experimental science driven by astrophysical observations. In addition to observations, describing and understanding our universe require an initial condition of the universe and a theory of fundamental fields/particles and interactions, such as general relativity for gravity and the standard model of particle physics for the others.

The modern version of the cosmic story told by observations is interesting and surprising. It involves the following unexpected characters.

- **Special initial condition.** The cosmic background was rather flat, homogeneous and isotropic; the primordial perturbations were rather adiabatic, scale invariant and Gaussian. The inflation scenario is doing a great job in giving such initial condition.
- **Extra gravity.** The extra attractive gravity is needed to help the cosmic structure formation. A favorite scenario of extra gravity is invoking dark matter.
- **Anti-gravity.** The repulsive gravity is needed to drive the accelerating expansion of the present universe. It invites the consideration of the energy source of anti-gravity dubbed "dark energy" and gives strong motivation for modifying gravity.

In the scenario with dark matter and dark energy, the two unknown dark components contribute 95% of the energy of the present universe, presenting us the greatest enigma in fundamental science at all times.

Anti-gravity is particularly mysterious. It may be caused by dark energy or modified gravity. The simplest candidate of dark energy is a positive cosmological constant. It was firstly introduced by Einstein and later abandoned as his biggest blunder. It is so far so consistent with the observational results and therefore has been widely considered. Nevertheless, the smallness of the cosmological constant and the coincidence problem (why the cosmic expansion just starts to accelerate recently or why the present matter and dark energy densities are comparable) in this model require fine-tuning and make this model look unnatural. The fine-tuning stems from the constant nature of the cosmological constant. To avoid the fine-tuning, a necessary condition is the time variation of the dark energy density. This invites the consideration of a scalar field as a simple phenomenological realization of dark energy with a time-varying energy density.

As to modified gravity, although there is no evidence of such modification, the above three surprising characters give strong motivation for modifying gravity. Although general relativity (GR) so far can pass all the local tests, it is still open for the cosmological tests. As an essential requirement from the success of GR in passing the local tests, in any viable modified gravity model the GR limit must exist and be stable, not just at the action level, but particularly at the solution level.

In addition to dark matter and dark energy, the recent cosmic microwave background (CMB) observations suggest one more dark component called "dark radiation" that represents the additional relativistic degree(s) of freedom. It is expected to give important effects in CMB and in Big-Bang nucleosynthesis (BBN).

The remainder of this paper will provide a simple personal view of three relevant issues: the cosmological constant problem, dark radiation, and the stability of the GR limit in modified gravity.

2. Cosmological Constant Problem

Observations have given an upper bound to the energy density of a cosmological constant: $\rho_\Lambda \lesssim 3 \times 10^{-11}\,\mathrm{eV}^4$, i.e. its energy scale $\lesssim 10^{-3}\,\mathrm{eV}$. This upper bound is much smaller than the expected contribution from the quantum vacuum, leading to the long-standing, notorious cosmological constant problem.[1]

In the framework of quantum field theory the vacuum energy can contribute to dark energy of the same form as a cosmological constant. Its size may be designated by the high-energy cut-off scale of the quantum field theory that, either the Planck scale, the electroweak scale or some other scale involved in the standard model of particle physics, is much larger than $10^{-3}\,\mathrm{eV}$. One may think this problem unrealistic because the physics, particularly that of gravity, around the cut-off scale is not well tested and the correct way of assessing the gravitational effect of the vacuum energy around the cut-off scale is not clear. Let us put aside the unclear high-energy cut-off scale and consider the low-energy scales such as the eV scale, i.e. the micron length scale. Even the quantum fluctuations of the eV scale can provide too large vacuum energy and ruin our universe, while the physics at such scale is well known and has been tested thoroughly. That tells the genuineness and the severeness of the cosmological constant problem.

One naive way of surviving the vacuum energy crisis is to compensate the vacuum energy with a bare cosmological constant that might be introduced at the very beginning of the universe. The size of the bare cosmological constant needs to be delicately chosen to balance the vacuum energy of quantum fields. Another naive way is to make the vacuum energies of different quantum fields cancel each other via carefully choosing the field contents and finely tuning the very details of the field theory. These two naive ways are so fine-tuning that one can hardly believe they can be a part of the grand design in nature.

Even if such fine-tuning is invoked at the beginning of the universe, the later phase transition(s) associated with spontaneous symmetry breaking (SSB), such as the electroweak phase transition, would ruin the initial fine-tuning. During a SSB phase transition, the vacuum energy may drop by an amount on the energy scale of the phase transition (e.g. $\sim 300\,\mathrm{MeV}$ for the electroweak phase transition), thereby ruining the perfect cancellation in the initial design. If one insists to invoke the brute-force cancellation, the design would be as tedious as the following sentence: It is necessary to foresee all possible SSB phase transitions and know the very details of the vacuum energy change during each of them, as detailed as $10^{-3}\,\mathrm{eV}$ at least, and then make the earlier cancellation imperfect, with the energy deficit on the scale of the phase transition and with the precision $10^{-3}\,\mathrm{eV}$ or better.

A good job of solving the vacuum energy crisis should not be as tedious as the brute-force cancellation. A satisfactory solution to the cosmological constant problem is yet to be found and the appropriate scenario for the solution is also not yet clear. The final solution may be associated with the reconciliation between gravity and quantum, while such ultimate paradigm is still in the mist.

3. Dark Radiation

Dark radiation is the additional relativistic degree(s) of freedom suggested by the recent CMB observations. It may be the only surprise so far in the 21st century in cosmology. In the 20th century there were several salient surprises in cosmology, such as the cosmic acceleration, dark energy, dark matter, etc. In contrast, in the 21st century the ΛCDM model fits the observational results so far so well, except the observational indication of dark radiation.

Conventionally cosmologists invoke the following fitting formula of the radiation energy density to fit data.

$$\rho_{\mathrm{rad}} = \left[1 + \frac{7}{8} \left(\frac{4}{11} \right)^{4/3} N_{\mathrm{eff}} \right] \rho_\gamma . \tag{1}$$

Here N_{eff} can be regarded as the effective number of the neutrino species, i.e, the number of the degrees of freedom of the relativistic neutrino-like particles (weakly interacting or even non-interacting fermions), or, phenomenologically, it parametrizes the energy density of the relativistic degrees of freedom additional to the CMB photons. Radiation is important in the early universe. With different amount of radiation, i.e. with different N_{eff}, the early universe has different looks, particularly regarding the CMB spectra and the BBN prediction of the light element abundance such as the ^{4}He abundance. Accordingly, the CMB and the BBN-related observations can give essential constraints on N_{eff}.

In the standard model of particle physics the contribution from neutrinos to N_{eff} is close to 3. In contrast, the recent CMB observations, together with the observations of large-scale structures and the measurements of the Hubble parameter, suggest 1 or 2 more degrees of freedom, i.e. $N_{\mathrm{eff}} = 4$–5, and the standard model value 3 is 2σ away from the best fit (see Refs. 2–4). As to BBN, the BBN theory with $N_{\mathrm{eff}} = 4$–5 is consistent with the observational results of the light element abundance (see Ref. 5). In the future the Planck observation of CMB is expected to give more precise information about N_{eff} with the precision $\delta N_{\mathrm{eff}} \simeq 0.26$.

Thus, in addition to dark matter and dark energy that contribute 95% of the energy density of the present universe, we may need to invoke one more dark component, dark radiation, that changes the early-time expansion history, thereby helping to explain the CMB and BBN data related to the early universe. Although dark radiation and dark energy provide very different functions, they provide the functions at two different epochs: one modifies the early-time expansion history and the other drives the late-time acceleration. Therefore, it is possible to combine them, i.e. with one single energy source that behaves like dark radiation at early times and like dark energy at late times. In this scenario the dark energy information may also be encoded in the early-time events such as CMB and BBN, in addition to the late-time events such as type Ia supernovae and structure formation. This distinct feature makes this possibility particularly worthy of further investigations.

4. Stability of the GR Limit in Modified Gravity

Since GR passes all the local tests, a viable model of modified gravity should behave very similar to GR at the local scales, particularly in the solar system. In addition, since the standard cosmology based on GR fits the observational results about CMB and BBN so far so well, people expect a viable modified gravity model should mimic GR at the early times relevant to CMB and BBN. Thus, the GR limit should exist and should be stable in modified gravity at the local scales and at the early times.

People may explore the existence of the GR limit at the action level. However, this is not good enough. The more essential is the existence and the stability at the solution level, because it is the solution but not the action that directly describes our universe. Around the GR limit people may treat GR as a good approximation of the modified gravity theory. Nevertheless, this may not be true when the GR limit is not stable.

This issue is particularly serious in the modified gravity theories with higher-order derivatives such as the $f(R)$ theory. The gravitational field equations of such theories are higher-order (higher than 2) differential equations while the Einstein equations in GR are 2nd-order differential equations. In this case, using GR to approximate modified gravity is to utilize the 2nd-order differential equations to approximate the higher-order differential equations, the validity of which is doubtful.

In this approximation a significant portion of the solution space is abandoned, and the remaining solution space is approximated by another simplified solution space. To verify the validity of this approximation, people need to show that the abandoned solution space is not important and the simplified solution space is truly a good approximation of the remaining solution space. Unfortunately this is not always true. In many cases the abandoned solution space is not negligible but may play an important role, and the simplified solution space as an approximation may be no good in a long run. That is, even if at the beginning the real solution is in the neighborhood of the simplified solution space, later it may leave away from the simplified solution space and even go deeply into the abandoned solution space. (For more details and for a heuristic demonstration of this issue, see Ref. 6.)

Thus, in addition to the existence, the stability of the GR limit at the solution level needs to be carefully examined for any modified gravity model to be viable.

5. Summary

Anti-gravity and dark-essences of the universe have been strongly suggested by astrophysical observations. They are the most mysterious in physics and cosmology. Their nature and origin are the most important unsolved puzzles in the 21st century. This paper presents a simple personal view of several relevant issues, particularly the cosmological constant problem, dark radiation, and the GR limit of modified gravity.

The cosmological constant problem may guide us to the final reconciliation between gravity and quantum. Dark radiation may be the early-time manifestation of dark energy, with which the nature of dark radiation indicated by the CMB and BBN observations can provide important information about dark energy. As to modified gravity, the need of anti-gravity in cosmology gives a strong motivation and the cosmological observations provide important tests. In addition to performing the tests, the even more essential is to examine not only the existence of the GR limit at the action level but also its stability at the solution level.

The clarification of these issues may help to solve the puzzle about the need of anti-gravity and dark-essences in cosmology. Hopefully the solution to this great puzzle will lead us to a new revolution in physics in the 21st century and bring us an unprecedentedly complete picture of our universe.

Acknowledgments

We thank the Dark Energy Working Group of the Leung Center for Cosmology and Particle Astrophysics (LeCosPA) at the National Taiwan University for the helpful discussions. We thank the National Center for Theoretical Sciences in Taiwan for the support. Gu is supported by the Taiwan National Science Council (NSC) under Project No. NSC 98-2112-M-002-007-MY3.

References

1. Je-An Gu, "Dark energy crisis," *Nucl. Phys. A* **844**, 245C (2010).
2. E. Komatsu *et al.* [WMAP Collaboration], "Seven-Year Wilkinson Microwave Anisotropy Probe (WMAP) Observations: Cosmological Interpretation," *Astrophys. J. Suppl.* **192**, 18 (2011) [arXiv:1001.4538 [astro-ph.CO]].
3. J. Dunkley, R. Hlozek, J. Sievers, V. Acquaviva, P. A. R. Ade, P. Aguirre, M. Amiri and J. W. Appel *et al.*, "The Atacama Cosmology Telescope: Cosmological Parameters from the 2008 Power Spectra," *Astrophys. J.* **739**, 52 (2011) [arXiv:1009.0866 [astro-ph.CO]].
4. M. Archidiacono, E. Calabrese and A. Melchiorri, "The Case for Dark Radiation," *Phys. Rev. D* **84**, 123008 (2011) [arXiv:1109.2767 [astro-ph.CO]].
5. Y. I. Izotov and T. X. Thuan, "The primordial abundance of 4He: evidence for non-standard big bang nucleosynthesis," *Astrophys. J.* **710**, L67 (2010) [arXiv:1001.4440 [astro-ph.CO]].
6. Je-An Gu, "Can f(R) gravity mimic general relativity?" *Int. J. Mod. Phys. Conf. Ser.* **10**, 63 (2012).

LOCAL INHOMOGENEITIES AND THE VALUE OF THE COSMOLOGICAL CONSTANT

ANTONIO ENEA ROMANO

Department of Physics, National Taiwan University, Taipei 10617, Taiwan, R.O.C.
Leung Center for Cosmology and Particle Astrophysics, National Taiwan University, Taipei 10617, Taiwan, R.O.C.
Instituto de Fisica, Universidad de Antioquia, A.A.1226, Medellin, Colombia
E-mail: are@phys.ntu.edu.tw

Supernovae observations strongly support the presence of a cosmological constant, but its value, which we will call apparent, is normally determined assuming that the Universe can be accurately described by a homogeneous model. Even in the presence of a cosmological constant we cannot exclude nevertheless the presence of a small local inhomogeneity which could affect the apparent value of the cosmological constant. Neglecting the presence of the inhomogeneity can in fact introduce a systematic misinterpretation of cosmological data, leading to the distinction between an apparent and the true value of the cosmological constant. But is such a difference distinguishable? Recently we set out to model the local inhomogeneity with a ΛLTB solution and computed the relation between the apparent and the true value of the cosmological constant. In this essay we reproduce the essence of our model with the emphasis on its physical implications.

Keywords: cosmological constant; large scale inhomogeneities.

1. Introduction

High redshift luminosity distance measurements[1–6] and the WMAP measurements[7,8] of cosmic microwave background (CMB) interpreted in the context of standard FLRW cosmological models strongly disfavor a matter dominated universe and strongly support a dominant dark energy component, which gives rise to an accelerated expansion of the universe.

One of the main assumptions of standard cosmology is that the metric describing space time on a sufficiently large scale is homogeneous, but this is more a simplifying theoretical hypothesis than an actual observational conclusive result. All the cosmological parameters whose apparent value is estimated under this homogeneity assumption may have different true values, if the Universe is actually inhomogeneous. The value of the cosmological constant for example could be different from the one which is obtained from fitting data with a homogeneous FLRW metric as it is common practice with the ΛCDM models, even in presence of relatively small large scale inhomogeneities. This type of effect would be more important for local inhomogeneities which surround the observer, and the first step towards taking them into account is to consider the effect of spherically symmetric inhomogeneities. A more general treatment would involve to include the effects of less symmetric cases, such as not central observers or not spherically symmetric spaces.

This type of space time geometry has already received a lot of attention in a cosmological context. As an alternative to dark energy, it has in fact been proposed[9,10] that we may be at the center of an inhomogeneous isotropic universe without cosmological constant, as described by a Lemaitre-Tolman-Bondi (LTB) solution of Einstein's field equations. Interesting analysis of observational data in inhomogeneous models without dark energy and of other theoretically related problems are given, for example, in[11–35]

Recently we have adopted a different approach.[36] We considered a Universe that consists of a cosmological constant and matter with some local large scale inhomogeneity. We modeled this by a ΛLTB solution. In this essay we will reproduce the essence of this model with the emphasis on its implications. For simplicity we will assume that we are located at the center of this local inhomogeneity. In this regard, this can be considered a first attempt to model local large scale inhomogeneities in the presence of the cosmological constant or, more generally, dark energy.

After calculating the null radial geodesics for a central observer we then compute the luminosity distance and compare it to that of ΛCDM model, finding the relation between the two different cosmological constants appearing in the two models, where we call apparent the one in the ΛCDM and true the one in ΛLTB. Our calculations show that the corrections to Ω_Λ^{app}, which is the value of the cosmological constant obtained from analyzing supernovae data by assuming homogeneity, can be important and should be taken into account.

2. LTB Solution with a Cosmological Constant

The LTB solution can be written as[37–39]

$$ds^2 = -dt^2 + \frac{\left(R_{,r}\right)^2 dr^2}{1 + 2\,E(r)} + R^2 d\Omega^2 \,, \tag{1}$$

where R is a function of the time coordinate t and the radial coordinate r, $E(r)$ is an arbitrary function of r, and $R_{,r} = \partial_r R(t,r)$.

The Einstein equations with dust and a cosmological constant give

$$\left(\frac{\dot{R}}{R}\right)^2 = \frac{2E(r)}{R^2} + \frac{2M(r)}{R^3} + \frac{\Lambda}{3} \,, \tag{2}$$

$$\rho(t,r) = \frac{2M_{,r}}{R^2 R_{,r}} \,, \tag{3}$$

with $M(r)$ being an arbitrary function of r, $\dot{R} = \partial_t R(t,r)$ and $c = 8\pi G = 1$ is assumed throughout this essay.

The general analytical solution for a FLRW model with dust and cosmological constant was obtained by Edwards[40] in terms of the elliptic functions. Inspired by the FLRW case, we can construct a general solution of the partial differential equation Eq.(2). First, we introduce a new coordinate $\eta = \eta(t, r)$ and a variable a by

$$\left(\frac{\partial \eta}{\partial t}\right)_r = \frac{r}{R} \equiv \frac{1}{a}\,, \tag{4}$$

and introduce new functions by

$$\rho_0(r) \equiv \frac{6M(r)}{r^3}\,, \quad k(r) \equiv -\frac{2E(r)}{r^2}\,. \tag{5}$$

Then Eq.(2) becomes

$$\left(\frac{\partial a}{\partial \eta}\right)^2 = -k(r)a^2 + \frac{\rho_0(r)}{3}a + \frac{\Lambda}{3}a^4\,, \tag{6}$$

where a is now regarded as a function of η and r; $a = a(\eta, r)$. It should be noted that the coordinate η, which is a generalization of the conformal time in a homogeneous FLRW universe, has been only implicitly defined by Eq.(4). The actual relation between t and η can be obtained by integrating $t = \int a \, d\eta$ once $a(\eta, r)$ is known.

Inspired by the construction of the solution for the FLRW case, we can now set

$$a(\eta, r) = \frac{\alpha}{3\phi(\frac{\eta}{2L}; g_2, g_3) + kL^2}\,, \tag{7}$$

which leads to the Weierstrass differential equation for the choice of the parameters given by

$$\alpha = \rho_0(r)L^2\,, \quad g_2 = \frac{4}{3}k(r)^2 L^4\,, \quad g_3 = \frac{4}{27}\left(2k(r)^3 - \Lambda\rho_0(r)^2\right)L^6\,, \tag{8}$$

where we have introduced the length L for dimensional consistency. $\phi(x; g_2, g_3)$ is the Weierstrass elliptic function satisfying the differential equation,

$$\left(\frac{d\phi}{dx}\right)^2 = 4\phi^3 - g_2\phi - g_3\,. \tag{9}$$

We finally get

$$a(\eta, r) = \frac{\rho_0(r)L^2}{3\phi\left(\frac{\eta}{2L}; g_2(r), g_3(r)\right) + k(r)L^2}\,. \tag{10}$$

In this essay we will set $L = (a_0 H_0)^{-1}$ and choose the so called FLRW gauge, i.e. the coordinate system in which $\rho_0(r)$ is constant.

3. Calculating the Luminosity Distance

We adopt the same method developed in[41] to solve the null geodesic equation written in terms of the coordinates (η, r). The luminosity distance for a central observer in the LTB space-time as a function of the redshift z is expressed as

$$D_L(z) = (1 + z)^2 R\left(t(z), r(z)\right) = (1 + z)^2 r(z) a\left(\eta(z), r(z)\right)\,, \tag{11}$$

where $(t(z), r(z))$ or $((\eta(z), r(z))$ is the solution of the radial geodesic equation as a function of z. Using the analytical solution we can derive the geodesics equations:

$$\frac{d\eta}{dz} = -\frac{\partial_r t(\eta, r) + F(\eta, r)}{(1+z)\partial_\eta F(\eta, r)} \equiv p(\eta, r) \,, \tag{12}$$

$$\frac{dr}{dz} = \frac{a(\eta, r)}{(1+z)\partial_\eta F(\eta, r)} \equiv q(\eta, r) \,, \tag{13}$$

where

$$F(\eta, r) \equiv \frac{R_{,r}}{\sqrt{1 + 2E(r)}} = \frac{1}{\sqrt{1 - k(r)r^2}} \left[\partial_r(a(\eta, r)r) - a^{-1}\partial_\eta(a(\eta, r)r)\, \partial_r t(\eta, r) \right] \,. \tag{14}$$

It is important that the functions p, q, F have explicit analytical forms.

In order to obtain the luminosity distance as a function of the redshift, we to use the following expansions:

$$k(r) = k_0 + k_1 r + k_2 r^2 + ... \tag{15}$$

$$t(\eta, r) = b_0(\eta) + b_1(\eta)r + b_2(\eta)r^2 + ... \tag{16}$$

Since we are interested in the effects due to the inhomogeneities, we will neglect k_0 in the rest of the calculation because this corresponds to the homogeneous component of the curvature function $k(r)$. Following the same approach given in,[31] we take local Taylor expansion in redshift for the geodesic equations, and find the luminosity distance as follows:

$$D_L^{\Lambda LTB}(z) = (1+z)^2 r(z) a^{\Lambda LTB}(\eta(z), r(z)) = D_1^{\Lambda LTB} z + D_2^{\Lambda LTB} z^2 + D_3^{\Lambda LTB} z^3 + .. \tag{17}$$

$$D_1^{\Lambda LTB} = \frac{1}{H_0},$$

$$D_2^{\Lambda LTB} = \frac{1}{36H_0(\Omega_\Lambda^{true} - 1)}(54B_1(\Omega_\Lambda^{true} - 1)^2 + 18B_1'(\Omega_\Lambda^{true} - 1) - 18h_{0,r}(\Omega_\Lambda^{true})^2$$
$$+ 30h_{0,r}\Omega_\Lambda^{true} + -12h_{0,r} + 6K_1\Omega_\Lambda^{true} - 10K_1 + 27(\Omega_\Lambda^{true})^2 - 18\Omega_\Lambda^{true} - 9), \tag{18}$$

where we have introduced the dimensionless quantities $K_0, K_1, B_1, B_1', h_{0,r}$ according to

$$H_0 = \left(\frac{\partial_t, a(t,r)}{a(t,r)}\right)^2\Bigg|_{t=t_0,r=0} = \left(\frac{\partial_\eta a(\eta,r)}{a(\eta,r)^2}\right)^2\Bigg|_{\eta=\eta_0,r=0}, \tag{19}$$

$$K_0 = k_0(a_0 H_0)^{-2}, \tag{20}$$

$$K_1 = k_1(a_0 H_0)^{-3}, \tag{21}$$

$$B_1(\eta) = b_1(\eta)a_0^{-1}, \tag{22}$$

$$B_1 = b_1(\eta_0)a_0^{-1}, \tag{23}$$

$$B_1' = \frac{\partial B_1(\eta)}{\partial \eta}\Bigg|_{\eta=\eta_0} (a_0 H_0)^{-2}, \tag{24}$$

$$h_{0,r} = \frac{1}{a_0 H_0}\frac{\partial_r a(\eta,r)}{a(\eta,r)}\Bigg|_{\eta=\eta_0,r=0}, \tag{25}$$

$$t_0 = t(\eta_0, 0), \tag{26}$$

and used the Einstein equation at the center ($\eta = \eta_0, r = 0$) with

$$1 = \Omega_k^0(0) + \Omega_M^0 + \Omega_\Lambda, \tag{27}$$

$$\Omega_k^0(r) = -\frac{k(r)}{H_0^2 a_0^2}, \tag{28}$$

$$\Omega_M^0 = \frac{\rho_0}{3H_0^2 a_0^3}, \tag{29}$$

$$\Omega_\Lambda = \frac{\Lambda}{3H_0^2}. \tag{30}$$

In order to put the formula for the luminosity distance in this form it is necessary to manipulate appropriately the elliptic functions and then reexpress everything in terms of physically meaningful quantities such as H_0.

For a FLRW space time we can calculate the luminosity distance using the following relation, which is only valid assuming flatness.

$$D_L^{\Lambda CDM}(z) = (1+z)\int_0^z \frac{dz'}{H^{\Lambda CDM}(z')} = D_1^{\Lambda CDM} z + D_2^{\Lambda CDM} z^2 + D_3^{\Lambda CDM} z^3 + \tag{31}$$

From which we can get

$$D_1^{\Lambda CDM} = \frac{1}{H_0}, \tag{32}$$

$$D_2^{\Lambda CDM} = \frac{3\Omega_\Lambda^{app} + 1}{4H_0}. \tag{33}$$

4. Apparent and True Values of the Cosmological Constant

So far we have calculated the first two terms of the redshift expansion of the luminosity distance for ΛLTB and ΛCDM models. Since we know that the latter provides a good fitting for supernovae observations, we can now look for the ΛLTB

models which give the same theoretical prediction. In order to find the relation between the apparent and the true value of the cosmological constant, we need in fact to match the terms in the redshift expansion, i.e.,

$$D_i^{\Lambda CDM} = D_i^{\Lambda LTB} \quad , \quad 1 \le i \le 2 \,. \tag{34}$$

From the above relations we find

$$H_0^{\Lambda LTB} = H_0^{\Lambda CDM} \,, \tag{35}$$

$$\Omega_\Lambda^{app} = \frac{1}{27(\Omega_\Lambda^{true} - 1)} \left[54B_1(\Omega_\Lambda^{true})^2 - 108B_1\Omega_\Lambda^{true} + 54B_1 + 18B_1'\Omega_\Lambda^{true} - 18B_1' \right.$$
$$-18h_{0,r}(\Omega_\Lambda^{true})^2 + 30h_{0,r}\Omega_\Lambda^{true} - 12h_{0,r} + 6K_1\Omega_\Lambda^{true} - 10K_1$$
$$\left. +27\Omega_\Lambda^{true}(\Omega_\Lambda^{true} - 1) \right] \,, \tag{36}$$

$$\Omega_\Lambda^{true} = -\frac{1}{6(6B_1 - 2h_{0,r} + 3)} \left[\left((36B_1 - 6B_1' - 10h_{0,r} - 2K_1 + 9\Omega_\Lambda^{true} + 9)^2 + \right.\right.$$
$$\left. -4(6B_1 - 2h_{0,r} + 3)(54B_1 - 18B_1' - 12h_{0,r} - 10K_1 + 27\Omega_\Lambda^{true}) \right)^{1/2} - 36B_1$$
$$\left. +6B_1' + 10h_{0,r} + 2K_1 - 9(\Omega_\Lambda^{true} - 1) \right] \,. \tag{37}$$

We can also expand the above exact relations by assuming that all the inhomogeneities can be treated perturbatively with respect to ΛCDM, i.e., $\{K_1, B_1, B_1'\} \propto \epsilon$, where ϵ stands for a small deviation from the $FLRW$ solution:

$$\Omega_\Lambda^{true} = \Omega_\Lambda^{app} - \frac{2}{27(\Omega_\Lambda^{app} - 1)}(27B_1(\Omega_\Lambda^{app} - 1)^2 + 9B_1'(\Omega_\Lambda^{app} - 1) - 9h_{0,r}(\Omega_\Lambda^{app})^2$$
$$+15h_{0,r}\Omega_\Lambda^{app} - 6h_{0,r} + 3K_1\Omega_\Lambda^{app} - 5K_1) + O(\epsilon^2) \,. \tag{38}$$

As expected, all these relations reduce to

$$\Omega_\Lambda^{true} = \Omega_\Lambda^{app}, \tag{39}$$

in the limit in which there is no inhomogeneity, i.e. when $K_1 = B_1 = B_1' = h_{0,r} = 0$.

5. Conclusions

We have derived for the first time the correction due to local large scale inhomogeneities to the value of the apparent cosmological constant inferred from low redshift supernovae observations. This analytical calculation shows how the presence of a local inhomogeneity can affect the estimation of the value of cosmological parameters, such as Ω_Λ. This effect should be properly taken into account both theoretically and observationally.

While this should be considered only as the first step towards a full inclusion of the effects of large scale inhomogeneities in the interpretation of cosmological observations, it is important to emphasize that we have introduced a general definition of the concept of apparent and true value of cosmological parameters, and shown the general theoretical approach to calculate the corrections to the apparent values obtained under the standard assumption of homogeneity.

Acknowledgments

This research is supported by Taiwan National Science Council under project No. NSC 97-2112-M-002-026-MY3 and by US department of Energy under Contract No. DE-AC03-76SF00515, and the CODI and Dedicacion exclusiva program of the University of Antioquia.

References

1. S. Perlmutter *et al.* [Supernova Cosmology Project Collaboration], "Measurements of Omega and Lambda from 42 High-Redshift Supernovae," Astrophys. J. **517**, 565 (1999) [arXiv:astro-ph/9812133].
2. A. G. Riess *et al.* [Supernova Search Team Collaboration], Astron. J. **116**, 1009 (1998) [arXiv:astro-ph/9805201].
3. J. L. Tonry *et al.* [Supernova Search Team Collaboration], Astrophys. J. **594**, 1 (2003) [arXiv:astro-ph/0305008].
4. R. A. Knop *et al.* [The Supernova Cosmology Project Collaboration], Astrophys. J. **598**, 102 (2003) [arXiv:astro-ph/0309368].
5. B. J. Barris *et al.*, Astrophys. J. **602**, 571 (2004) [arXiv:astro-ph/0310843].
6. A. G. Riess *et al.* [Supernova Search Team Collaboration], Astrophys. J. **607**, 665 (2004) [arXiv:astro-ph/0402512].
7. C. L. Bennett *et al.*, Astrophys. J. Suppl. **148**, 1 (2003) [arXiv:astro-ph/0302207];
8. D. N. Spergel *et al.*, arXiv:astro-ph/0603449.
9. Y. Nambu and M. Tanimoto, arXiv:gr-qc/0507057.
10. T. Kai, H. Kozaki, K. i. nakao, Y. Nambu and C. M. Yoo, Prog. Theor. Phys. **117**, 229 (2007) [arXiv:gr-qc/0605120].
11. A. E. Romano, Phys. Rev. D **75**, 043509 (2007) [arXiv:astro-ph/0612002].
12. D. J. H. Chung and A. E. Romano, Phys. Rev. D **74**, 103507 (2006) [arXiv:astro-ph/0608403].
13. C. M. Yoo, T. Kai and K. i. Nakao, Prog. Theor. Phys. **120**, 937 (2008) [arXiv:0807.0932 [astro-ph]].
14. S. Alexander, T. Biswas, A. Notari and D. Vaid, "Local Void vs Dark Energy: Confrontation with WMAP and Type Ia Supernovae," arXiv:0712.0370 [astro-ph]. CITATION = ARXIV:0712.0370;
15. H. Alnes, M. Amarzguioui and O. Gron, Phys. Rev. D **73**, 083519 (2006) [arXiv:astro-ph/0512006].
16. J. Garcia-Bellido and T. Haugboelle, JCAP **0804**, 003 (2008) [arXiv:0802.1523 [astro-ph]].
17. J. Garcia-Bellido and T. Haugboelle, JCAP **0809**, 016 (2008) [arXiv:0807.1326 [astro-ph]].
18. J. Garcia-Bellido and T. Haugboelle, JCAP **0909**, 028 (2009) [arXiv:0810.4939 [astro-ph]].

19. S. February, J. Larena, M. Smith and C. Clarkson, Mon. Not. Roy. Astron. Soc. **405**, 2231 (2010) [arXiv:0909.1479 [astro-ph.CO]].

20. J. P. Uzan, C. Clarkson and G. F. R. Ellis, Phys. Rev. Lett. **100**, 191303 (2008) [arXiv:0801.0068 [astro-ph]].

21. M. Quartin and L. Amendola, Phys. Rev. D **81**, 043522 (2010) [arXiv:0909.4954 [astro-ph.CO]].

22. C. Quercellini, P. Cabella, L. Amendola, M. Quartin and A. Balbi, Phys. Rev. D **80**, 063527 (2009) [arXiv:0905.4853 [astro-ph.CO]].

23. C. Clarkson, M. Cortes and B. A. Bassett, JCAP **0708**, 011 (2007) [arXiv:astro-ph/0702670].

24. A. Ishibashi and R. M. Wald, Class. Quant. Grav. **23**, 235 (2006) [arXiv:gr-qc/0509108].

25. T. Clifton, P. G. Ferreira and K. Land, Phys. Rev. Lett. **101**, 131302 (2008) [arXiv:0807.1443 [astro-ph]].

26. M. N. Celerier, K. Bolejko, A. Krasinski arXiv:0906.0905 [astro-ph.CO].

27. A. E. Romano, Phys. Rev. D **76**, 103525 (2007) [arXiv:astro-ph/0702229].

28. A. E. Romano, JCAP **1001**, 004 (2010) [arXiv:0911.2927 [astro-ph.CO]].

29. A. E. Romano, JCAP **1005**, 020 (2010) [arXiv:0912.2866 [astro-ph.CO]].

30. A. E. Romano, Phys. Rev. D **82**, 123528 (2010) [arXiv:0912.4108 [astro-ph.CO]].

31. A. E. Romano, M. Sasaki and A. A. Starobinsky, arXiv:1006.4735 [astro-ph.CO].

32. N. Mustapha, C. Hellaby and G. F. R. Ellis, Mon. Not. Roy. Astron. Soc. **292**, 817 (1997) [arXiv:gr-qc/9808079].

33. M. N. Celerier, Astron. Astrophys. **353**, 63 (2000) [arXiv:astro-ph/9907206].

34. C. Hellaby, PoS **ISFTG**, 005 (2009) [arXiv:0910.0350 [gr-qc]].

35. A. E. Romano, arXiv:1105.1864 [astro-ph.CO].

36. A. E. Romano and P. Chen, arXiv:1104.0730 [astro-ph.CO].

37. G. Lemaitre, Annales Soc. Sci. Brux. Ser. I Sci. Math. Astron. Phys. A **53**, 51 (1933).

38. R. C. Tolman, Proc. Nat. Acad. Sci. **20**, 169 (1934).

39. H. Bondi, Mon. Not. Roy. Astron. Soc. **107**, 410 (1947).

40. D. Edwards, Monthly Notices of the Royal Astronomical Society, 159, 51 (1972).

41. A. E. Romano and M. Sasaki, arXiv:0905.3342 [astro-ph.CO].

CONSTRAINING A MODEL OF VARYING ALPHA WITH PARITY AND CHARGE PARITY VIOLATION

DEBAPRASAD MAITY[1,2*] and PISIN CHEN[1,2,3]

[1] *Department of Physics and Center for Theoretical Sciences, National Taiwan University,*
Taipei 10617, Taiwan
[2] *Leung Center for Cosmology and Particle Astrophysics,*
National Taiwan University, Taipei 106, Taiwan
[3] *Kavli Institute for Particle Astrophysics and Cosmology,*
SLAC National Accelerator Laboratory, Menlo Park, CA 94025, U.S.A.
** E-mail: debu.imsc@gmail.com*

In this article we will study our new phenomenological model of parity and charge parity (PCP) violating varying fine structure constant α. We will be particularly focusing on the effect of PCP violation in various observable phenomena. We qualitatively estimate the effect of PCP violation on the variation of α during cosmological evolution and its observational constraints. However, to place more precise constraint on our model, we consider various laboratory measurements on the optical rotation and ellipticity and also Sunyaev-Zel'dovich (SZ) measurement of CMB passing through a galaxy cluster medium. We find for the PCP violating parameter $\beta = 0$, scale of α variation ω should be $> 6.4 \times 10^9$ GeV which is consistent with the previous studies. In general, as the absolute value of β increases, lower bound on ω also increases. We also discuss about our prediction and the possible experimental issues on the CMB polarization.

Keywords: Varying alpha theory; Parity violation; Optical rotation; SZ-effect.

1. Introduction

Variation of fine structure constant (α) has been the subject of interest for the last several years. A consistent theory of varying alpha was first introduce by Bekenstein in.[1] After that very little has been studied on this subject until an evidence of variation of alpha has been claimed in the absorption spectra of Quasers.[2,3] Subsequently this subject has been studied extensively.[4–6] According to the studies, the value of α was calculated to be lower in the past at the cosmological time scale, with $\Delta\alpha/\alpha = -0.72 \pm 0.18 \times 10^{-5}$ for redshift $z \approx 0.5 - 3.5$.

Recently we have extended the model of varying alpha to include parity (P) and charge-parity (CP) violation.[7] Thanks to the present era of high precession cosmological as well as laboratory experiments such as PLANCK and LHC respectively which in addition to test the prediction of standard models of physics, gives us some hope to observe new physics beyond. One of the simplest new physics that has been haunting physicists for a long time is Parity violation in the gauge sector. This is one of our main motivations that leads us to consider PCP violation in the photon sector which naturally arises in the framework of varying α. In the subsequent sections we will show how this leads to various interesting observable phenomena.

We organize this article as follows: in Sec.2 we discuss about our extension of PCP violating varying alpha model. In sec.2 we will discuss about the alpha variation during cosmological evolutions and it observational aspects. In the subsequent sections we concentrate on constraining our model using number of experimental observations such as vacuum birefringence measurement in the laboratory and SZ measurements on the CMB photon passing through the intra-cluster medium (ICM). Concluding remarks and future prospects are provided in Sec.6.

2. Parity Violating Varying-Alpha Theory

In the framework of the varying alpha theory, the variation of α was constructed by simply requiring the electric charge to be spacetime varying as $e = e_0 e^{\phi(x)}$, where e_0 denotes the coupling constant of a charged particle and $\phi(x)$ is a dimensionless scalar field. The fine-structure constant is therefore $\alpha = e_0^2 e^{2\phi(x)}$[1,4,5]. There is an arbitrariness involved in the definition of $\phi(x)$ under $\phi \to \phi + c$. Because of explicit breaking of charge conservation, $U(1)$ gauge invariance is lost. Keeping these in mind, most general action that one can write down is

$$\mathcal{L} = M_p^2 R - \frac{\omega}{2}\partial_\mu \phi \partial^\mu \phi - \frac{1}{4}e^{-2\phi}F_{\mu\nu}F^{\mu\nu} + \frac{\beta}{4}e^{-2\phi}F_{\mu\nu}\tilde{F}^{\mu\nu} + \mathcal{L}_m, \tag{1}$$

where new gauge transformation would be

$$e^\phi A_\mu \to e^\phi A_\mu + \chi_{,\mu}. \tag{2}$$

corresponding field strength would be looking like

$$F_{\mu\nu} = (e^\phi A_\nu)_{,\mu} - (e^\phi A_\mu)_{,\nu}. \tag{3}$$

In the above action and for the rest of this paper we set $e_0 = 1$ for convenience. The action is also clearly invariant under the shift symmetry of ϕ. Here the coupling constant ω characterizes an inherent length scale of the theory above which the Coulomb force law is valid. From the present experimental constraints the scale has to be above a few tens of MeV to avoid conflict with experiments.

Parity violating parameter is β. $\tilde{F}^{\mu\nu} = \epsilon^{\mu\nu\sigma\rho}F_{\sigma\rho}$ is the Hodge dual of the Electromagnetic field tensor. As we have explained in the introduction, at the present level of experimental accuracy PCP violation in the electromagnetic sector may not be ruled out, and if the PCP in this EM sector is indeed violated, then there should be some interesting consequences which we will be discussing about.

3. Varying α Cosmology

The effect of cosmic evolution on the variation of the fine structure constant in the framework of the variation of a scalar field $\phi(x)$ has been extensively studied.[4–6] This has been referred to as the Bekenstein-Sandvik-Barrow-Magueijo (BSBM) theory. Here we discuss on the variation of α during cosmological evolution particularly focusing on the effect of PCP violating parameter β.

Assuming a usual FRW metric with expansion scale factor $a(t)$,

$$ds^2 = -dt^2 + a(t)^2(dx^2 + dy^2 + dz^2), \tag{4}$$

we obtain the Friedmann equation

$$\left(\frac{\dot{a}}{a}\right)^2 = \frac{1}{3M_p^2}\left[\rho_m\left\{1 + e^{-2\phi}\zeta_m\right\} + e^{-2\phi}\rho_r + \rho_\phi\right] + \frac{\Lambda}{3} \tag{5}$$

where Λ is a constant cosmological vacuum energy density and $\rho_\phi = \frac{1}{2}[\dot{\phi}^2 + V(\phi)]$. For the scalar field we get

$$\ddot{\phi} + 3H\dot{\phi} = \frac{e^{-2\phi}}{\omega}[-2\zeta_m\rho_m + \frac{4}{a^3}\langle\mathbf{E}\cdot\mathbf{B}\rangle], \tag{6}$$

where $H \equiv \dot{a}/a$. ζ_m is a new unknown constant which parametrises the nature of dark matter in our universe. The standard parametrization of $\zeta_m = -(1/4)F_{\mu\nu}F^{\mu\nu}/\rho_m.\rho_m$ is the dark matter energy density. The conservation equations for the non-interacting radiation and matter densities ρ_r and ρ_m, respectively, are

$$\dot{\tilde{\rho}}_m + 3H\tilde{\rho}_m = 0, \tag{7}$$
$$\partial_t(e^{-2\phi}\rho_r) + 4He^{-2\phi}\rho_r = 0, \tag{8}$$

where ρ_r is the radiation energy density. Equations (5-8) govern the Friedmann universe with a time-varying fine-structure constant $\alpha(t)$. They depend on the choice of the parameters ζ_m/ω and β/ω^2. In general it is difficult to solve the Eqs.(5,6). Since the effect of the new scalar field is expected to be very small on the background cosmological evaluation, we will try to solve the scalar field evolution equation to the leading order approximation with the standard Hubble expansion.

3.1. *Variation of alpha during different cosmological era*

- We showed that during the radiation era there exits non-trivial variation of α due to PCP violating term. The solution is

  $\alpha \sim \exp\left[\phi_0 + 2\mathcal{C}_1 t^{-\alpha_+} + 2\mathcal{C}_2 t^{-\alpha_-}\right]$, for $a(t) \sim t^{1/2}$

  where $\mathcal{C}_{1,2}$ are the integration constants. $\alpha_\pm = 1/4\left(1 \pm \sqrt{1 - 8\mathcal{A}}\right)$; $\mathcal{A} = 8\beta^2\langle\mathbf{B}\cdot\mathbf{B}\rangle/\omega$

  So, the change of α is controlled by the average energy density of the radiation, $\mathcal{A}$ and β.

- The behaviour of alpha during matter dominated universe comes out to be[4,5]

  $\alpha \simeq 1 - \frac{\zeta_m}{4\pi G\omega^2}\log(a(t))$ for $a(t) \sim t^{3/2}$.

  During matter domination alpha varies slowly as logarithm in time

- The behaviour of alpha during cosmological constant dominated universe comes out to be[4,5]

$$\alpha \simeq 1 + \frac{\zeta_m}{4\pi G\omega^2}\left(\frac{8\pi G\rho_m}{3H}\right) Hte^{-3Ht} \text{ for } a(t) \sim e^{\sqrt{\Lambda/3}t}$$

where $H = \sqrt{\Lambda/3}$

alpha quickly tends to a constant value during Λ dominated universe.

Let us now see the following recent observational constraint on the variation of alpha

- Oklo Natural Reactor in Gabon (2 Gyrs old, $z \sim 0.1 - 0.15$) gives[8] $\frac{\delta\alpha}{\alpha} \simeq = (8.8 \pm 0.7) \times 10^{-8}$.
- From BBN latest bound comes out to be[9] $-0.007 \leq \frac{\delta\alpha}{\alpha} \leq 0.017$ at 95%.
- WMAP 7-year data study gives[10] $-0.005 \leq \frac{\delta\alpha}{\alpha} \leq 0.008$ at 95%
- The latest astrophysical constraints turns out to be[11] $\frac{\delta\alpha}{\alpha} \simeq = (.61 \pm 0.2) \times 10^{-5}$ for $z > 1.8$

If we consider the above constraints one gets $\frac{\delta\alpha}{\alpha} \simeq \frac{\zeta_m}{4\pi G\omega^2} \simeq (.3\pm0.4)\times 10^{-7}$ from Oklo measurement. On the other hand the constraint coming from the astrophysical sources at $z > 1.8$ becomes $\frac{\zeta_m}{4\pi G\omega^2} \simeq (.6 \pm 0.2) \times 10^{-6}$. As one sees it is difficult to get constraint on β. Naively difference between BBN and WMAP observation may give us the constraints about $-0.002 \leq \frac{\delta\alpha_{rad}}{\alpha} \leq 0.009$. Although this does not provide direct constraint on β. The appropriate boundary condition for fixing $\mathcal{C}_{1,2}$ is important which we defer for our future study. In the following sections we will consider other classes of experimental results to constrain our model parameters.

4. Constraining through Laboratory Experiments

In this section we will talk about some laboratory experimental such as BFRT, PVLAS, Q & A,.[12–14] In all those experiments, main goal is to indirectly detect a new scalar fields which has non-trivial coupling with a photon in the strong background magnetic field.

Because of the non-trivial scalar-photon coupling, state of a polarized photon changes after passing through the strong background magnetic field. The physical quantities are optical rotation and ellipticity. Interested reader can look our paper for the explicit expressions of those physical quantities.[15] In this article we will just show the results

After considering those experimental results given in Tab.1, one gets constraints summarized in Tab. 2.

5. Effect on CMB through Galaxy Cluster Medium: SZ-Like Effect

In this section we will discuss about the effect of our model on the CMB photon passing through the magnetized medium of a galaxy-cluster. As we know in our

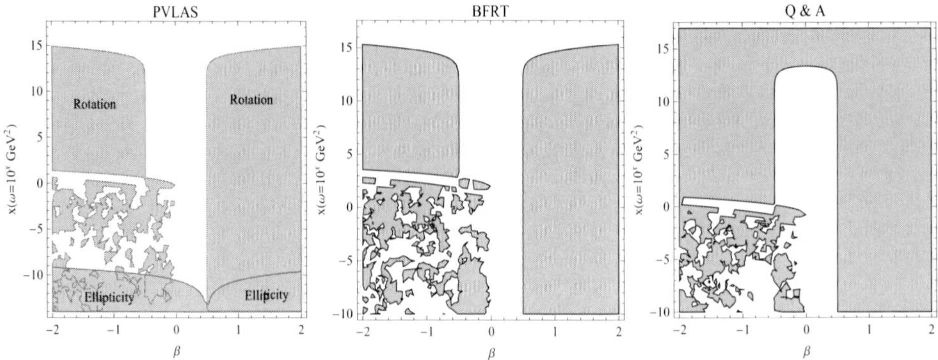

Fig. 1. Bounds on β and ω using different experimental results for the rotation and the ellipticity.

Table 1. Laboratory bounds on rotation and ellipticity.

Experiment	$\lambda(nm)$	$\mathbf{B}_0(T)$	$L(m)$	N	Rotation/Ellipticity
BFRT	514	3.25	8.8	250	$3.5 \times 10^{-10}/1.4 \times 10^{-8}$
PVLAS	1064	2.3	1	45000	1.0×10^{-9}
Q & A	1064	2.3	0.6	18700	$(-0.375 \pm 5.236) \times 10^{-9}$

Table 2. Laboratory constraints on β and ω.

Range of ω in GeV	Bound on β	Experiment
$10^{-5} \lesssim \omega \lesssim 10$	$0 \leq \beta \leq 0.5$	PVLAS
$\omega \lesssim 10$		BFRT, Q&A
$10 \lesssim \omega \lesssim 3.3 \times 10^6$	$-0.5 \leq \beta \leq 0.5$	PVLAS, BFRT, Q&A
$\omega \gtrsim 10^7$	$|\beta| \geq 1$	PVLAS, BFRT

universe magnetic field is omnipresent at all scales from Galactic to Intra-galactic to even huge cosmological scale with the intensity of the order of Gauss(G) to micro Gauss (μG). The states of the CMB photon, therefore, will get affected after passing through the magnetized plasma in the intra-galactic medium (ICM). All these effects can be written in terms of photon-to-scalar conversion probability. In order to calculate this probability we assume a particular model of ICM magnetized plasma called power spectrum model in which the magnetic field $\mathbf{B}$ and the electron density ρ_e are written as

$$\mathbf{B} = \mathbf{B}_0 + \delta\mathbf{B} \quad ; \quad \rho_e = \rho_{e0} + \delta\rho_e$$

The two point correlation function of those approximately Gaussian fluctuations $\delta\mathbf{B}_i$ and $\delta\rho_e$ are defined as

$$< \delta\mathbf{B}_i(y)\delta\mathbf{B}_i(x+y) >= \frac{1}{12\pi}\delta_{ij} \int d^3k P_{\mathbf{B}}(k)e^{i\mathbf{k}\cdot\mathbf{x}},$$

$$< \delta\rho_e(y)\delta\rho_e(x+y) >= \frac{1}{4\pi} \int d^3k P_e(k)e^{i\mathbf{k}\cdot\mathbf{x}}.$$

Various observations indicate that both the power spectra behave like a power law in momentum modes

$$k^2 P_{\mathbf{B}}(k) = \mathcal{P}_{\mathbf{B}} k^\gamma \quad ; \quad k^2 P_e(k) = \mathcal{P}_e k^\gamma,$$

where $\mathcal{P}_{\mathbf{B}}$ and $P_e(k)$ are the normalization constants. According to the observations value of γ changes within $[-1, -2]$.

Once we assume the above model for ICM, one can calculate the change of intensity of the CMB photon after traversing a distance z^{16} as follows

$$I(z) \simeq I(0)(1 - P_{\gamma \to \phi})$$

$$\bar{P}_{\gamma \to \phi} = \bar{P}_{\gamma \to \phi}^{reg} + \bar{P}_{\gamma \to \phi}^{ran} + \frac{8\beta^2}{3} \left(\int_{\bar\Delta}^\infty + \int_{\bar\Delta'}^\infty \right) k dk \mathcal{F}_{(1)}^k, \tag{9}$$

where $P_{\gamma \to \phi}$ is total photon-to-scalar conversion probability. $I(0)$ is the initial intensity of the CMB photon. $\bar{P}_{\gamma \to \phi}^{reg}$ and $\bar{P}_{\gamma \to \phi}^{ran}$ are the contribution coming from regular and random part the magnetic field respectively. The explicit expression of the above equation can be found in Ref. 16. The total conversion probability induces an additional temperature anisotropy in the CMB.

$$\frac{\delta T}{T_0} \approx \frac{(e^{-\mu\varpi} - 1)}{\mu\varpi} \bar{P}_{\gamma \to \phi}(L).$$

The Boltzmann factor $\mu = \frac{1}{k_B T_0}$.

For the Coma cluster the constraint on $P_{coma}(204 GHz)$ has been calculated[17] to be $< 6.2 \times 10^{-5}$. We use this bound to constrain our model parameters showed in the left panel of Fig. 2. On the right panel of Fig. 2, we compare our results coming from the both laboratory and CMB measurements. It is, therefore, clear that CMB measurement gives us tighter constraints on our model parameters.

5.1. *Comment on the induced polarization of the CMB photon*

In this section we discuss about the induced photon polarization and some issues on there experimental measurements. So far the existing experiments have not reached the required label of sensitivity to measure the CMB photon polarization induced by the ICM magnetized plasma. According to our results the induced polarization are

$$\bar{V}(L) \simeq -\beta I(0) \bar{P}_{\gamma \to \phi}^{ran},$$

$$\bar{Q}(L) \simeq I(0) \bar{P}_{\gamma \to \phi}^{reg} (\cos 2\theta - 4\beta \sin 2\theta),$$

$$\bar{U}(L) \simeq I(0) \bar{P}_{\gamma \to \phi}^{reg} (\sin 2\theta + 4\beta \cos 2\theta).$$

If we use the bound on ω coming from the Coma galaxy cluster for non-zero β, naively the bound on linear polarization becomes $\bar{P}_{\gamma \to \phi}^{reg}(L) \leq (0.89 - 1.51) \times 10^{-10}$ and that for the circular polarization becomes $\bar{P}_{\gamma \to \phi}^{ran}(L) \leq 6.2 \times 10^{-5} \gg \bar{P}_{\gamma \to \phi}^{reg}(L)$.

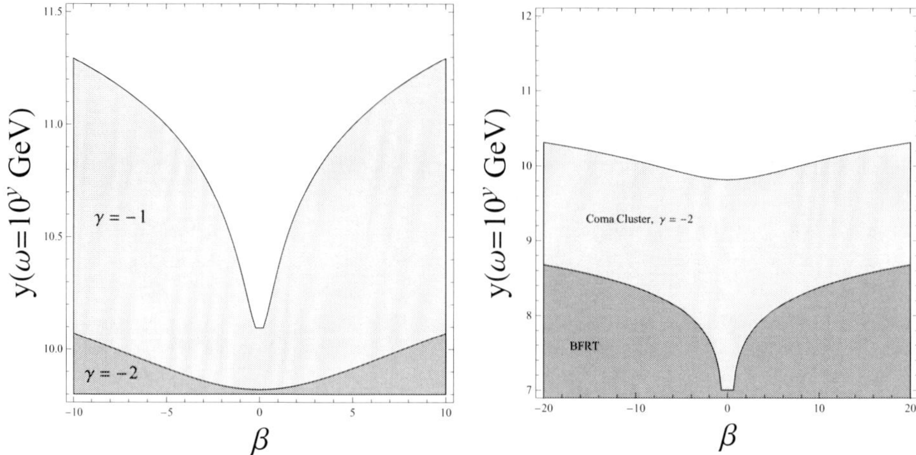

Fig. 2. Left panel: Bounds on β and ω from SZ-measurement on Coma cluster. Right panel: Combined constraint from Laboratory and CMB measurement. Sheded regions are excluded

We know that the amount of intrinsic fractional polarizations of the CMB photon itself are $\langle Q(0)^2 \rangle^{1/2}/I(0), \langle U(0)^2 \rangle^{1/2}/I(0) \sim \mathcal{O}(10^{-7}), \langle V(0)^2 \rangle^{1/2}/I(0) \ll \mathcal{O}(10^{-7})$. So if one naively compares our prediction of the amount of circular polarization, probably β should be very small. But before we come to the above conclusion, it is important to remember

- The required angular resolution of the experiment should be $\theta < 0.03$ arcmin (considering the known parameters of Coma Cluster).
- Spectral resolution should be $\delta\lambda/\lambda < 10^{-6} - 10^{-7}$
- Where as the typical value of those resolutions in the existing experiments are: several arcmin for angular resolution and spectral resolution $\delta\lambda/\lambda < 10^{-4} - 10^{-2}$ for CMB experiments.

There exist few earth-based experiments, such as SPT-Pol, ALMA, POLAR, which are either on- going or under development, that have detectors to measure the polarization of CMB also on the low scale. All these experiments with high angular resolution could in principle shed some light on the parity-violating effect on the polarization of the CMB photon.

6. Conclusions

We have studied a PCP violating varying alpha model in the light of various laboratory as well as CMB measurements. Although until now there has been no positive observational evidence of this parity-violating effect, near future experiments with higher precession may hopefully give some hints beyond the standard model. In the cosmological context our model predicts a new contribution to the fine-structure constant variation in the radiation dominant era. Because of that, BBN (Big Bang

Nucleosynthesis) becomes the main observational window to constrain the evolution of the PCP violating varying fine structure constant. We defer this for our future study.

In this article we have mainly focused on two different classes of experiments to constrain our model. We have considered the laboratory measurements on the optical rotation and ellipticity and also consider the SZ measurement of the CMB photon passing through the Coma cluster. Using those results we found $\omega \geq (0.66 - 4.04) \times 10^{10}$ GeV for $\beta = 0$ which corresponds to the standard dilaton-photon system. Furthermore if we increase β, lower bound on ω also increases. With present experimental input parameters and accuracy, we further concluded that the laboratory experiments will not be able to detect the signal for optical rotation and ellipticity in near future. In regard to the polarization of the CMB photon future experiments such as ALMA can shed some light on it.

Acknowledgments

This research is supported by Taiwan National Science Council under Project No. NSC 97-2112-M-002-026-MY3, by Taiwan's National Center for Theoretical Sciences (NCTS), and by US Department of Energy under Contract No. DE-AC03-76SF00515.

References

1. J.D. Bekenstein, Phys. Rev. **D25**, 1527 (1982).
2. M.T.Murphy et. al, MNRAS, **327**, 1208 (2001) .
3. J.K. Webb, V.V. Flambaum, C.W. Churchill, M.J. Drinkwater & J.D. Barrow, Phys. Rev. Lett. 82, 884 (1999); J.K. Webb *et al*, Phys. Rev. Lett. **87**, 091301 (2001).
4. H. B. Sandvik, J. D. Barrow and J. Magueijo, Phys. Rev. Lett. **88**, 031302 (2002), Phys. Rev. D **65** , 123501 (2002), Phys. Rev. D **66**, 043515 (2002), Phys. Lett. B **541**, 201 (2002), J.D. Barrow and D. Mota, Class. Quant. Grav. **19**, 6197 (2002).
5. J. D. Barrow, H. B. Sandvik and J. Magueijo, Phys. Rev. D **65**, 063504 (2002).
6. K. A Olive and Maxim Pospelov, hep-ph/0110377; T. Chiba and K. Kohri, Prog. Theor. Phys. **107**, 631 (2002); J. P. Uzan, Rev. Mod. Phys. **75**, 403 (2003); D. S. Lee, W. Lee and K. W. Ng, Int. J. Mod. Phys. D **14**, 335 (2005).
7. D. Maity and Pisin Chen, Phys.Rev. **D83**,083516 (2011).
8. Y. Fujii, Lect.Notes, Phys.648:167-185,2004, hep-ph/0311026
9. T. Dent, S. Stern and C. Wetterich, Phys. Rev. **D76**, 063513 (2007).
10. S. J. Landau and C. G. Scoccola, arXiv:1002.1603 [astro-ph.CO]
11. J. K. Webb *et al* 1008.3907
12. Y. Semertzidis *et al.* [BFRT Collaboration], Phys. Rev. Lett. **64**, 2988 (1990), R. Cameron *et al.* [BFRT Collaboration], Phys. Rev. D. **47**, 3707 (1993).
13. E. Zavattini *et al.* [PVLAS Collaboration], Phys. Rev. Lett. **96** (2006) 110406, [arXiv:hep-ex/0507107].
14. S. J. Chen, H. H. Mei and W. T. Ni [Q & A Collaboration], arXiv:hep-ex/0611050.
15. D. Maity and Pisin Chen, Phys.Rev. **D84**,026008 (2011).
16. D. Maity and Pisin Chen, Phys.Rev. **D85**,043512 (2012).
17. A. C Davis, C. A. O. Schelpe and D. J. Shaw, Phys.Rev. **D80** 064016 (2009); *ibid* Phys.Rev. **D83** 044006 (2011).

Section IV

Particle Astrophysics

COSMIC CONNECTIONS: FROM COSMIC RAYS TO GAMMA RAYS, COSMIC BACKGROUNDS AND MAGNETIC FIELDS

ALEXANDER KUSENKO

*Department of Physics and Astronomy, University of California, Los Angeles,
CA 90095-1547, USA
and
Kavli IPMU, University of Tokyo, Kashiwa, Chiba 277-8568, Japan
E-mail: kusenko@ucla.edu*

Combined data from gamma-ray telescopes and cosmic-ray detectors have produced some new surprising insights regarding intergalactic and galactic magnetic fields, as well as extragalactic background light. We review some recent advances, including a theory explaining the hard spectra of distant blazars and the measurements of intergalactic magnetic fields based on the spectra of distant sources. Furthermore, we discuss the possible contribution of transient galactic sources, such as past gamma-ray bursts and hypernova explosions in the Milky Way, to the observed flux of ultrahigh-energy cosmic-rays nuclei. The need for a holistic treatment of gamma rays, cosmic rays, and magnetic fields serves as a unifying theme for these seemingly unrelated phenomena.

Keywords: cosmic rays; gamma rays; galactic and extragalactic magnetic fields.

1. Gamma Ray Astronomy of Cosmic Rays

Gamma rays from Active Galactic Nuclei (AGN) are studied extensively using ground-based atmospheric Cherenkov telescopes (ACT), as well as Fermi Space Telescope and other instruments. Their signals reveal important information about the sources, as well as about extragalactic background light (EBL) and intergalactic magnetic fields (IGMF) along the line of sight. The same sources are expected to accelerate cosmic rays, although it is more difficult to associate cosmic rays with their sources because the local, galactic magnetic fields alter the arrival directions of cosmic rays.

1.1. *Secondary gamma rays from the line-of-sight interactions of cosmic rays*

It was recently proposed that the hardness (and uniform redshift-dependent shape) of gamma-ray spectra of distant blazars can be naturally explained by the line-of-sight interactions of cosmic rays accelerated in the blazar jets.[1-8] The cosmic rays with energies below $10^{17} - 10^{18}$ eV can cross large distances with little loss of energy and can generate high-energy gamma rays in their interactions with cosmic background photons relatively close to the observer. Such *secondary* gamma rays

232

can reach the observer even if their energies are well above TeV. In the absence of cosmic-ray contribution, some unusually hard intrinsic spectra[9] or hypothetical new particles[10,11] have been invoked to explain the data.

As long as the IGMFs are smaller than ~ 10 femtogauss, secondary gamma rays come to dominate the signal from a sufficiently distant source. One can see this from the way the flux scales with distance for primary and secondary gamma rays:[3]

$$F_{\text{primary},\gamma}(d) \propto \frac{1}{d^2} e^{-d/\lambda_\gamma} \tag{1}$$

$$F_{\text{secondary},\gamma}(d) \propto \frac{\lambda_\gamma}{d^2}\left(1 - e^{-d/\lambda_\gamma}\right) \tag{2}$$

$$\sim \begin{cases} 1/d, & \text{for } d \ll \lambda_\gamma, \\ 1/d^2, & \text{for } d \gg \lambda_\gamma. \end{cases} \tag{3}$$

Obviously, for a sufficiently distant source, secondary gamma rays must dominate because they don't suffer from the exponential suppression as in Eq. (1). The predicted spectrum turns out to be similar for all the distant AGN, depending only on their redshift. These predictions are in excellent agreement with the data.[1–3]

One can see the transition from primary to secondary gamma rays in Fig. 1, which shows the spectral index difference for blazar spectra as a function of their redshifts. At small redshifts, the data confirm the Stecker – Scully relation,[12] but, at redshifts 0.15 and beyond, there is clearly a new population of blazars, whose observed spectral index shows only a weak dependence on the redshift. The nearby population is obviously the blazars from which primary gamma rays are observed. The distant blazars are observed in secondary gamma rays, which are produced in line-of-sight cosmic ray interactions. These

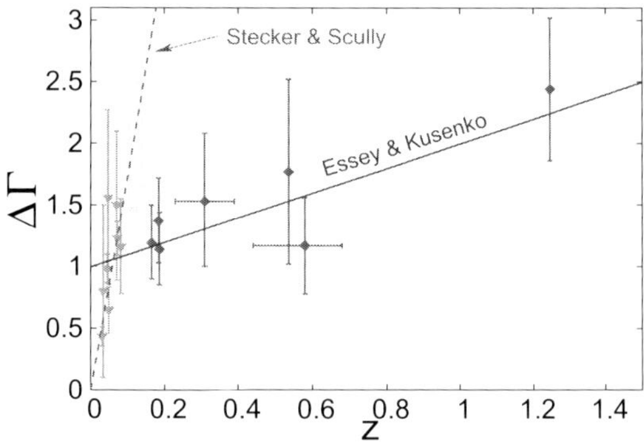

Fig. 1. Spectral index change $\delta\Gamma = \Gamma_{\text{GeV}} - \Gamma_{\text{TeV}}$ as a function of redshift. While the low-redshift blazars agree with the Stecker–Scully relation,[12] the data indicate the existence of an additional, distinct population with a weak redshift dependence at redshifts 0.15 and beyond.[5] In particular, the recently measured redshift[13] of PKS 0447-439 is in agreement with the trend.

secondary gamma rays are produced relatively close to the observer, regardless of the distance to the source. Hence, their redshift dependence is much weaker.[5] Finally, there is an intermediate population around redshift 1.2 which is composed of some blazars seen in primary gamma rays and some seen in secondary gamma rays.

A recent redshift measurement of PKS 0447-439 redshift[13] further strengthens our interpretation. Gamma rays with energies above 1 TeV have been observed from this blazar by HESS.[14] The spectral properties agree with the trend (Fig. 1). Furthermore, there is no way for primary gamma rays to reach Earth from such a distant source, while secondary gamma rays provide a consistent explanation of the PKS 0447-439 spectrum.[15]

This motivates future observations by ACT of blazars with known large redshifts. Secondary gamma rays with TeV and higher energies can be observed even from some sources located at cosmological ($z \sim 1$) distances.[15]

The spectral slope of protons and the level of EBL do not have a strong effect on the spectrum of secondary photons, as one can see from Fig. 2. However, for the same photon flux, the neutrino flux varies depending on the maximal energy $E_{\max}$ to which the protons are accelerated. Indeed, there are two competing processes that generate secondary photons: $p\gamma_{EBL} \to p\pi^0 \to p\gamma\gamma$ and $p\gamma_{CMB} \to pe^+e^-$. For smaller $E_{\max}$, a larger fraction of photons come from the hadronic channel, which is accompanied by production of neutrinos via $p\gamma_{EBL} \to n\pi^\pm$ followed by the decays of charged pions and the neutron. Neutrino observations can help determine this parameter.[2]

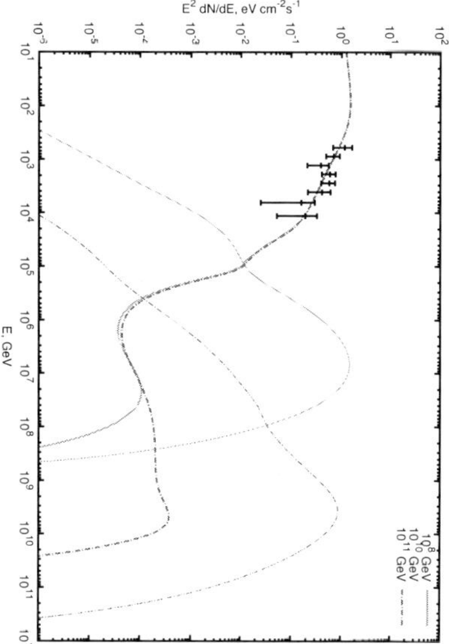

Fig. 2. Photon (low energy) and neutrino (high energy) spectra[2] expected from an AGN at $z = 0.14$ (such as 1ES0229+200), normalized to HESS data points (shown),[16] for $E_{\max} = 10^8$GeV, 10^{10}GeV, and 10^{11}GeV shown by the solid, dashed, and dash-dotted lines, respectively.

1.2. IGMFs and EBL

The success of this picture lends support to the hypothesis of cosmic ray acceleration in AGN. Furthermore, one can use the spectral gamma ray data to study EBL and IGMFs. The predicted spectra depend to some extent on the EBL model, as shown in Fig. 3, although this dependence is too weak to distinguish between different models.[3] IGMFs, however, can have a strong effect on the goodness of fit. Based on the spectra of several distant blazars, one can set both upper and lower limits on IGMF:[4]

$$10^{-17}\mathrm{G} < B < 3 \times 10^{-14}\mathrm{G}.$$

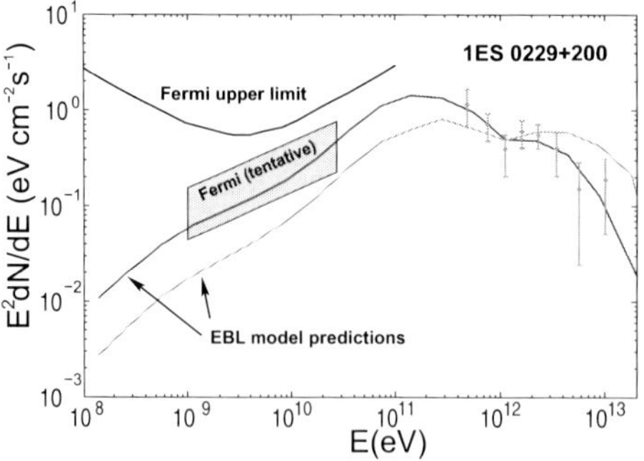
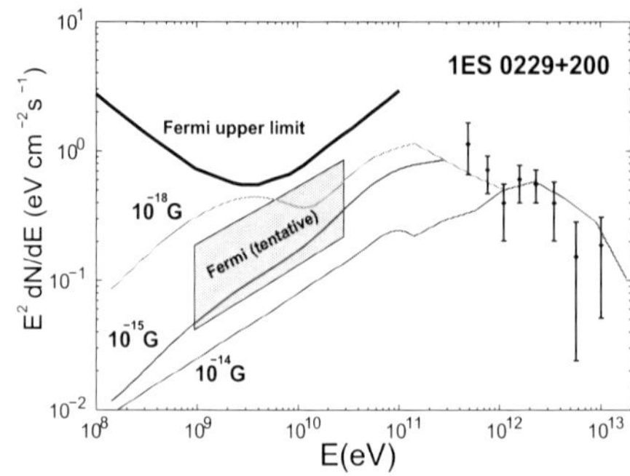

Fig. 3. Sensitivity of secondary gamma-ray spectra to the model of EBL and to the average value of IGMFs. *Fermi* upper limit and tentative detection[17] for blazar 1ES 0229+200 are shown at lower energy, and HESS data points at high energy. The predictions of two models of EBL are shown in the left panel according to Essey *etal.*,[3] and the effects of intergalactic magnetic fields are shown in the right panel.[4]

1.3. *Time variability*

An important property of secondary gamma rays is the lack on short-scale time variability.[8] For $E > 1$ TeV and $z > 0.15$, one expects the signal to be dominated by secondary photons, and any time variability on short scales should be erased by delays in the propagation of protons and electromagnetic cascades. Fig. 4 shows the time delays as a function of the proton energy.

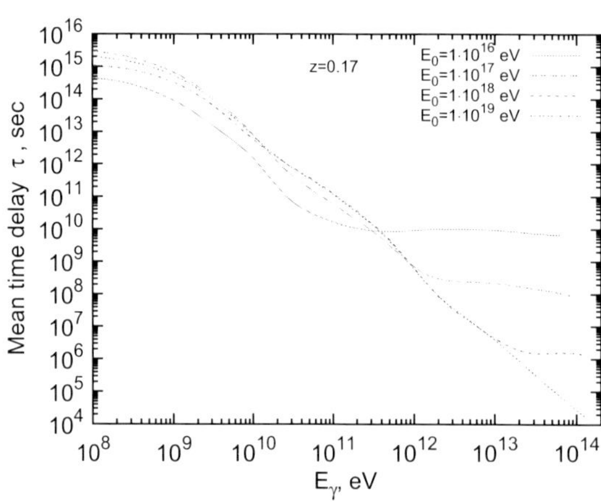

Fig. 4. Time delays of gamma rays emitted at redshift $z = 0.17$ for different proton energies E_0 in a femtogauss random IGMF with a correlation length of 1 Mpc.

The present data are not yet sufficient to probe time variability at the relevant energies and redshifts because the data points are too few. While time variability has been observed for *nearby* TeV blazars at TeV energies, as well as for distant TeV blazars at energies above a few hundred GeV, no variability has been reported so far for *distant* TeV blazars at *TeV energies*. One can infer from Fig. 1 how distant the source has to be for the secondary signals to dominate. It is evident that the secondary component takes over for redshifts beyond 0.15.

2. Composition of UHECR and Past Transient Phenomena in the Milky Way

Let us now turn to another phenomenon related to cosmic rays and magnetic fields, only this time we will concentrate on the magnetic fields inside the galaxy and their effect on the observed fluxes of ultrahigh-energy nuclei.[18,19] There is a growing evidence that long GRBs are caused by a relatively rare type(s) of supernovae, while the short GRBs probably result from the coalescence of neutron stars with neutron stars or black holes. Compact star mergers undoubtedly take place in the Milky Way, and therefore short GRBs should occur in our Galaxy.

Although there is some correlation of long GRBs with star-forming metal-poor galaxies,[20] many long GRBs are observed in high-metallicity galaxies as well,[21-23] and therefore one expects that long GRBs should occur in the Milky Way. Less powerful hypernovae, too weak to produce a GRB, but can still accelerate UHECR,[24] with a substantial fraction of nuclei.[25,26]

If the observed cosmic rays originate from past explosions in our own Galaxy, PAO results have a straightforward explanation.[18]

GRBs have been proposed as the sources of extragalactic UHECR,[26-28] and they have also been considered as possible Galactic sources.[29-31] It is believed that GRBs happen in the Milky Way at the rate of one per $t_{\mathrm{GRB}} \sim 10^4 - 10^5$ years.[32-36] Such events have been linked to the observations of positrons.[37-40]

If local sources, such as past GRBs, hypernovae, and other stellar explosions in the Milky Way, produce a small fraction of heavy nuclei,[41] the observed fraction of UHE nuclei is greatly amplified by diffusion. This is because the galactic magnetic fields are strong enough to trap and contain nuclei but not protons with energies above EeV. This observation leads to a simple explanation of the composition trend observed by PAO.

As illustrated in Fig. 5, diffusion depends on rigidity, and, therefore, the observed composition can be altered by diffusion.[18,42] Changes in composition due to a magnetic fields have been discussed in connection with the spectral "knee",[42] and also for a transient source of UHECR.[43] The "knee" in the spectrum occurs at lower energies than those relevant PAO, and at higher energies the cosmic rays effectively probe the spectrum of magnetic fields on greater spatial scales, of the order of 0.1 kpc.[44]

One can use a simple model[18] to show how diffusion affects the observed spectrum of the species "i" with different rigidities. Let us suppose that all species are produced with the same spectrum $n_i^{(\mathrm{src})} = n_0^{(\mathrm{src})} \propto E^{-\gamma}$ at the source located in the center of the Milky Way and examine the observed spectra altered by the energy dependent diffusion and by the trapping in the Galactic fields.

In diffusive approximation, the transport inside the Galaxy can be described by the equation:

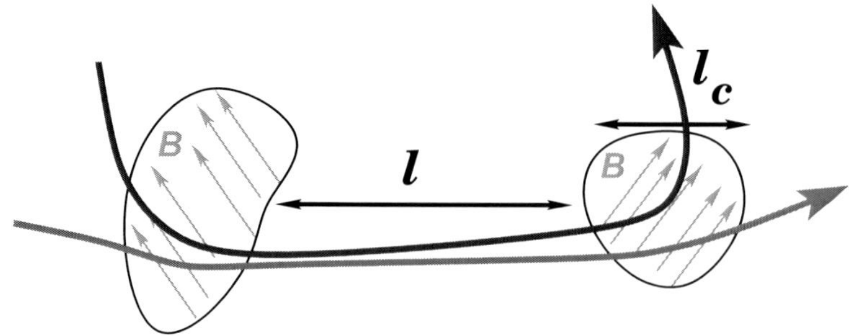

Fig. 5. For each species, there is a critical energy $E_{0,i}$ for which the Larmor radius R_i is equal to the magnetic coherence length l_c. For $E \ll E_{0,i}$, the mean free path of the diffusing particle is $l \sim l_0$, and $D_i(E) = l_c/3$. For $E \gg E_{0,i}$, the particle is deflected only by a small angle $\theta \sim l_0/R_i$, and, after k deflections, the mean deflection angle squared is $\bar{\theta^2} \sim k(l_0/R_i)^2$. The corresponding diffusion coefficient is $D_i(E) \propto (\frac{E}{E_{0,i}})^2$, for $E \gg E_{0,i}$.

$$\frac{\partial n_i}{\partial t} - \vec{\nabla}(D_i \vec{\nabla} n_i) + \frac{\partial}{\partial E}(b_i n_i) =$$
$$Q_i(E, \vec{r}, t) + \sum_k \int P_{ik}(E, E') n_k(E') dE'.$$

Here $D_i(E, \vec{r}, t) = D_i(E)$ is the diffusion coefficient, which we will assume to be constant in space and time. The energy losses and all the interactions that change the particle energies are given by $b_i(E)$ and the kernel in the collision integral $P_{ik}(E, E')$. For energies below GZK cutoff, one can neglect the energy losses on the diffusion time scales.

The diffusion coefficient $D(E)$ depends primarily on the structure of the magnetic fields in the Galaxy. Let us assume that the magnetic field structure is comprised of uniform randomly oriented domains of radius l_0 with a constant field B in each domain. The density of such domains is $N \sim l_0^{-3}$. The Larmor radius depends on the particle energy E and its electric charge $q_i = e Z_i$:

$$R_i = l_0 \left(\frac{E}{E_{0,i}} \right), \text{ where } E_{0,i} = E_0 Z_i, \tag{4}$$

$$E_0 = 10^{18} \text{eV} \left(\frac{B}{3 \times 10^{-6} \, \text{G}} \right) \left(\frac{l_0}{0.3 \, \text{kpc}} \right). \tag{5}$$

The spatial energy spectrum of random magnetic fields inferred from observations suggests that $B \sim 3\mu G$ on the 0.3 kpc spatial scales, and that there is a significant change at $l = 1/k \sim 0.1 - 0.5$ kpc.[44] This can be understood theoretically because the turbulent energy is injected into the interstellar medium by supernova explosions on the scales of order 0.1 kpc. This energy is transferred to smaller scales by direct cascade, and to larger scales by inverse cascade of magnetic helicity. Single-cell-size models favor ~ 0.1 kpc scales as well.[44]

As explained in the caption of Fig. 1, diffusion occurs in two different regimes depending on whether the Larmor radius is small or large in comparison with the correlation length. As a result, the diffusion coefficient changes its behavior dramatically at $E = E_{0,i}$:

$$D_i(E) = \begin{cases} D_0 \left(\frac{E}{E_{0,i}} \right)^{\delta_1}, & E \leq E_{0,i}, \\ D_0 \left(\frac{E}{E_{0,i}} \right)^{(2-\delta_2)}, & E > E_{0,i}. \end{cases} \tag{6}$$

Here the two parameters $0 \leq \delta_{1,2} \leq 0.5$ are different from zero if the magnetic domains are not of the same size. The exact values of these parameters depend on the power spectrum of turbulent magnetic fields.

The approximate solution of the transport equation in our simple model yields

$$n_i(E, r) = \frac{Q_0}{4\pi r \, D_i(E)} \left(\frac{E_0}{E} \right)^{\gamma}. \tag{7}$$

Since diffusion depends on rigidity, the composition becomes energy dependent. Indeed, at critical energy $E_{0,i}$, which is different for each nucleus, the solution (7) changes from $\propto E^{-\gamma}$ to $\propto E^{-\gamma-2}$ because of the change in $D_i(E)$, as discussed in the caption of Fig. 1. Since the change occurs at a rigidity-dependent critical energy $E_{0,i} = eE_0 Z_i$, the larger nuclei lag behind the lighter nuclei in terms of the critical energy and the change in slope. If protons dominate for $E < E_0$, their flux drops dramatically for $E > E_0$, and the heavier nuclei dominate the flux. The higher Z_i, the higher is the energy at which the species experiences a drop in flux.

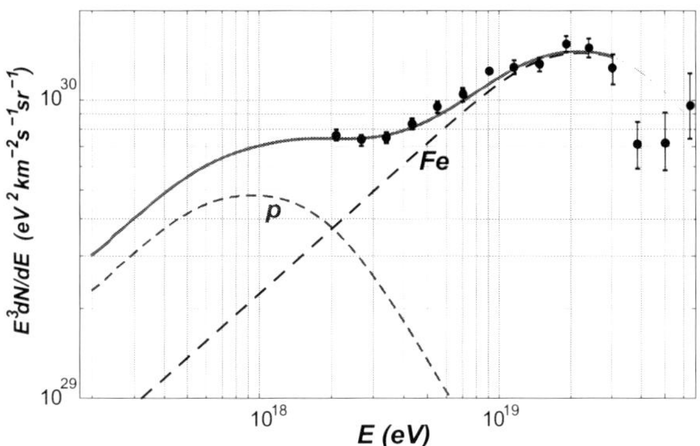

Fig. 6. UHECR spectra,[18] for the magnetic field $\sim 10\mu$G, coherent over $l_0 = 100$ pc domains. The power and the iron fraction were adjusted to fit the PAO data.[45]

The model[18] provides a qualitative description of the data (see Fig. 6). To reproduce the data more accurately, it must be improved. First, one should use a more realistic source population model. Second, one should include the coherent component of the Galactic magnetic field. Third, one should not assume that UHECR comprise only two types of particles, and one should include a realistic distribution of nuclei. Finally, one should include the extragalactic component of UHECR produced by distant sources, such as active galactic nuclei (AGN) and GRBs (outside the Milky Way). A recent realization that very high energy gamma rays observed by Cherenkov telescopes from distant blazars are likely to be secondary photons

238

produced in cosmic ray interactions along the line of sight lends further support to the assumption that cosmic rays are copiously produced in AGN jets.[1,2] For energies $E > 3 \times 10^{19}$ eV, the energy losses due to photodisintegration, pion production, pair production and interactions with interstellar medium become important and must be included. The propagation distance in the Galaxy exceeds 10 Mpc, so that the Galactic component should exhibit an analog of GZK suppression in the spectrum. Extragalactic propagation can also affect the composition around 10^{18} eV.[46]

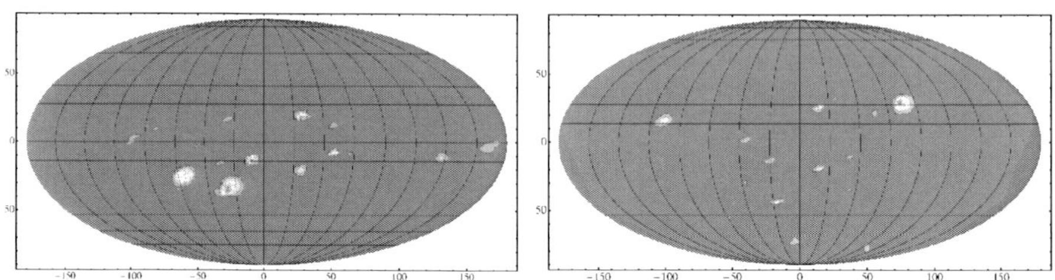

Fig. 7. Monte Carlo simulation of a typical set of "hot spots" in the directions of the closest, recent GRBs or hypernovae, in the model of Calvez *et al.*[18] The distribution of transient sources was modeled for long GRBs (left) and short GRBs (right).

Galactocentric anisotropy for a source distribution that traces the stellar counts in the Milky Way is small.[18] Although the anisotropy in protons is large at high energies, their contribution to the total flux is small, so the total anisotropy was found to be $< 10\%$, consistent with the observations. The latest GRBs do not introduce a large degree of anisotropy, as it would be in the case of UHE protons, but they can create "hot spots" and clusters of events (Fig. 7).

The model[18] leads to the following prediction for the highest-energy cosmic rays. Just as the protons of the highest energies escape from our Galaxy, they should escape from the host galaxies of remote sources, such as AGN. Therefore, UHECR with $E > 3 \times 10^{19}$ eV should correlate with the extragalactic sources. Moreover, these UHECR should be protons, not heavy nuclei, since the nuclei are trapped in the host galaxies. If and when the data will allow one to determine composition on a case-by-case basis, one can separate $E > 3 \times 10^{19}$ eV events into protons and nuclei and observe that the protons correlate with the nearby AGN. This prediction is one of the non-trivial tests of our model: at the highest energies the proton fraction should exist and should correlate with known astrophysical sources, such as AGN. The microgauss magnetic fields in the Milky Way cause relatively small deflections for the highest-energy protons. As for the intergalactic magnetic fields, there are reasons to believe that they are relatively weak, of the order of a femtogauss,[4] and, therefore, they should not affect the protons significantly on their trajectories outside the clusters of galaxies.

If local, Galactic GRBs are the sources of UHECRs, the energy output in cosmic rays should be of the order of 10^{46} erg per GRB. This is a much lower value than what would be required of extragalactic GRBs to produce the same observable flux.

Indeed, in our model the local halo has a much higher density of UHECR than intergalactic space, and so the overall power per volume is much smaller. The much higher energy output required from extragalactic GRBs[26-28] in UHECR has been a long-standing problem. The same issue does not arise in our case because it seems quite reasonable that a hypernova or some other unusual supernova explosion would generate 10^{46} erg of UHECR with energies above 10 EeV.

3. A Gamma-Ray Signature of Cosmic-Ray Nuclei

A spectral feature, namely an "iron shoulder" at 5-10 GeV can help identify cosmic nuclear accelerators.[47] Nuclei are likely to come out of acceleration regions unstable because they can lose a nucleon or a few nucleons to photodisintegration in the high-density photon environments accompanying some accelerators.[41] An unstable nucleus decays, and most of such decays are β-decays. With a probability of order one, the β-decay electron is captured by the Coulomb potential of the fully ionized atom.[48] Hence, a non-negligible fraction of nuclei come out of astrophysical accelerators in the form of one-electron ions.

In a narrow energy range, CMB photons have energies $\approx$ 7 keV in the rest frame of the ion. Such photons can excite the ion, which later emits a 7 keV photon (in the ion's rest frame). Multiple excitations and de-excitations can take place resulting in emission of gamma rays, which have energies of $5 - 10$ GeV in the laboratory frame. The spectral feature around 8 GeV (Fig. 8) can be used for identi-

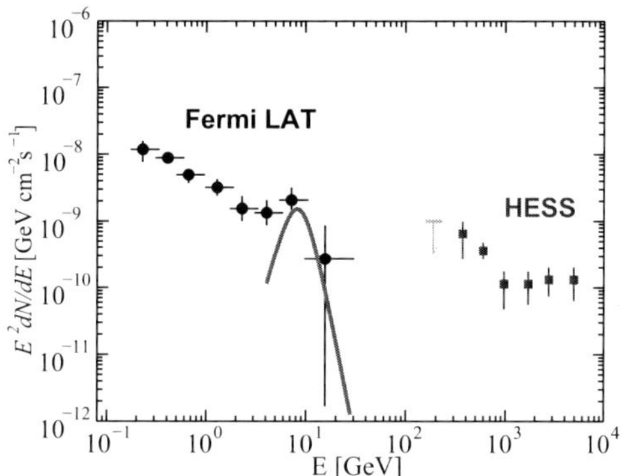

Fig. 8. Expected signature of nuclear emission for Cen A (solid line), normalized to total observed flux of iron, and the data from Fermi[49] and HESS.[50,51]

fying astrophysical sources of nuclei, or (in the case of non-detection) for setting the upper limits on nuclear acceleration.[47]

4. Conclusions

Based on the recent data, one can make several remarkable inferences about the ultrahigh-energy cosmic rays and magnetic fields inside the Milky Way and in the intergalactic space. Gamma-rays detected from most distant blazars are most likely dominated by the secondary photons produced in line-of-sight interactions of cosmic rays. This interpretation allows one to set both upper and lower bounds on intergalactic magnetic fields, $10^{-17}\mathrm{G} < B < 3 \times 10^{-14}\mathrm{G}$.[4]

Furthermore, the energy dependent composition of UHECR, with heavier nuclei at high energy, points to a non-negligible contribution from Galactic sources.[18] Diffusion in turbulent Galactic magnetic field traps the nuclei more efficiently than protons, leading to an increase in the nuclear fraction up to the energy at which iron escapes (~ 30 EeV). At higher energies, the extragalactic protons should dominate the flux of UHECR, and their arrival directions should correlate with locations of the known sources.

If and when the neutrino telescopes, such as IceCube,[52] detect point sources, one can learn about the cosmic-ray sources and photon backgrounds by comparing the neutrino flux to the photon flux. Neutrino and gamma-ray observations can help distinguish the local Galactic sources from extragalactic sources of UHE nuclei.[53–55] These inferences open exciting new opportunities for multi-messenger photon, charged-particle, and neutrino astronomy.

This work was supported by DOE Grant DE-FG03-91ER40662.

References

1. W. Essey and A. Kusenko, *Astropart. Phys.* **33**, 81 (2010).
2. W. Essey, O. E. Kalashev, A. Kusenko and J. F. Beacom, *Phys. Rev. Lett.* **104**, p. 141102 (2010).
3. W. Essey, O. Kalashev, A. Kusenko and J. F. Beacom, *Astrophys. J.* **731**, p. 51 (2011).
4. W. Essey, S. Ando and A. Kusenko, *Astropart. Phys.* **35**, 135 (2011).
5. W. Essey and A. Kusenko, *Astrophys.J.* **751**, p. L11 (2012).
6. K. Murase, C. D. Dermer, H. Takami and G. Migliori, *Astrophys. J.* **749**, p. 63 (2012).
7. S. Razzaque, C. D. Dermer and J. D. Finke, *Astrophys. J.* **745**, p. 196 (2012).
8. A. Prosekin, W. Essey, A. Kusenko and F. Aharonian, *arXiv:1203.3787* (2012).
9. E. Lefa, F. M. Rieger and F. Aharonian, *Astrophys. J.* **740**, p. 64 (2011).
10. A. De Angelis, O. Mansutti and M. Roncadelli, *Phys.Rev.* **D76**, p. 121301 (2007).
11. D. Hooper and P. D. Serpico, *Phys. Rev. Lett.* **99**, p. 231102 (2007).
12. F. W. Stecker and S. T. Scully, *Astrophys. J.* **652**, L9 (2006).
13. H. Landt, *arXiv:1203.4959* (2012).
14. A. Zech *et al.*, *PoS* **TEXAS2010**, p. 200 (2010).
15. F. Aharonian, W. Essey, A. Kusenko and A. Prosekin, *arXiv:1206.6715* (2012).
16. F. Aharonian *et al.*, *Nature* **440**, 1018 (2006).
17. M. Orr, F. Krennrich and E. Dwek, *Astrophys. J.* **733**, p. 77 (2011).
18. A. Calvez, A. Kusenko and S. Nagataki, *Phys. Rev. Lett.* **105**, p. 091101 (2010).
19. A. Kusenko, *Nucl.Phys.Proc.Suppl.* **212-213**, 194 (2011).
20. A. S. Fruchter *et al.*, *Nature* **441**, 463 (2006).
21. S. Savaglio, *New J. Phys.* **8**, p. 195 (2006).
22. A. J. Castro-Tirado *et al.*, *arXiv:0708.3043* (2007).

23. E. M. Levesque, L. J. Kewley, J. F. Graham and A. S. Fruchter, *Astrophys. J.* **712**, L26 (2010).
24. X.-Y. Wang, S. Razzaque, P. Meszaros and Z.-G. Dai, *Phys. Rev.* **D76**, p. 083009 (2007).
25. X.-Y. Wang, S. Razzaque and P. Meszaros, *Astrophys. J.* **677**, 432 (2008).
26. K. Murase, K. Ioka, S. Nagataki and T. Nakamura, *Phys. Rev.* **D78**, p. 023005 (2008).
27. E. Waxman, *Phys. Rev. Lett.* **75**, 386 (1995).
28. M. Vietri, *Astrophys. J.* **453**, 883 (1995).
29. C. D. Dermer and J. M. Holmes, *Astrophys. J.* **628**, L21 (2005).
30. P. L. Biermann, S. Moiseenko, S. V. Ter-Antonyan and A. Vasile (2003).
31. P. L. Biermann, G. A. Medina-Tanco, R. Engel and G. Pugliese, *Astrophys. J.* **604**, L29 (2004).
32. M. Schmidt, *ApJ* **523**, L117 (1999).
33. D. A. Frail *et al.*, *Astrophys. J.* **562**, p. L55 (2001).
34. S. R. Furlanetto and A. Loeb, *ApJ* **569**, L91.
35. R. Perna, R. Sari and D. Frail, *Astrophys. J.* **594**, 379 (2003).
36. X.-H. Cui, J. Aoi and S. Nagataki, *AIP Conf.Proc.* **1279**, 136 (2010).
37. G. Bertone, A. Kusenko, S. Palomares-Ruiz, S. Pascoli and D. Semikoz, *Phys. Lett.* **B636**, 20 (2006).
38. E. Parizot, M. Casse, R. Lehoucq and J. Paul (2004).
39. K. Ioka, *Prog. Theor. Phys.* **123**, 743 (2010).
40. A. Calvez and A. Kusenko, *Phys. Rev.* **D82**, p. 063005 (2010).
41. S. Horiuchi, K. Murase, K. Ioka and P. Meszaros, *arXiv:1203.0296* (2012).
42. S. D. Wick, C. D. Dermer and A. Atoyan, *Astropart. Phys.* **21**, 125 (2004).
43. K. Kotera *et al.*, *Astrophys. J.* **707**, 370 (2009).
44. J.-L. Han, K. Ferriere and R. N. Manchester, *Astrophys. J.* **610**, 820 (2004).
45. J. Abraham *et al.*, *arXiv:0906.2189* (2009).
46. C. T. Hill and D. N. Schramm, *Phys. Rev.* **D31**, p. 564 (1985).
47. A. Kusenko and M. B. Voloshin, *Phys. Lett.* **B707**, 255 (2012).
48. J. N. Bahcall, *Phys. Rev.* **124**, 495 (1961).
49. A. A. Abdo *et al.*, *Astrophys. J.* **719**, 1433 (2010).
50. F. Aharonian *et al.*, *Astrophys. J. Lett.* **695**, L40 (2009).
51. F. Aharonian *et al.*, *Astron. Astrophys.* **441**, 465 (2005).
52. F. Halzen and S. R. Klein, *Rev. Sci. Instrum.* **81**, p. 081101 (2010).
53. K. Murase and J. F. Beacom, *Phys. Rev.* **D81**, p. 123001 (2010).
54. K. Murase and J. F. Beacom, *Phys. Rev.* **D82**, p. 043008 (2010).
55. D. Hooper, A. M. Taylor and S. Sarkar, *Astropart. Phys.* **34**, 340 (2011).

Fermi – LARGE AREA TELESCOPE: ACCOMPLISHMENTS AND CHALLENGES

TUNEYOSHI KAMAE

KIPAC and SLAC, Stanford University, Menlo Park, California 94025, USA
E-mail: kamae@slac.stanford.edu

Fermi Gamma-ray Space Telescope (*FGST*) has been making many exciting discoveries in astrophysics since its launch in June 2008. One main motivation for the *Fermi/GLAST* LAT proposal in early 1990s was to promote research in particle astrophysics through close collaboration between particle physics and astrophysics communities. We have accomplished this goal by bring the two communities into one collaboration and by winning funds from their respective funding agencies. The success in finding many exciting phenomena and making high precision measurements have brought many new challenges in interpreting on fundamental physics. I will select a few such discoveries and explain the challenges.

Keywords: Gamma-ray astronomy; Cosmic-ray; Cosmic-ray acceleration; Dark matter.

1. Introduction

As of February 2012, *Fermi* collaboration has 189 papers published or accepted for publication in refereed journals. Out of many topics described in these papers I select the topics listed below and present brief presentations. All publications and PhD theses can be seen at `http://www-glast.stanford.edu/cgi-bin/pubpub`.

- *GLAST/Fermi* Large Area Telescope (LAT) project: A strong international collaboration has been formed involving particle physics, space astrophysics, and ground-based astronomy communities.
- LAT is a miniature e^+e^- collider experiment operated in space.
- Intense diffuse gamma-ray emissions produced by cosmic-rays (CR) interacting with Galactic gas and star light have been studied; more than 1400 point-like or extended gamma-ray sources detected.
- Cosmic-ray spectra have been measured: *Fermi* is providing the largest exposure among the past and planned experiments in GeV to ~ 1 TeV energy range.
- Very High Energy and Ultra High Energy cosmic-ray accelerators have been searched: Supernova remnants (SNRs) are GeV proton sources; Crab Nebula accelerates e^-e^+ to $\sim$ PeV; A special class of active galactic nuclei (AGN) may be possible UHE proton sources.
- Particle dark matter signals have been searched in various parts of sky:

Line gamma-rays from extra-galactic space; continuum gamma-rays from dwarf galaxies around Milky-Way.

2. *GLAST*/*Fermi* Large Area Telescope Collaboration

2.1. *History*

- 1990-1992: Discussion began amon Atwood, Bloom, and Michelson about a proposal for a Si detector-based upgrade of EGRET and brought to SLAC Director, Dr. Richter. Interaction with DOE and NASA began.
- 1992-2001: Large Area Telescope collaboration formed.
- 1996: First beam test at SLAC with a simple Si detector array.
- 2000: LAT coll. won the award for NASA AO 99-055-03.
- 2001: Balloon flight with one prototype tower in Texas.
- 2000-2006: LAT subsystems constructed in the collaborating institutions in Italy, UCSC, SLAC, Japan (Tracker); NRL, Sweden, France (Calorimeter); GSFC and SLAC (ACD); SLAC (DAQ, system integration). Subsystems tested at SLAC and GSI beams.
- 2006: LAT assembled and tested at SLAC (Fig. 1a) and NRL; Beam tests with spare towers at CERN PS and SPS.
- 2007: LAT integrated with the satellite.
- June 11, 2008: Launch at Kennedy Flight Center (Fig. 1b).
- August 4, 2008: Science operation started.
- 2010-2011: Fermi Source Catalog published with $\sim$ 1400 sources (Fig. 1c and [1,2]).

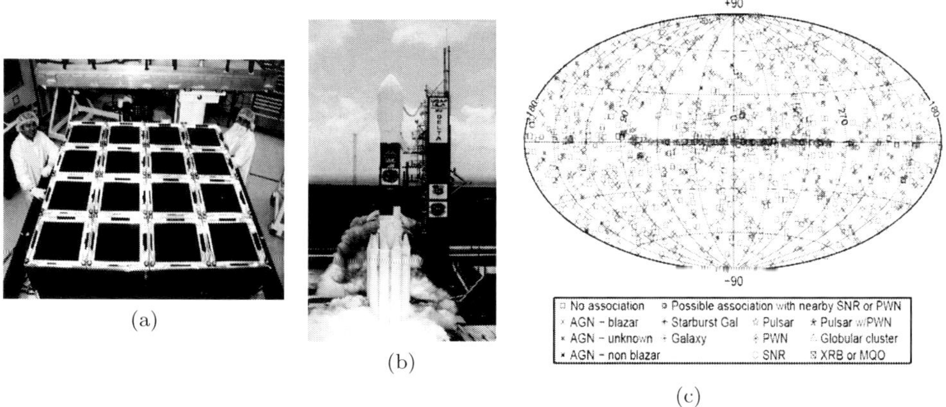

Fig. 1. (a) 16 Tracker Towers mounted onto the structure at SLAC. (b) Launch at KSC. (c) Sources in the 2nd Fermi Source Catalog.[2]

2.2. *Synergy among physics, astrophysics, and astronomy*

Since early 1990s, a successful model of cosmology has given a backbone based on particle physics, astrophysics and astronomy. Most fields of astronomy and astrophysics are now integrated into the standard cosmic evolution paradigm and accelerator-based particle physics experiments, laboratory dark-matter searches, space-based observations, and ground-based observations are interconnected in some way. Astronomical observations are interpreted in the context of the standard cosmic evolution paradigm from the Big Bang to the present universe.

The *GLAST/Fermi* collaboration has been supported in multiple scientific communities: our financial supports come from diverse government agencies such as DOE and NASA in US; MEXT and ISAS in Japan; INFN and ASI in Italy; CNRS and CEA in France; Wallenberg foundation, SRC, and SNSB in Sweden. The collaboration has been mandated to disseminate data to the international science community including particle physicsists and astrophysicists. The diversity of authorship and topics seen in the 800+ publications by non-collaboration and collaboration scientists on the *Fermi* data testifies to the success of the *FGST* project.

3. *Fermi* LAT: Instruments and Science Operation

3.1. *Instruments*

Fermi Large Area Telescope has a modular structure by design and consists of 16 identical Towers and an anti-coincidence detector covering all the Towers (see Fig. 2a). A gamma-ray is detected after being converted to an e^+e^- pair in a tungsten (W) foil in a Tower and the charged particles are traced by silicon (Si) stripe detectors and the CsI calorimeter as shown schematically in Fig. 2b. The gamma-ray direction is reconstructed based on the e^+e^- tracks recorded in Si stripe detectors and its energy based on the e^+e^- tracks and energy distribution in the the CsI calorimeter as shown in Fig. 3. Absence of signal in the anti-coincidence detector tells us that the particle is neutral and the pattern of tracks and energy depositions in the Si stripe detectors and CsI calorimeter allow us to select gamma-rays. Details on the LAT instrument are given in Atwood *et al.* [3].

The *Fermi* Space Gamma-ray Telescope consists of two instruments as shown in the upper panel of Fig. 4: the primary instrument, LAT, to detect gamma-rays from ~ 20 MeV to ~ 300 GeV and the auxiliary instrument, GBM, to detect gamma-rays from ~ 8 keV to ~ 40 MeV. The GBM detector consists of 12 NaI scintillation counters and 2 BGO scintillation counters and serves primarily to detect gamma-ray bursts (GRBs) and rapid transient events [4,5].

The LAT has an effective area, ~ 0.7 m^2 and a field-of-view, ~ 2.7 sr in its prime energy range, $1 - 3$ GeV after the standard event selection (P7Source_V6 class). Its ability to measure the gamma-ray direction (point-spread-function, PSF) and energy (energy resolution) depend on the gamma-ray energy: they are ~ 0.6 deg (68% containment) and ~ 9 % (68% containment) in the prime energy range, respectively. Technical details are updated as the science analysis

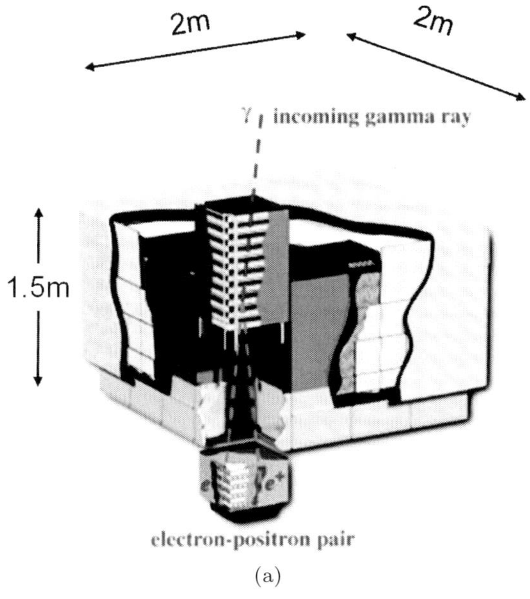

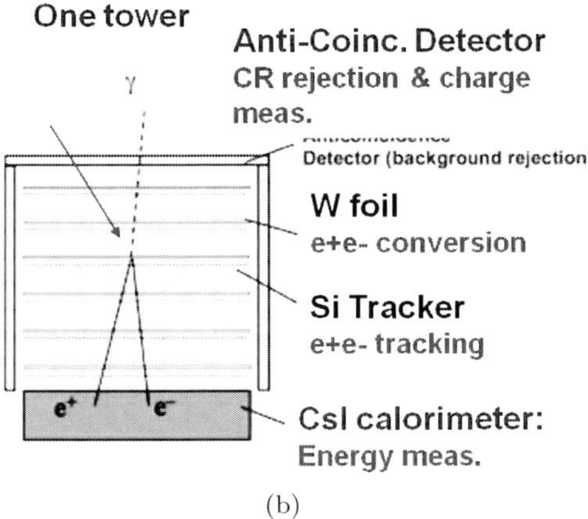

Fig. 2. (a) *Fermi* Large Area Telescope. (b) Schematic drawing of one of the 16 Tracker Towers in LAT: it is a pair-conversion gamma-ray spectrometer consisting of an Anti-Coincidence Detector, a Silicon Tracker consisting of tungsten (W) foils and Silicon (Si) strip detectors, and a CsI calorimeter.

tools evolve at `http://www-glast.stanford.edu/instrument.html`, and `http://www.slac.stanford.edu/exp/glast/groups/canda/lat_Performance.htm`.

3.2. *Science operation*

The *Fermi* mission is operated nominally in the sky survey mode which takes advantage of its large field of view. The LAT surveys almost the full sky every orbit of $\sim$ 90 min. To make the exposure more uniform over the entire sky, the LAT pointing is offset from the zenith by 30-50 deg northward or southward every other orbit. The integrated exposure is shown by color in the lower 4 panels of Fig. 4. In two orbits or in about 3 hours, the sky coverage becomes quite uniform.

4. Gamma-Ray Sky: What Do We See?

The gamma-ray sky seen by LAT is shown in the Galactic coordinate in Fig. 5. The emission represented by the thin yellow band originates mostly from the H_2 gas distribution in the inner Galaxy (distance from the Earth = a few -10kpc); the emission from the wide pink band is from H_2 clouds and H gas near the solar system (dist < 1 kpc) [6]. Emission from atomic H gas in the Galactic halo and the outer Galaxy overlaps to the above structures and barely visible in Fig. 5. And the extra-Galactic diffuse emission covers all sky.

Nearly 1400 point-like or extended-but-confined sources have been detected in the entire sky [2]. They includes some well-known sources and the sources to be discussed in this presentation (circled): supernova remnants (SNRs) and pulsar wind nebulae (PWN) in yellow; and active galactic nuclei (AGN) in black.

5. Cosmic Ray Spectra

Fermi-LAT is the largest space cosmic-ray experiment in the integrated exposure, the product of the effective area and the live time, in the energy range between sub-GeV to TeV [7]. However it does not have a magnet and hence cannot differentiate

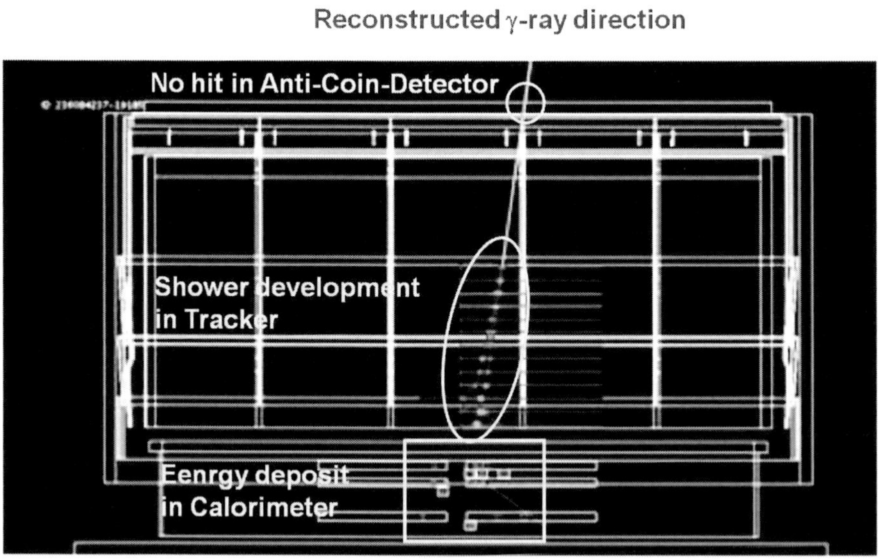

Fig. 3. Reconstruction of a gamma-ray event based on the hits in the Si tracker and the energy deposited in the CsI calorimeter.

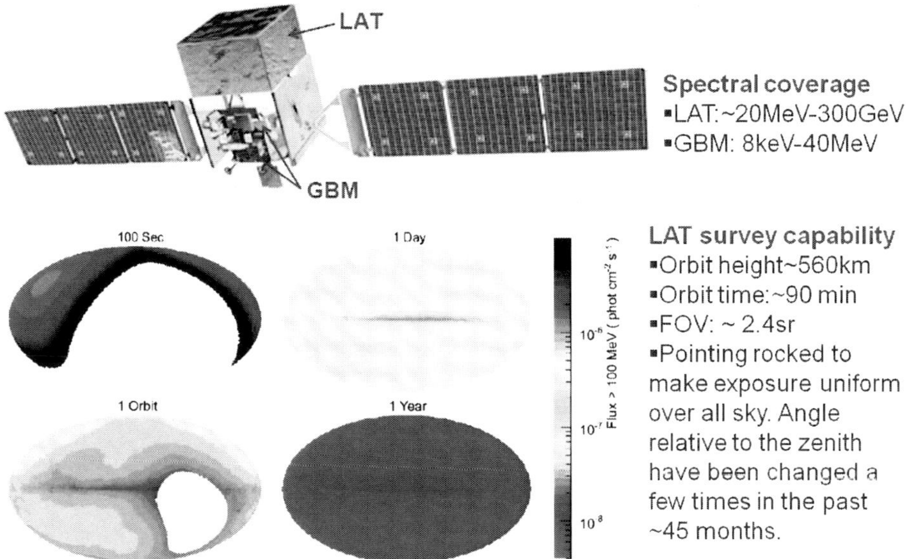

Fig. 4. (Top) The *Fermi* space gamma-ray telescope consisting of the LAT and the GBM. (Bottom) The sky coverage in the Galactic Coordinate for various time intervals.

positively charge particles from negatively charged particles nor can measure the mass of particles, except for some limited cases.

The limited cases in which *Fermi* can differentiate particles are: in the non-

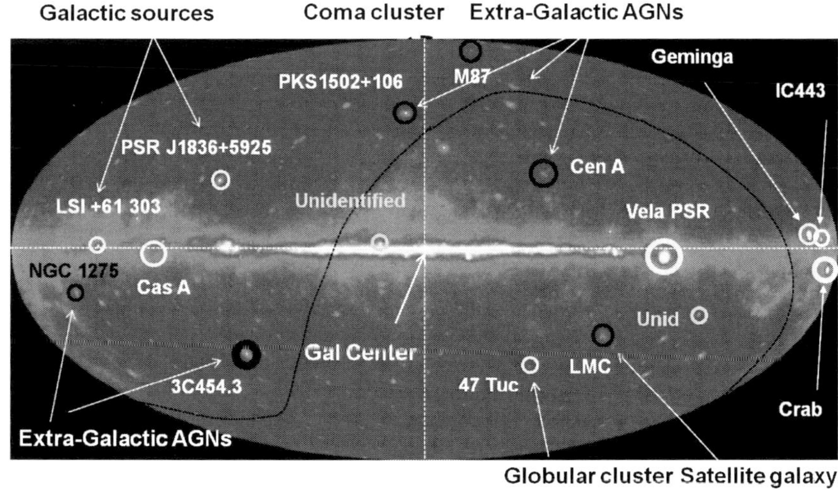

Fig. 5. Gamma-ray sky seen by the *Fermi* LAT: some sources are marked.

relativistic energy range, p, α, and heavy ions can be differentiated by dE/dx in the plastic scintillator in ACD [3]; in the GeV range, electrons and positrons can be differentiated from nuclear particles by the electromagnetic showers [3]; in the energy range between a few 10 GeV and a few 100 GeV positively charged and negatively charged cosmic-rays can be differentiated by using the Earth's magnetic field. Since *Fermi* has a large field-of-view and surveys the entire sky several times a day, it can measure anisotropy in the cosmic-ray arrival direction. We show 2 examples of such measurements by *Fermi*.

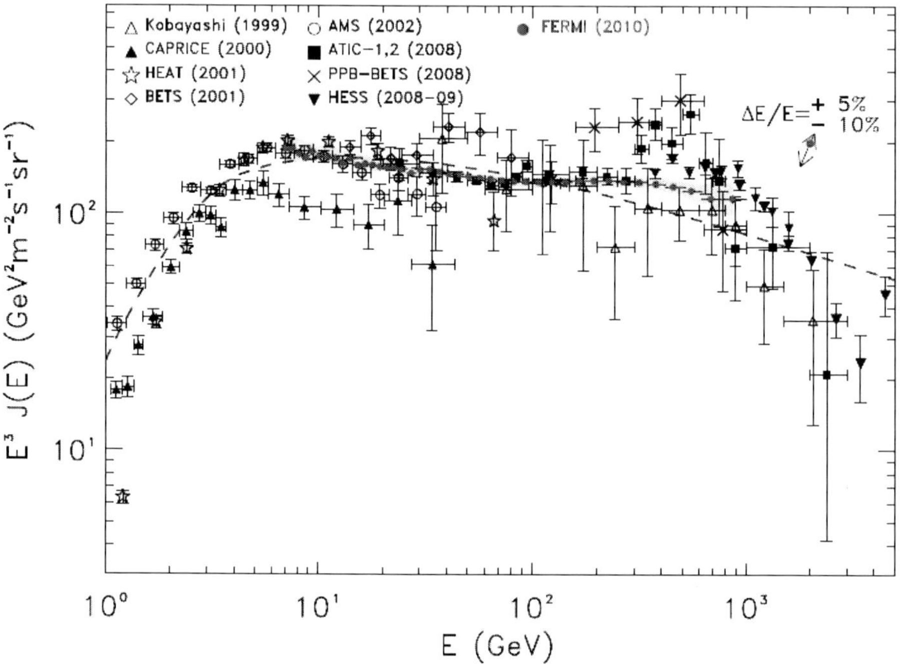

Fig. 6. Spectrum of electrons plus positrons measured by *Fermi* (red data points) compared with those reported by other experiments. From [9].

The first is the measurements on the spectrum of electrons plus positrons extending to TeV shown in Fig. 6 [8,9]. What makes the *Fermi* measurements stand out among others is the precision and the breadth of the energy coverage. These features come from the fact that e^-, e^+, and gamma-rays are detected, reconstructed, and calibrated with the same set of detectors, programs, and calibration procedures which have been designed and built for high precision measurements.

The second example is the positron-to-electron ratio measured using the Earth's magnetic field [10]. Fig. 7a shows the distribution of the direction of the charged particles identified as e^- or e^+ in the LAT. Electrons and positrons are reachable from outside of the Earth's magnetosphere to the crescent-shaped areas shown in the left and right columns, respectively. The areas shrink as the particle energy

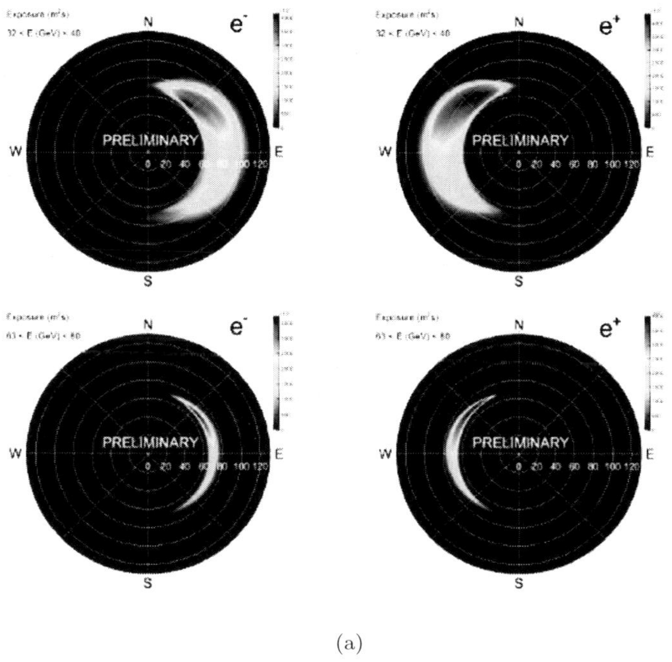

(a)

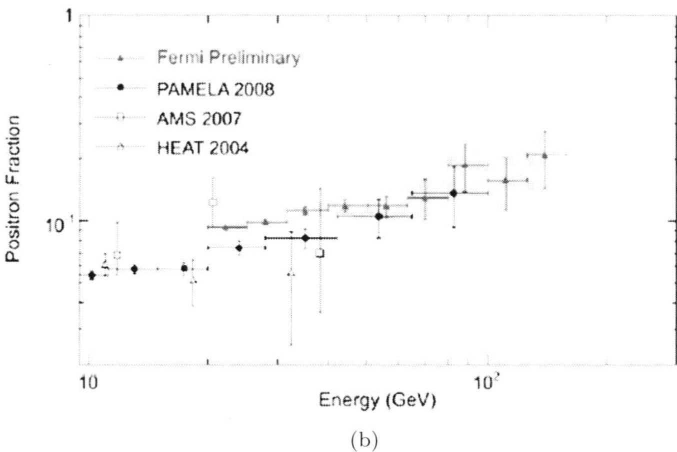

(b)

Fig. 7.   (a) Distribution of arrival directions of e^- and e^+ relative to the Earth. Upper-left for $32 - 40$GeV e^-; Lower-left for $63 - 80$GeV e^-; Upper-right for $32 - 40$GeV e^+; Lower-right for $63 - 80$GeV e^+; (b) Positron fraction, $e^+/(e^- + e^+)$, measured by $Fermi$ compared with that by PAMELA. From [10].

increases: $32 - 40$ GeV for the upper crescent areas and $63 - 80$ GeV for the lower crescent areas in Fig. 7a. The positron fraction has been calculated on the ratio of the counts in these areas and shown in Fig. 7b. *Fermi* has confirmed the rising trend seen by PAMELA and extended the data points to ~ 150 GeV [10].

6. Very High Energy Cosmic-Ray Accelerators

Although many Galactic and extra-Galactic sources show signs of particle acceleration, evidence has been lacking for proton acceleration especially in the VHE range extending to the "Knee" region ($E \sim 10^{15}$ eV). *Fermi* has studied many potential cosmic-ray sources and gathered an evidence for acceleration of protons (incl. nuclei) in some SNRs and very-high-energy (VHE) electrons (incl. positrons), in Crab Nebula, as will be discussed here. In addition, we will bring to your attention a speculative possibility that protons are accelerated to EeV in an "extreme" blazar.

6.1. *SNR*

The gamma-ray spectra observed from some supernova remnants are particularly intense and compatible as coming from the molecular clouds interacting with the forward shock as shown in Fig. 8 [11–15]. Younger SNRs and SNRs located away

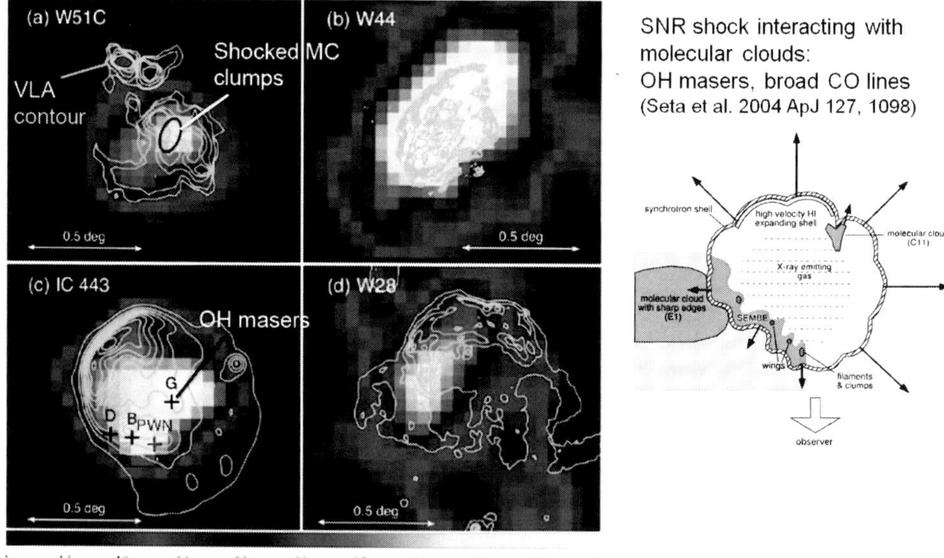

Fig. 8. Supernova remnants interacting with molecular clouds observed by *Fermi*. Color maps are for gamma-ray intensity and the contours are for radio continuum: Upper-left, W51C; upper-right, W44; lower-left, IC443; lower-right: W28. The right panel shows the locations of the forward shock and the molecular clouds in W44 measure by Seta *et al.* [21]

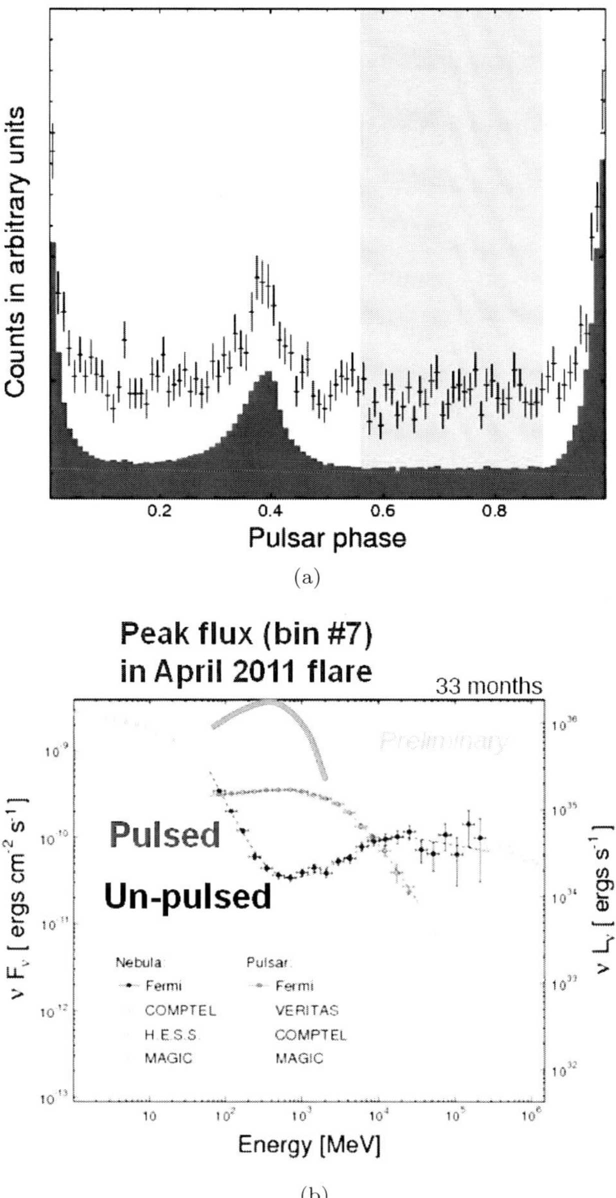

Fig. 9. (a) Gamma ray counts folded at the Crab Pulsar phase. The data points are during the April flare and the red profile is the average over 33 months. (b) The observed gamma-ray spectra for (green) the non-pulsed phase in the gray band in (a) during the April flare, (red) the pulsed phase during the April flare, and (black) the averaged non-pulsed emission. From [23].

from the Galactic Plane are less likely to interact with molecular clouds and their gamma-ray emission is primarily due to electrons [16,17]. However we found some evidence for proton interaction with diffuse Galactic gas in some SNRs on the Galactic Plane [18–20].

6.2. *Giant GeV Flares in Crab Nebula: Evidence for a PeV accelerator*

Fermi and *AGILE* have detected a few giant flares from Crab Nebula at a rate about 1 per year since their launch [22–24]. The *Fermi* collaboration has been monitoring the gamma-ray counts from Crab in two pulsar phases, one in the gray band (non-pulsed) and the rest (pulsed) of Fig. 9a. So far 3 flares have been detected. The most spectacular of them is the one occurred in April 2011 and its light curve is shown in Fig. 10. The rise-time of the highest peak is hour-scale. There is no such non-pulsed flare has been seen in Crab. The gamma-ray spectrum observed during the April 2011 flare is shown in Fig. 9b together with the average (normal) non-pulsed and pulsed spectra. The spectral shape is consistent for being a synchrotron spectrum produced by VHE electrons (incl. positrons). Their energy depends on the magnetic field strength of the emission region but likely be much higher than 100TeV and in the PeV range [23].

6.2.1. *How and where are VHE e+e- accelerated?*

Many papers have discussed about particle acceleration near or at the termination shock of the relativistic striped wind, especially that of Crab Nebula. However non of them has predicted hour-scale acceleration to ∼PeV energy scale. Because of this time-scale, fast magnetic reconnection is considered as a strong candidate.

Particle acceleration associated with magnetic reconnection has been observed and analyzed in the solar flares and in the interaction of the solar wind and Earth

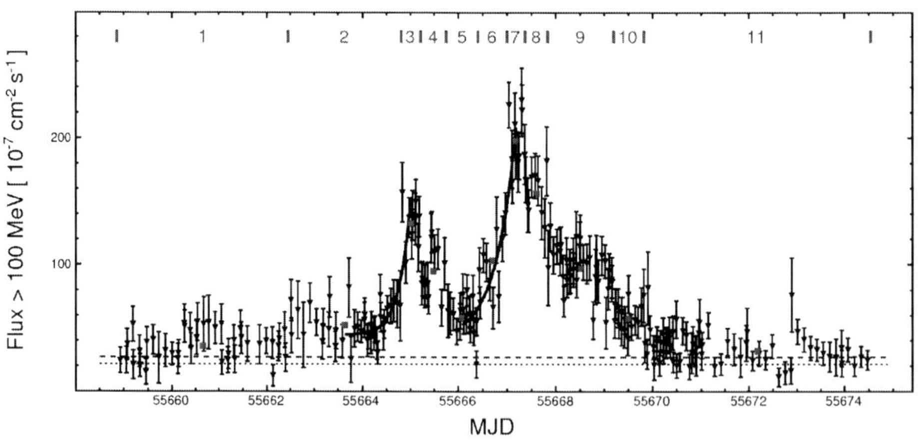

Fig. 10.   Detailed time profile of non-pulsed gamma-ray flux during the April flare. From [23].

magnetosphere. The former is reviewed well by Benz [25] (observations) and by Shibata and Magara [26] (theoretical analyses). An analysis of satellite observations for the latter is reported in Oeierose *et al.* [27]. The particle energy scale is, however, GeV or less for the solar flares and a few 100keV or less for the latter. Particles involved are electrons and protons (ions) and their speed is non-relativistic.

In Crab Nebula, the particles involved are most likely very high energy electrons and positrons. Magnetic field is quite strong because it comes from the neutron star. So the parameter space for the GeV flares in Crab Nebula is very different from that for the solar flares and the solar wind. Many numerical simulations have been carried out in the past 10 years. They can be grouped into 2 categories: the first is a large scale 3D simulation of the entire pulsar wind (e.g. Crab Nebula) like the works by Spitkovsky [28] and by Kalapotharakos and Contopoulos [29]; magnetic reconnection in a limited region is simulated in the second group. The second group is divided further into 2 sub-categories: one is two-fluid MHD simulations [30,31]; the other, Particle-In-Cell simulations of $e^- e^+$ plasma in a scale smaller than the entire PWN [32–34]. These studies generally find that particles are accelerate quickly but not necessarily to very high energy [34,35].

Several possible scenarios for the observed gamma-ray flares and the acceleration mechanism behind the emission have been proposed. They all have some issues to become a convincing model for the observed GeV flares.

- Komissarov & Lyutikov [36]: The emission comes from an inner knot on the arch shock 1-2 light-day away from the PSR.
- Bykov *et al.* [37]: Intermittency in B-field fluctuation generated by ion cyclotron instability in the wisp region.
- Bedranek & Idec [38]: Inverse Compton scattering of CMB photons by a few $\times 10^{14}$ eV electrons.
- Cerutti, Uzdensky,& Begelman [39]: Aligned reconnection regions with a focusing effect.

6.3. *Proton Acceleration in AGN Jets?*

Theorists have been arguing since early 1990s that protons must exist in the AGN jets by comparing the power emitted as the electromagnetic wave, the power in the Poynting flux, and the kinetic energy of the bulk plasma containing electrons. Ghisellini and Tavecchio estimate the p/e ratio must exceed 10% for the 3C454.3 flares observed with *Fermi*, UV, and X-ray [40–42]. It is to be noted that protons can carry the majority of power in the AGN jet as shown for PKS 122+210 in an analysis by Tavecchio *et al.* [43].

Although various observations mentioned above indicate that protons are accelerated in AGN jets, there is no observational indications for these protons reaching very high energy hadron as reviewed by Reimer [44]. Even if protons reach very high energy, they have to hit dense target media or photon field and develop electromagnetic cascade to produce the observed synchrotron spectrum. Several blazar

emission models based on hadronic interaction have been proposed as reviewed by
Reimer [44]. While they add more freedom in parameter adjustment, they yet lack
compelling reason to replace the leptonic scenario for general AGN jets.

Recently, observations on an "extreme blazar" IES0229+200 are found to be
challenging for the leptonic scenario. The broad-band spectrum of the blazar is
shown in Fig. 11 [45]. The synchrotron spectrum extending past 100keV is similar
to that seen in flares of some BL Lac (e.g., Mkn501). If the emission is due to
very high energy electrons as in Mkn501 flares, we expect the X-ray flux to vary
by a large factor in short time scales. Observations in the X-ray band found the
flux stable over several years to a few % level [46]. Another abnormality is in the
higher-energy part of the emission, presumably due to inverse Compton scattering
(IC): the spectrum is considered to be unusually hard (photon index $\sim$ 1-1.5) after
correcting for absorption by EBL [45,46].

These unusual features have led to a speculation that the TeV gamma-rays
are produced by ultra high energy cosmic rays (UHECR) interacting inter-galactic
photons [47], in which case ultra high energy neutrinos would be observed. Another
interpretation is that the TeV gamma-rays are synchrotron radiations produced by
UHECR protons [48]. So far we only have one such sample: should there be more,
these speculations can be seriously tested.

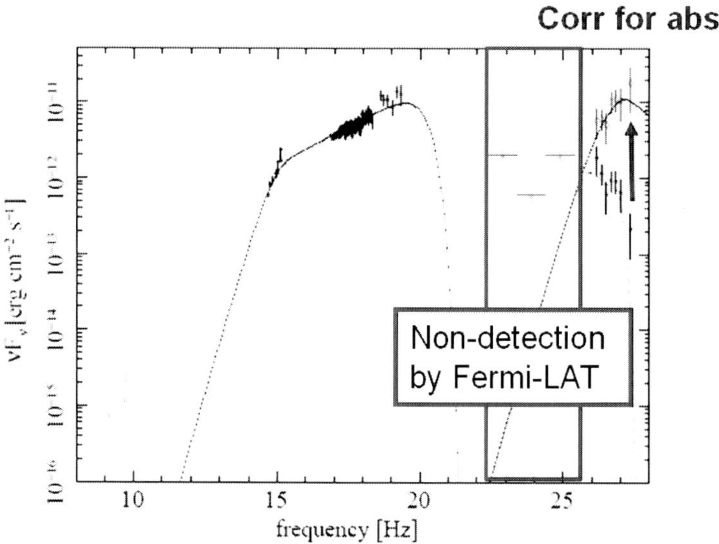

Fig. 11.   Broad-band spectrum for 1ES0029+200. From [46].

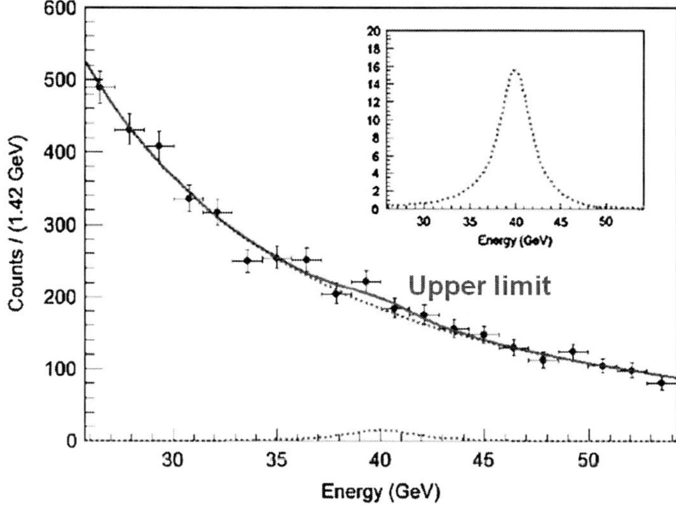

Fig. 12. Typical fit to obtain an upper limit to the line spectrum: (data points) Counts in 11 months of data in the search region; (red) the best fit with a line signal at the assumed energy of 40GeV; (dot) the background counts. From [49].

6.4. *Challenge: Sources of VHE and UHE protons?*

Fermi has detected gamma-ray emission from ~ 10 SNRs and obtained evidence for intense proton acceleration in the mid-aged SNRs interacting with molecular clouds. However the implied proton spectra are soft beyond a few 100 GeV. Gamma-ray spectra from some other SNRs are consistent being of pionic origin and having harder power-law index but the counts beyond ~ 10 GeV are too low to be confirmed as VHE cosmic-ray accelerators.

We discovered GeV gamma-ray flares from Crab Nebula implying that $e^- e^+$ are accelerated to $\sim PeV$. However the observed spectrum shows no hint of proton acceleration.

Theoretical analyses of AGN flares strongly suggest that protons are carrying the majority of power in the jet. However no direct evidence for proton acceleration to VHE in the observed spectra. Analyses of the extreme blazar 1ES0229+200 have led to interesting speculations that UHECR may be accelerated there.

7. Search for Dark Matter Signals

Signals for the dark matter has been searched in several locations in the sky: the Galactic Center region; Dwarf satellite galaxies; the Galactic Halo; Extra-galactic space; Clusters of galaxies. The spectral features looked for are emission lines and broad enhancements consistent with annihilation of dark matter particles.

The most straightforward is the search for emission lines [49]. The search region

256

is set to the high Galactic latitude sky excluding the Galactic plane and the Galactic bulge where the known gamma-ray emissions add strong background to the search. We found no evidence for the line feature and set an upper limit as shown in Fig. 12 [49].

The most sensitive search for minimal SUSY dark matter has been done at

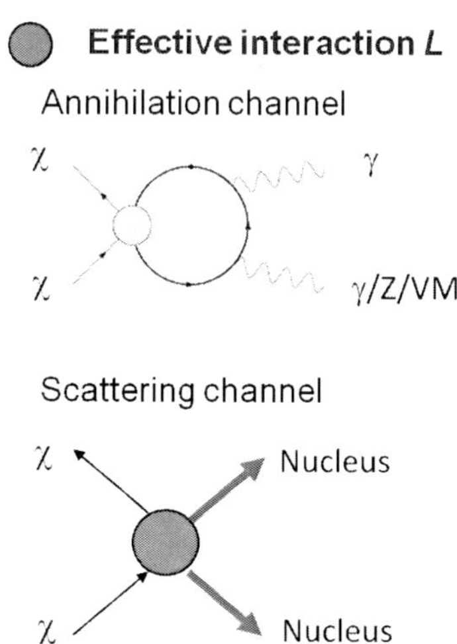

Fig. 13. Diagram important in the "indirect search" (upper) and in the "direct search" (lower).

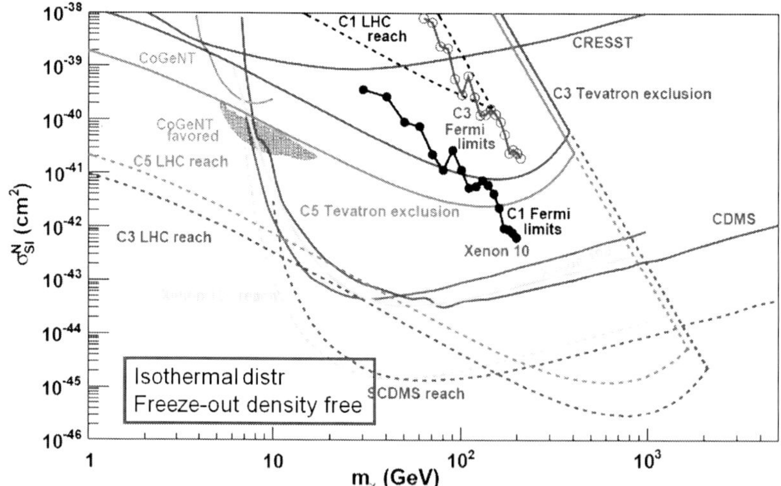

Fig. 14. Upper limits set by "direct search" experiments, "indirect search" experiments, and expected reach of LHC experiments for two operator products. From [54].

the gravitational center of 10 satellite dwarf galaxies [50–52]. An upper limit of $\sim 1 - 7 \times 10^{-26} \mathrm{cm}^3\mathrm{s}^{-1}$ has been obtained for the SUSY WIMP decaying to $b\bar{b}$ in the mass range between 1 and $100 \mathrm{GeV}/c^2$ [52].

7.1. *Challenge: How to compare with the direct search results?*

Dark matter particles have been searched in the accelerator experiments and the underground scattering experiments (the "direct search") as well as in the gamma-ray spectra from the sky (the "indirect search"). The limits set by these experiments are interrelated. It is possible the upper limit set by a direct search experiment may rule out possibility to find any dark matter annihilation signal in the sky.

This topics has been addressed in several publications in the past 10 years, for example by Baltz *et al.* [53]. Production of DM particles in Large Hadron Collider involves many processes and highly model dependent. The cross-sections for DM scattering with normal matter ("direct" searches) and for DM annihilation into gamma-rays ("indirect searches") can be related simply by rotating the Feynman diagram in the 4-dim space as shown in Fig. 13. The blobs in the upper and lower panels represent the same Lorentz-covariant operator product. The work by Goodman *et al.* [54] is among the first to exploit this relation and compare the upper limits in the direct and indirect searches on the effective theory formalism. One example is shown in Fig. 14 where the limits by the direct search experiments (CDMS, Xenon10, and CREST), *Fermi* limits from the gamma-ray line search, and reach of LHC experiments are compared for two operator products, C3 and C5 in [54].

The number of possible operator products is manageable if the underlying interaction is described by the minimum SUSY. There are several other dark matter models proposed in the past decades including Extra-Dimension and Little Higgs. They can be casted into some effective theories consisting of operator products but they will be larger in number and the true interaction may not be represented by one or two operator products. How will then the limits or discoveries in any one of the 3 types of experiments constrain the signals in the other 2 types? Findings by LHC experiments and limits set by "direct searches" will have to be incorporated in the *Fermi* analyses assuming possible underlying theories.

8. Conclusions

Fermi-LAT has been making many exciting discoveries. However their interpretation is not straightforward. Well-focused collaboration with broad science community in observations theories and numerical analyses. We have high expectation that LeCosPA will nurture such collaborations.

Acknowledgments

I thank the hospitality offered by Pisin Chen and LeCosPA personnels. Assistances given by the *Fermi* collaboration, especially by A. Reimer, R. Buehler, and Y. Uchiyama are appreciated.

References

1. Abdo, A. A. *et al.* 2010, ApJS, 188, 405
2. Nolan, P. L. *et al.* 2011, ApJS, 199, 31
3. Atwood, W. B. *et al.* 2009, ApJ, 697, 1071
4. Meegan, C. A., *et al.* 2009 ApJ 702, 791
5. Bissaldi,E., *et al.* 2009 Experimental Astronomy, 24, 47
6. Ackermann, M., *et al.* 2012 arXiv:1202.4039
7. Thompson, D. J., Baldini, L., & Uchiyama, Y., 2012 arXiv1201.0988
8. Abdo, A. A. *et al.* 2009, Phys. Rev. Lett., 102, 181101
9. Ackermann, M. *et al.* 2010, Phys. Rev. D, 82, 092004
10. Ackermann, M. *et al.* 2012, Phys. Rev. Lett., 108, 011103
11. Abdo, A. A. *et al.* 2010, ApJ, 718, 348
12. Abdo, A. A. *et al.* 2010, ApJ, 712, 459
13. Abdo, A. A. *et al.* 2010, Science, 327, 1103
14. Abdo, A. A. *et al.* 2009, ApJL, 706, L1
15. Abdo, A. A. *et al.* 2010, ApJ, 722, 1303
16. Abdo, A. A. *et al.* 2010, ApJL, 710, L92
17. Abdo, A. A. *et al.* 2011, ApJ, 734, 28
18. Giordano, F. *et al.* 2012, ApJL, 744, L2
19. Ajello, M. *et al.* 2012, ApJ, 744, 80
20. Katagiri, H. *et al.* 2011, ApJ, 741, 44
21. Seta *et al.*, AJ 127, 1098
22. Abdo, A. A. *et al.* 2011, Science, 331, 739
23. Buehler, R., Scargle, J. D., Blandford, R. D., *et al.* 2011, ApJ (in press); arXiv:1112.1979
24. Tavani M. *et al.* 2011, Science 331, 736
25. Benz, A. O. 2008 Living Rev. Solar Phys. 5, 1
26. Shibata, K. and Magara, T. 2011, Living Rev. Solar Phys. 8, 6
27. Øieroset, *et al.* 2001, Nature, 412, 414
28. Spitkovsky, A. 2006 ApJL 648, L51
29. Kalapotharakos, C. and Contopoulos, I 2009 A&A 496, 495
30. Zenitani, S., Hesse, M., & Klimas, A. 2009, ApJ, 696, 1385
31. Zenitani, S., Hesse, M., & Klimas, A. 2009, ApJ, 705, 907
32. Zenitani, S., & Hoshino, M. 2007, ApJ, 670, 702
33. Zenitani, S., & Hoshino, M. 2008, ApJ, 677, 530
34. Liu, W., Li, H., Yin, L., *et al.* 2011, Physics of Plasmas, 18, 052105
35. Sironi, L. and Spitkovsky, A. 2011 ApJ 741, 39
36. Komissarov and Lyutikov, 2011, MNRAS 414, 2017
37. Bykov, *et al.* 2012, MNRAS, 421, L67
38. Bednarek, W., & Idec, W. 2011, MNRAS 414, 2229
39. Cerutti, B., Uzdensky, D. A., & Begelman, M. C. 2012, ApJ 746, 148
40. Ghisellini, G., & Tavecchio, F. 2010, MNRAS, 409, L79
41. Ackermann, M., *et al.* 2010 ApJ 721, 1383
42. Bonnoli, G., *et al.* 2011, MNRAS, 410, 368
43. Tavecchio, F., Becerra-Gonzalez, J., Ghisellini, G., *et al.* 2011, A&A 534, A86
44. Reimer, A. 2011 Workshop on Beamed and Unbeamed γ-rays from Galaxies (Muonio, 2011).
45. Aharonian, F., Akhperjanian, A. G., Barres de Almeida, U., *et al.* 2007, A&A, 475, L9
46. Kaufmann, S., Wagner, S. J., Tibolla, O., & Hauser, M. 2011, A&A, 534, A130
47. Murase *et al.* 2012 ApJ 746, 164
48. Zacharopoulou, O., *et al.* 2011 ApJ 738, 157
49. Abdo, A. A. *et al.* 2010, Phys. Rev. Lett., 104, 091302
50. Abdo, A. A. *et al.* 2010, ApJ, 712, 147
51. Ackermann, M. *et al.* 2011, Phys. Rev. Lett., 107, 241302
52. Ackermann, M., *et al.* 2012 ApJ 747, 121
53. Baltz, E. A., Battaglia, M., Peskin, M. E., & Wizansky, T. 2006, PRD, 74, 103521
54. Goodman, J., *et al.* 2011 Nucl. Phys. B 844, 55

ULTRA-FAST FLASH OBSERVATORY (UFFO) FOR OBSERVATION OF EARLY PHOTONS FROM GAMMA RAY BURSTS

I.H. Park[1*], S. Ahmad[2], P. Barrillon[2], S. Brandt[3], C. Budtz-Jorgensen[3], A.J. Castro-Tirado[4], P. Chen[5], Y.J. Choi[6], P. Connell[7], S. Dagoret-Campagne[2], C. Eyles[7], B. Grossan[8], M.–H.A. Huang[9], A. Jung[1], S. Jeong[1], J.E. Kim[1], M.B. Kim[1], S.-W. Kim[10], Y.W. Kim[1], A.S. Krasnov[11], J. Lee[1], H. Lim[1], E.V. Linder[1,8], T.–C. Liu[5], N. Lund[3], K.W. Min[6], G.W. Na[1], J.W. Nam[5], M.I. Panasyuk[11], J. Ripa[1], V. Reglero[7], J.M. Rodrigo[7], G.F. Smoot[1,8], J.E. Suh[1], S. Svertilov[11], N. Vedenkin[11], M.–Z. Wang[5], I. Yashin[11]

[1]*Ewha Womans University, Seoul, Korea*
[2]*University of Paris-Sud 11, Orsay, France*
[3]*Technical University of Denmark, Copenhagen, Denmark*
[4]*Instituto de Astrofisica de Andalucia - CSIC, Granada, Spain*
[5]*National Taiwan University, Taipei, Taiwan*
[6]*Korea Advanced Institute of Science and Technology, Daejeon, Korea*
[7]*University of Valencia, Valencia, Spain*
[8]*University of California, Berkeley, USA*
[9]*National United University, Miao-Li, Taiwan*
[10]*Yonsei University, Seoul, Korea*
[11]*Moscow State University, Moscow, Russia*
**E-mail: ipark@ewha.ac.kr*

One of the least documented and understood aspects of gamma-ray bursts (GRB) is the rise phase of the optical light curve. The Ultra-Fast Flash Observatory (UFFO) is an effort to address this question through extraordinary opportunities presented by a series of space missions including a small spacecraft observatory. The UFFO is equipped with a fast-response Slewing Mirror Telescope (SMT) which uses rapidly moving mirror or mirror arrays to redirect the optical beam rather than slewing the entire spacecraft to aim the optical instrument at the GRB position. The UFFO will probe the early optical rise of GRBs with a sub-second response, for the first time, opening a completely new frontier in GRB and transient studies, the only GRB system which can point and measure on these time scales. Its fast response measurements of the optical emission of dozens of GRB each year will provide unique probes of the burst mechanism, shock breakouts in core-collapse supernovae, tidal disruptions around black holes, test Lorentz violation, be the electromagnetic counterpart to neutrino and gravitational wave signatures of the violent universe, and verify the prospect of GRB as a new standard candle potentially opening up the z>10 universe. As a first step, we employ a motorized slewing stage in SMT which can point to the event within 1s after X-ray trigger, in the UFFO-pathfinder payload onboard the Lomonosov satellite to be launched in 2012. The pathfinder was a small and limited, yet remarkably powerful micro-observatory for rapid optical response to bright gamma-ray bursts, the first part of our GRB and rapid-response long-term program. We describe the early photon science, the space mission of UFFO pathfinder, and our plan for the next step.

Keywords: Gamma Ray Burst; fast transient flash; slewing mirror telescope; optical afterglow; X-ray; Lomonosov spacecraft; UFFO; coded mask.

1. Introduction

GRBs are the most luminous explosions in the universe, emitting the highest energy photons, and should be seen to the highest redshift of any object in the universe. These properties provide great leverage in time, wavelength, and information, and thus a unique opportunity to understand not only the nature of the universe but fundamental physics, to name a few, as listed in the following.

GRB allows the study of the evolution of stars and stellar populations from redshifts $z=0\sim15$ or over 98% of the age of the universe. The first stars and galaxies at $z=10\sim15$ would be identified with their formation history, and the reionization history at $z\sim7$ could be unveiled as GRB would be signposts of the faint galaxies providing the bulk of UV photons to reionization of the universe. GRB could be a potential candidate of the next generation cosmological standard candle. If such a standardization is realized, GRBs prove the early universe at red shift values of up to $z=15$. It would be nothing less than a revolution in extremely high-z cosmology, as other cosmological tools (SNIa, BAO, lensing) are much more limited in z range.

GRB would offer a great opportunity to explore the extreme universe. Among the many other avenues of research are: the origin of GRBs including short GRBs and dark GRBs; fast transient cosmic flash events such as primordial black holes; black hole event horizon; tidal disruption and shock break-out associated with extremely heavy gravitational objects. GRBs are believed to be sources of ultra-high energy cosmic rays (UHECR) and gravitational waves (GW), while the burst and afterglow spans some 9 orders of magnitude in photon energy. It enables synoptic astronomy and multi-messenger astrophysics which requires fast-response observations of photons aided with observations of UHECR particles through air showers (e.g. TA, Auger, TUS, JEM-EUSO), of neutrinos (e.g. ICECUBE, ANITA), and searches for GW (e.g. LIGO, LISA).

It is clear that thorough understanding of GRBs and their underlying physics will be possible with GRB observations not only in various probes including frequencies from X-rays to radio, but in wide time domain particularly early phase. Hundreds of GRBs UV-optical light curves have been measured since the discovery of optical afterglows. A prime example is the *Swift* spacecraft, which can detect simultaneously gamma ray, X-ray, UV, and optical signals[1], a spectacular success by any measure. However, even after nearly 7 years of operation of the *Swift*, the immediate aftermath of the explosion is scarcely observed in the optical or ultraviolet, despite a highly visible afterglow. *Swift* is the fastest high-sensitivity, space observatory for following the UV-optical afterglows, typically responds in ~100 seconds or more, while ground-based telescopes do occasionally respond faster, but only a handful of rapid detections have been produced to date. Less than a handful of short duration GRB have been detected to date in the UV-optical within the first minute after the gamma ray signal.

This lack of early observations and the blindness to the rise phase of the afterglow immediately following the explosion, or to other abrupt or highly variable transient sources, leaves fertile astrophysical territory and many important physical questions arising at the short time scales unexplored. On the astrophysical side, this includes the origin of short-hard type GRBs, rapid-rising GRBs associated with prompt emissions and progenitors, the nature of the burst energy release mechanism, the formation of the neutron star or black hole, the development of shocks and interaction with the surrounding medium, etc. Rapid data collection is also essential for tests of fundamental physics such as constraints on Lorentz violations and CPT from the time delay between different energy photons, or between photons and neutrinos. Coincident or successive observations of the explosion event as an electromagnetic counterpart to a neutrino observatory of gravitational wave observatory signal would revolutionize astronomy, and greatly improve our understanding of black holes, neutron stars, and strong field gravity.

We have developed methods, for the first time, for reaching sub-minute and sub-second time scales in a spacecraft observatory appropriate for launch even on small satellites. Rather than slewing the entire spacecraft to aim the optical instrument at the GRB position, we have proposed Slewing Mirror Telescope which employs rapidly moving mirror or mirror arrays to redirect our optical beam. We describe in the following the idea and development of a fast-response optical telescope, the science and the mission of the UFFO (Ultra-Fast Flash Observatory) project, the current status of the first mission UFFO-pathfinder onboard *Lomonosov* spacecraft to be launched in November 2012, and a proposed full-scale UFFO-100 as the next step.

2. Current Limits of Rapid Response Measurements

The *Swift* observatory produces UV-optical light curves by first serendipitously detecting the onset of a GRB within the very large field of the Burst Alert Telescope(BAT)[2]. The BAT then produces a crude sky position via a standard coded mask technique. After this, the entire observatory spacecraft slews to point the UV-optical telescope (UVOT) and other instruments at the GRB position. After slewing, a period of time is required for the pointing to stabilize, after which a series of UVOT exposures begins. Though a great success of this system has been made by producing numerous detections of optical afterglows associated with GRB, only a handful of responses have occurred in less than 60 s. Due to finite mission lifetime, *Swift* cannot be expected to significantly increase this number of sub-minute responses.

In fact, one should not think of the *Swift* system as being limited to the *Swift* spacecraft. The position calculated by *Swift* within 5~7 seconds is also broadcasted over the internet via the gamma-ray coordinate network (GCN). At this point, any instrument may respond and follow up the coordinates in the easily-machine-readable format of the GCN alerts, via email or socket connection. Although the response of some instruments on ground (to name a few: ROTSE-I-III, RAPTOR, PAIRITEL, Super-LOTIS, BOOTES) is extremely rapid, e.g. 25 sec for ROTSE-III, the sensitivity is far less than that of the *Swift* UVOT. Due to their small size, and to the limitations of ground-based observing including daytime and weather, together these instruments have managed only a handful of rapid detections [3]. Because of the ability of space-based telescopes to detect photons throughout the UV-optical band without scattering or absorption, the 30 cm aperture UVOT telescope compares favorably in sensitivity to a 4-m ground-based telescope [4], and such telescopes are not capable of sub-minute s response. The slower slew times of such larger terrestrial telescopes makes them uncompetitive for the sub-

262

1000 sec regime. The *Swift* limit of 60 sec response is therefore the practical minimum for sensitive UV-optical GRB studies for the near to mid-term future. In the following we propose a new technical solution to overcome the constraints of current instrumentation.

3. Slewing Mirror Telescope

After triggering, conventional GRB observatories in space or on the ground must reorient their entire spacecraft or telescope to aim their narrow field instruments at the GRB. The time to rotate the spacecraft to slew the UV-optical telescope is the limiting factor in the *Swift* response time; not only must the entire spacecraft mass be rotationally accelerated and decelerated, but after the movement some additional time is required for any vibrations to cease.

Our approach to accelerate the slew capabilities is to redirect the optical path at an astronomical telescope via relatively or much lightweight slewing mirror rather than move the entire payload or telescope [5]. The slewing system can be either a flat mirror or mirror arrays such as MEMS (Micro-Eletro-Mechanical Systems) mirror array (MMA), mounted on a gimbal platform. In either case, large field of view (FOV) is accessible without the aberration inherent in wide-field optical systems.

Figure 1 shows the concept of Slewing Mirror Telescope (SMT)[6]. The parallel rays are directed on-axis with respect to the fixed optics by the slewing mirror system. The net effect is to steer the UV-optical instrument beam, instead of moving the telescope or the spacecraft itself. The beam can be steered by two-axis rotation of the mirror plate, rotation of the individual MMA devices, or rotation of MMA and also gimbal afterward.

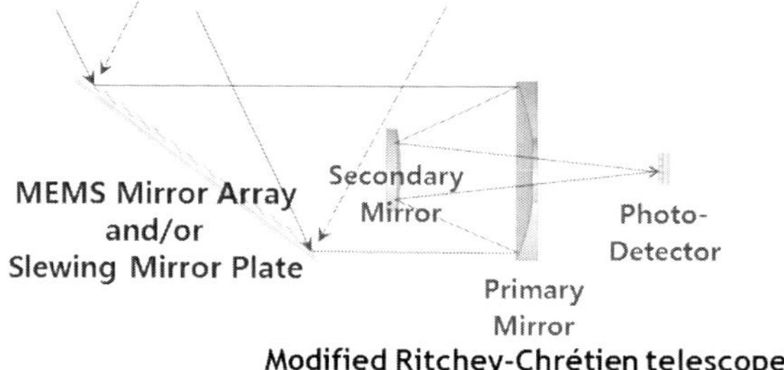

Fig. 1. A simplified schematic of the SMT showing the location of the MMA and rotating plate is given.

We find that various types of rotating mirrors move across entire field of view wider than 180° x 180°, point, and settle in less than 1 sec. In order to build a telescope with milliseconds slew speed both for *x*- and *y*-directions at a time, our lab consortium has produced small mirror arrays driven by MEMS devices. Resembling mirror segments mounted on two-axis gimbals, MEMS micormirrors are fabricated in arrays using advanced silicon and integrated circuit technologies. These MEMS mirror arrays, fabricated like other microelectronics devices, can move, point, and settle in less than a few msec with rotation angle ±15° off axis and thus FOV of 60° x 60°. Only voltages are applied to tiny electric actuators for rapid pointing to observe bursts. Such an extremely lightweight and low power device fits well to space applications, e.g. the platform of a microsatellite. A series of small prototype MMA system have been developed in our group since 2004 [7,8]. We fabricated a small prototype of 3 mm caliber telescope to demonstrate the idea of fast slewing or tracking. It was flown once in space on the ISS in 2008, and once on Tatiana-2 satellite in 2009, with excellent performance, both for nadir observation of transient luminous events occurring in the upper atmosphere [9,10].

Our simulations of our segmented MMA show that the point spread function (PSF) of SMT will have a FWHM of about 1 arcsec with the micromirrors at zero tilt. When the micromirrors are tilted, however, the PSF spreads to a FWHM of 2 arcmin due to the difference in beam path length created by tilt of elements. Therefore, we use the MMA to steer the beam to measure very early photons, starting $\sim 10^{-3}$ sec after trigger+location. At this ultra-fast mode, the mirror plate remains aimed at the middle of the field as shown in Figure 2. We perform high-resolution imaging of the source later, using the mirror plate to steer the beam (with the MMA at zero tilt). This fast mode motion takes much longer, ~ 1 sec. So, the ultra-fast mode with MMA minimizes response time, while the fast mode with rotating plate maximizes sensitivity.

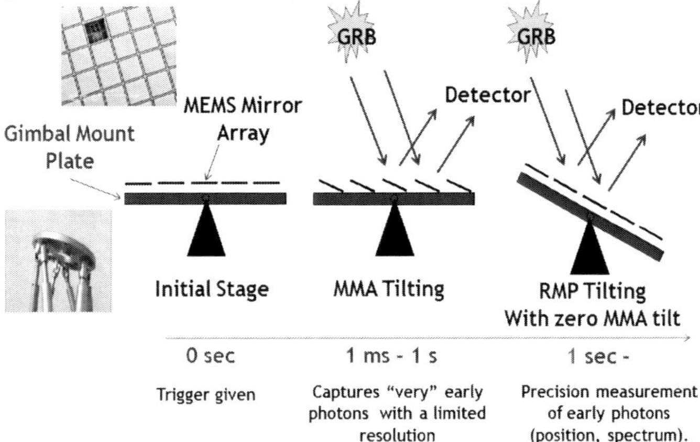

Fig. 2. In ultra-fast mode, the mirror plate remains aimed at the middle of the field while the MMA elements point at the target. In fast mode, the MMA elements are in standby or zero deflection mode, and the rotatable plate aims at the source. The first mode minimizes response time, the second maximizes sensitivity. The MMA provides ultra-fast response times, but at some cost in terms of PSF broadening. There is essentially no degradation in performance for the MMA in standby mode, and the full resolution (and therefore sensitivity) of the telescope is obtained in fast mode.

4. The Science Case for Prompt Response UV/Optical Observations

Beyond the possible physics with GRBs mentioned in the first section, the SMT offers a unique opportunity to probe a new, very early emission parameter space to thoroughly investigate the rise phase of GRB, which thus far has been observed only occasionally. A variety of rise time physics are as follows.

4.1. *Early rise of light curves*

The discovery of optical afterglows of GRB was a monumental event in modern astrophysics, ending the thirty-year mystery of the GRB distance scale. The study of GRB UV-optical afterglows and their host galaxies has led to knowledge of the origin of some types of GRB and the discovery of the most distant object known (GRB090423 at z=8.2). Much progress has been made in GRB science since the launch of the *Swift* observatory in 2004 [1]. The observations from *Swift* did not produce a simple picture of GRB, but rather documented the richness and complexity of this phenomenon. After some 370 UV-optical observations by *Swift* UV-optical telescope made to date, a huge variation in light curves has been observed, especially in the early rise time. There appear to be distinct classes of fast-rising ($t_{peak}<10^2$ sec) and slow-rising bursts [11]. Additionally, the light curves are complex, with decays, plateaus, changes in slope, and other features that are not yet understood.

It is claimed that the luminosity distribution of the fast-rising bursts at ~ 10^3 sec is quite narrow, and has promise as a kind of "standard candle" which would make GRBs useful as a cosmological probe of the very high redshift universe. In order to move this possible trend to the status of a refined tool, a larger sample of such objects is required, and in particular, better resolution is required at early times: The fastest bursts often have just one measurement in their rising phase – hardly enough to understand the physics in this regime – and many other bursts have no early measurements at all. Only less than 10 GRBs in this study were measured at less than 100 sec after their burst trigger and *not a single measurement* at less than 15 sec after trigger.

In this respect, several fundamental questions arise. Are there more features in the early light curve that are missed by such sparse sampling? Does any feature of the rise correlate with the luminosity or a particular aspect of the physics? How many bursts are misclassified because the rapid rise was missed? The need for earlier measurements (faster UV-optical response after the initial gamma-ray burst) is clear and compelling.

4.2. *Short duration GRB*

It has become apparent that there are several types of GRBs. First, GRBs have an obvious (but overlapping) separation on the spectral hardness vs. duration plane, so the basic taxonomy separates these into short and hard type (SHGRBs) less than two seconds in duration and long and soft type (LSGRBs), more than two seconds in duration. The short time scale of SHGRB emission, the associated lower luminosity and shorter time scale of the X-ray and optical afterglow lead to speculation that the two classes have fundamentally different physical origins. LSGRBs are thought to originate from the collapse of massive stars, e.g. the collapsar model[12], and SHGRBs from the merger of compact objects like neutron stars and black holes (for a review of SHGRBs, see e.g. [13]). Other types of classifications, including those with more of a physical than phenomenological motivation, have been proposed (e.g. [14]). Very short GRB (VSGRB) may originate from the evaporating Primordial Black Holes[15].

The recent progress in short-hard GRBs is extremely exciting. As of this writing, however, only about 10 short-hard GRBs had UV-optical measurements often with only one measurement above background, and thus suffer from poor time resolution in their light curves. Two measurements during the decay period are required to determine the most rudimentary decay time constant, assuming a power-law decay. The rise of SHGRB optical afterglow has never been observed. What is the shape of the rise? Is the shape homogeneous? The rise time may give rich information including the size of the system and the surrounding environment. The physical origin of this type of burst remains an outstanding mystery, so any hints as to this origin would be extremely valuable. Is there any prompt UV-optical emission from such events? What would we see if we observed more of these events in the sub-minute or sub-second regime? Are there ultra-short events on the accretion disk dynamical time scale of compact objects (that are beamed so we can see them)? Earlier observations would answer these questions and open a new window probing compact object structure, populations, and evolution.

4.3. *Dark GRB*

"Dark" GRB are those that stand out as having a very faint optical compared to X-ray afterglow. Only recently, extinction has been found to be the dominant source of dark GRB[16]. An alternative scenario, however, suggests that some Dark GRB are simply due to a faster decay for optical than X-ray emission. In this scenario, the optical emission fades in less than $\sim 10^2$ sec, so that most observations would not detect the optical afterglow. Better short time scale observations would shed light on this two-mechanism model.

4.4. *Physical time scales in compact objects*

In a more general sense, resolving the light curve peak time at any epoch gives a hint of the most important physical processes in that epoch. Coalescence of neutron star and black hole systems are features of a number of GRB models, particularly models for the less understood short GRB. The light crossing time of outer accretion disk bounds, the dynamic time scales of large accretion disk systems, and other time scales are in the sub-minute regime, requiring rapid response for their measurement. The time scales of jet formation or deceleration in these smaller systems may also be in this time regime.

4.5. *Association of emission processes by cross-correlations*

Another general tool that rapid-response observations afford is the correlation of light curves from different bands. If complex light curves in different bands have clear correlation, this is a very strong argument for a physical linkage between the processes of emission in the two wave bands. The delay between the light curves gives further information about both processes. The γ-ray light curves have structure at every time scale investigated, up to ~10's of minutes. X-ray light curves at greater times, however, don't show much structure; all the action is at early times. Again referring to the correlation of rapid-response light curves referred to earlier, it is intriguing that some of the most early UV-optical light curves show good correlation to their γ light curves, yet others do not. Is this a clue to additional processes, or a hint that the origin of these GRBs is quite different, a la SN Ia vs. Ib? What will we see if we can extend these correlations of early emission to SHGRB?

4.6. *Association of emission processes by spectral slope*

The emission spectra of GRB are featureless power laws. However, the spectra evolve in time. Chromatic and achromatic jet breaks are important predictions/distinguishing features of models. One feature is the well-known transition from relativistic to non-relativistic emission, the transition from "prompt" emission to afterglow. The change in spectral slope, and the time of this change, are therefore important diagnostics of the interaction of the jet and the surrounding medium, and/or injection of additional energy into the jet. The broad-band spectral slope itself is a discriminator of the electron energy distribution, magnetic field, and other features of the emission mechanism.

4.7. *Test of shock models with bulk Lorentz factor*

Measurement of early UV-optical emission can serve as a probe of the physical conditions in the GRB fireball at short times. A simple, nearly model-independent argument[17], shows that the bulk Lorentz factor depends on the time of the early UV-optical emission peak. Measurement of the peak will therefore provide a measurement of the bulk Lorentz factor.

4.8. *Identification of internal shock via fast variability*

Currently, UV-optical emission at early times in typical bursts is believed to come from external shocks, and predicted to have a smooth, monotonic rise (e.g. see [18] and references therein). Observation of an early time UV-optical light curve that more closely resembles a gamma-X light curve, jagged, and with multiple peaks, would clearly indicate the presence of internal shock produced prompt emission in this band. Sub-minute measurements would be required to learn more about such prompt emission. What are the prospects for pushing to shorter time scales in GRB measurements?

4.9. *Tie-in to compact objects and gravitational wave observations*

The upcoming generation of gravitational wave observatories should regularly detect the coalescence of binary compact systems, the favored scenario for SHGRBs. This will open an entirely new field of astronomy. In part because of its novel nature, corresponding UV-optical measurements will be highly important to interpret the astrophysics of the event. Moreover, while gravitational wave signals from binary system inspirals have the potential to yield highly accurate distance measurements, they alone cannot break the degeneracy in parameters to yield the redshift – this requires observation of an electromagnetic counterpart. Fast-response optical observations can test Lorentz violations from the time delay between different energy photons, or between photons and electromagnetic counterpart of neutrinos or early emission with GW. Such a fast-response would be essential for deep understanding of compact objects and cosmology [19].

4.10. *Time-Evolution of local dust*

Extinction is extremely important in GRB studies, and later we discuss untangling effects of extinction and low intrinsic luminosity. Through the association of dust, GRB are probes of star formation to the highest observable red shifts. However, can the modeled roles of circumstellar dust vs. non-local host galaxy associated dust be separated? In fact yes, if high time-resolution observations are available to observe dust destruction. This is predicted on ~100 sec time scales, and clearly shows the role of local dust.

5. The UFFO Program

The UFFO (Ultra-Fast Flash Observatory) is to pioneer the hitherto unexplored time domain nature of GRB by using the concept of SMT's fast or ultra-fast slewing mirror technology. It will respond to initial photons within a fraction of a second after the burst of GRB. The UFFO project will be carried out in a series of relatively light payloads to be adaptable readily to micro or small satellites. The first is a small payload, UFFO-Pathfinder, which will be flown aboard the *Lomonosov* spacecraft by November 2012. The next serious version, UFFO-100 with its payload mass of 120kg and 30cm telescope aperture, is foreseen to be in space in 2015. The UFFO-100 will extend its measurement capability to near-IR (NIR) and polarization using dichroic beam splitter on the SMT optics bench. We will demonstrate that such a small mass of payload is sufficient to make major advances in GRB science.

A peculiar property of GRBs makes relatively small X-ray trigger instruments possible, the GRB distribution in flux. Typical astrophysical sources have a number count vs. flux distribution close to a uniform Euclidian distribution, with a sharp cutoff caused by the sensitivity of the instrument. GRBs, in contrast, are so bright that they are limited by the volume of the observable universe since star formation, rather than the sensitivity of our largest instruments. This means that for simple location of a GRB (not detailed study of the X-ray through gamma-ray spectrum), a small instrument can locate a large fraction of those detected by much larger instruments. This property of GRB made the UFFO concept possible. It also makes possible very advanced missions improving on the UFFO-Pathfinder with relatively modest size. Indeed, the GRB detection rate of UFFO-Pathfinder will be about 60% of *Swift*, and the sub-minute measurements of the UV-optical emission of dozens of GRB each year provide the first detailed measurements of fast-rise GRB optical light curves, and help verify the prospect of GRB as a new standard candle.

In the next decade, *Swift* will likely cease operations, and the UFFO will continue to provide numerous GRBs for follow up studies by the optical ground-based multi-wavelength communities, providing future possibilities for understanding GRBs of all types, and giving hope for their use as extreme $-z$ cosmological probes. If other *Swift*-like GRB missions were to fly soon in the future, their sensitivity would complement UFFO's rapid response.

6. UFFO-Pathfinder

The system of the UFFO-Pathfinder was designed to (i) fit the constraints of the *Lomonosov* spacecraft, (ii) use all pre-proven technologies and (iii) to be available for fast delivery. The main constraints for inclusion in *Lomonosov* are 20 kg total instrument mass and 800 cm maximum length. The payload consists of two instruments: SMT (Slewing Mirror Telescope) for rapid coverage of UV-optical sky and UBAT (UFFO Burst Alert Trigger) for X-ray triggers. We have designed a small telescope to provide imaging measurements by employing a gimbal beam-steering system in SMT which can point to the event within 1 sec after localization of GRBs with X-rays. The X-ray trigger system, UBAT, is a wide-field coded mask camera and essentially scaled down version of the *Swift* BAT mainly to fit the available mass and size requirements.

The UFFO-Pathfinder was built and went through space environments test, including thermal, vacuum, shock, and vibrations, successfully at National Space Organization of Taiwan (NSPO) in August 2011. The final integration of the Flight Model (FM) to the *Lomonosov* spacecraft and space environments test is currently under way at a branch of Roscosmos (see Figure 3).

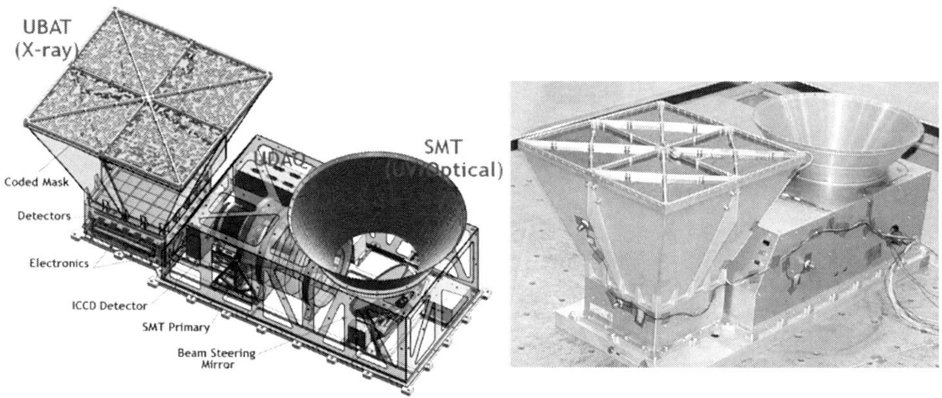

Fig. 3. A rendering of integrated UFFO Pathfinder (left) and fabricated flight model (right).

6.1. *SMT*

As a subsystem of UFFO, the Slewing Mirror Telescope is designed for fast observation of the prompt optical/UV photons from GRBs. The SMT uses a gimbal system which provides 1 sec response over the entire FOV of UBAT, $90° \times 90°$. Electric motors driving gimbal-mounted mirrors are a fundamentally simple and robust technology. For UFFO-Pathfinder we used off-the shelf encoders and motors with sealed bearing systems and have already obtained sub-arcsecond settling over +/- 90 degrees with t < 1 sec travel + settle time.

The SMT optics includes a Ritchey-Chretien telescope with a 100 mm diameter aperture. Its field of view is 17 x 17 arcmin^2 and f-number is 11.4. The primary and secondary mirrors were fabricated with the precision of about RMS 0.02 waves in wave front error (WFE) and 84.7% in average reflectivity over 200~650 nm. The entire SMT optics was aligned at the accuracy of RMS 0.05 waves in WFE at 632.8 nm.

The focal detector is an Intensified Charge-Coupled Device (ICCD) with a pixel size of 4 x 4 arcsec^2 and a wavelength sensitive to 200~650 nm. The ICCD operates in photon counting mode and observes faint objects up to 19.11 magnitude B-star in white light per 100 msec. The SMT has the readout rate of 4 msec and can take 250 frames per second.

One of non-trivial issues on UFFO-Pathfinder and SMT structures was the reduction of mass, while satisfying structural and functional requirements related to stiffness, strength, dimensional restriction and thermal conduction. The housing was made of carbon fiber. Mirrors and substructure were built through finite element analysis and lightweight engineering. Two identical flight models of SMT shown in Figure 4 has been built and delivered. More details of UFFO SMT is presented in [20].

Fig. 4. The SMT of the UFFO-Pathfinder. The opto-mechanics of 10 cm Ritchey-Chretien telescope, two-axis gimbal mirror, and readout electronics, are shown, together with UDAQ box and UBAT detector module. The settle time of the slewing mirror is 120 msec for < 1" point spread function.

6.2. *UBAT*

Numerous instruments have used coded-mask aperture shadow cameras (e.g. *Beppo-Sax*, *HETE-2*, *Integral*, BATSE, and *Swift*) to determine positions of GRBs. We use the most recently launched *Swift* observatory as the benchmark of our X-ray system. The *Swift* BAT can localize bursts at 90% probability to a region ~17' across, the FOV of its UV-optical camera. Note that, as with most GRB instrument figures of merit, the accuracy applies to some fraction of the brightest bursts, and will vary greatly with more faint or rapid bursts. Typically, a larger number of counts will be observed in a 48 msec time bin, tripping a "rate trigger" in the BAT. The instrument will acquire data for some additional time until enough counts are recorded to determine a position. Approximately 6~7 seconds of on-board computing, involving Fourier Transforms, are then performed to determine the position. At this time, the entire spacecraft is slewed to the calculated position, and then a series of exposures varying in time and choice of filter wheel are then performed.

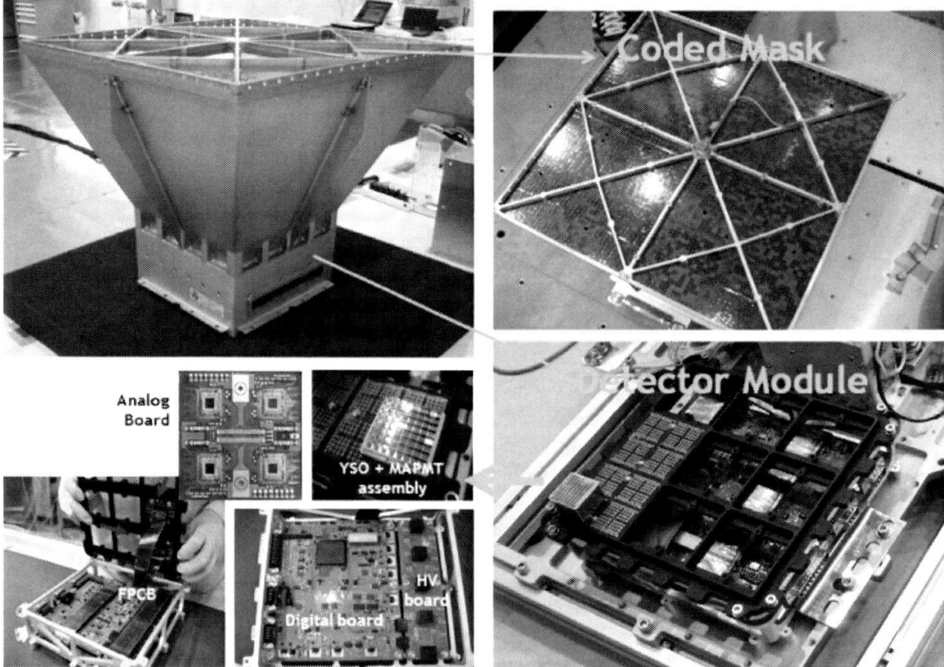

Fig. 5. Fabricated flight model of UBAT. It consists of coded mask(upper right figure), hopper and shielding (upper left), detector module (lower right). Part of YSO crystals, MAPMTs and readout electronics are shown in lower left figure.

Our science goals and objectives all stem from a large sample as much as possible of homogenously selected GRB. This translates to an X-ray instrument capable of detecting tens of GRB per year, and precise localization of GRBs and rapidly triggering the UV-IR observations. Though there are worlds of science in additional information from the high-energy emission of GRB, we make no specific requirements on energy resolution or flux measurements. We require that the instrument provide only rough flux and duration information for each burst. We anticipate that *Fermi* with extremely wide sky coverage and high sensitivity and other instruments like *Suzaku* will provide additional flux and duration information for a large fraction of bursts.

With the time constraint to meet the launch schedule as well as mass and power constraints for UBAT (only approximately 10 kg and 10 W), we adopted well established coded-mask technique similar to *Swift* BAT but scaled down for the localization of bright, transient X-ray sources. With only 196 cm^2 of detecting area, our collaboration has made a viable camera with which we expect to detect dozens of GRB/year. In order to respond over a wider energy range, e.g. 5~200 keV, however, we used pixellated YSO scintillating crystal red out by 36 64-ch multi-anode photo multipliers (MAPMTs) with 36 64-ch SPACIROC ASICs. The resulting sensitivity is 310 mCrab in 10 sec at 5 σ. Figure 5 shows the integrated UBAT system.

6.3. *UDAQ*

The UFFO Data Acquisition (UDAQ) is in charge of: central control of the payload not only with preset commands but upload commands from the ground; interfacing to the spacecraft; data collection from SMT and UBAT, storing in several NOR flash memories and transfer to the spacecraft. It is also responsible for: monitoring of all housekeeping parameters; calculation of the orbit and recognition of day and night with its photosensors; arbitration and prioritization of triggers from UBAT and BDRG (another *Fermi*-like payload of *Lomonosov*); power management, etc. All of these functions are implemented in a ACTEL field programmable gates arrays (FPGA) for the low power consumption and fast real-time processing. Entire trigger calculations with the data from UBAT, including rate trigger and imaging trigger, are also performed in a way of pipelining in another ACTEL FPGA, which reduces the latency significantly.

7. UFFO-100

Awaiting the completion and launch of the UFFO-Pathfinder, the UFFO collaboration has been exploring its next step, a more ambitious project: UFFO-100 (named indicating the mass of payload), based on the same design principle but with total mass larger than 100 kg.

The great instrumental challenge of the UFFO concept is to see changes in the optical light curve on short time scales which requires short exposures Therefore, the aperture size of the instrument is the fundamental limitation on both the total number of GRB that may be detected, and the time resolution. GRB gamma-ray light curves, even the longer-duration class, have high amplitude variability at every observed time scale. Comparison of the variability between the gamma-X bands and the optical bands can tell us a great deal about the emission physics at the source. Thus far, with the most rapid optical measurements available, it is not known whether gamma-X and optical emission correlates, has lags, or perhaps correlates only in certain types of bursts. There is simply not enough short time scale data. The UFFO-100 will answer such an intriguing question of "What would *Swift* have seen if it could have responded faster?", with the slewing mirror telescope of an aperture 30 cm as large as that of *Swift*, but with several enhancements to make it even more sensitive and productive, enabling detections at even shorter time resolution.

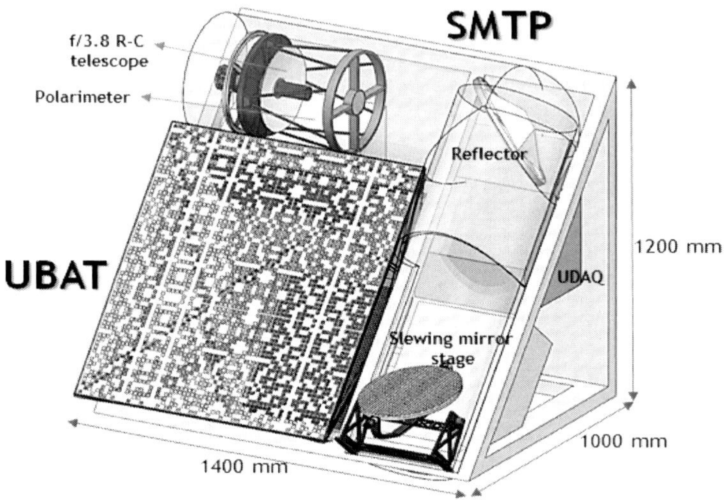

Fig. 6. Schematic of UFFO-100.

Though some enhancements may be restricted by the precise restrictions of payload, UFFO-100 would afford a 1024 cm^2 X-ray camera but improved detector technology. The goal is to finally integrate the MMA technology with the motorized slewing mirror and to add an NIR-sensitive camera and polarimeter to detect the distinguished bursts. The UV/optical and NIR cameras, both with 17' fields, use the incoming beam from the SMT after being split from a dichroic. Much of the instrumentation, particularly the electronics, will be built on the heritage of UFFO-pathfinder. The pathfinder basic telescope design, fast-mode beam steering, spacecraft bus interface, and data acquisition system architecture will be shared with UFFO-100. The key components and the dimensions of UFFO-100 are shown in Figure 6. We expect UFFO-100 to be flown as one of the scientific payloads of the Russian Resurs-P3 satellite in 2015.

8. Summary

We propose space missions of Ultra-Fast Flash Observatory (UFFO) for the investigation of a new area of gamma-ray burst phase space both quantitatively and qualitatively. The

UFFO equipped with Slewing Mirror Telescope (SMT) has an extraordinary capability by permitting the first ever systematic study of GRB UV-optical-NIR emission far earlier than 1 sec after trigger. Our fundamental science objective is to use our ability to probe this new, very early emission parameter space to make measurements of and thoroughly investigate the rise phase of GRB, which is thus far only occasionally observed. In the time domain, this improves on *Swift*'s response by several orders of magnitude. In the spectral domain, we will improve on *Swift*'s sensitivity by ~2.5 mag, and we expect to detect afterglow components that are invisible to *Swift* because of extinction.

The UFFO-Pathfinder has now entered the final stage of completion, heading for launch onboard *Lomonosov* satellite in November 2011. The pathfinder is a small and limited, yet remarkably powerful micro-observatory for rapid optical response within 1 sec after X-ray trigger to bright gamma-ray bursts. Its far sub-minute measurements of the optical emission of dozens of GRB each year will result in a more rigorous test of current internal shock models, probe the extremes of bulk Lorentz factors, provide the first early and detailed measurements of fast-rise GRB optical light curves, and test the prospect of GRB as extreme z cosmological probes. We foresee not only its exciting findings but the proof-of-principle for this new approach to future GRB telescopes.

References

1. N. Gehrels *et al.*, ApJ **611,** 1005 (2004).
2. S.D. Barthelmy, Proc. SPIE **5165**, 175 (2004).
3. C. Akerlof *et al.*, Nature **398**, 400 (1999).
4. P. Roming, *et al.*, Space Sci. Rev. **120**, 95 (2005).
5. I.H. Park, Nucl. Phys. B Supp. Proc. **134**, 196 (2004).
6. I. H. Park *et al.*, arXiv:0912.0773.
7. J. H. Park *et al.*, Optics Express **16,** 25, 20249 (2008).
8. M. Kim *et al.*, J. Micromech. Microeng **19**, 035014 (2009).
9. S. Nam *et al.*, Nucl. Instrum. Meth. A **588**, 197 (2008).
10. B.W. Yoo *et al.*, Optics Express **17**, 5, 3370 (2009).
11. A. Panaitescu and W. Vestrand, MNRAS **387**, 497 (2008).
12. A. I. MacFadyen and S. E. Woosley, ApJ **524**, 262 (1999).
13. E. Nakar, Physics Reports 442, 166–236 (2007). arXiv:astro-ph/0701748.
14. Norris & Bonnell, ApJ 643, 266 (2006).
15. D.B. Cline, S. Otwinowski, B. Czerny, A. Janiuk, arXiv:1105.5363.
16. D.A. Perley *et al.*, AJ 138, 1690 (2009).
17. E. Molinari *et al.*, A&A 469, 13 (2007).
18. T. Piran, Reviews of Modern Physics **76**, 1143 (2004).
19. L. Stoldosky, Phys. Lett. B. **473**, 61 (2000).
20. J.W. Nam *et al.*, this proceedings.

THE UFFO SLEWING MIRROR TELESCOPE FOR EARLY OPTICAL OBSERVATION FROM GAMMA RAY BURSTS

JIWOO NAM

Graduate Institute of Astrophysics & LeCosPA Center, National Taiwan University, 1 Roosevelt Road Taipei, 10617, Taiwan
E-mail: jwnam@phys.ntu.edu.tw

For The UFFO Collaboration

S. AHMAD[1], K. AHN[2], P. BARRILLON[1], S. BRANDT[3], C. BUDTZ-JØRGENSEN[3], A.J. CASTRO-TIRADO[4], S.-H. CHANG[5], C.-R. CHEN[5], P. CHEN[6], Y.J. CHOI[7], P. CONNELL[8], S. DAGORET-CAMPAGNE[1], C. EYLES[8], B. GROSSAN[9], M.-H. A. HUANG[10], J.-J. HUANG[6], S. JEONG[11], A. JUNG[11], J.E. KIM[11], S.H. KIM[2], Y.W. KIM[11], J. LEE[11], H. LIM[11], C.-Y LIN[5], E.V. LINDER[10,11], T.-C. LIU[6], N. LUND[3], K.W. MIN[7], G.W. NA[11], M.I. PANAYUK[12], I.H. PARK[11], J. RIPA[11], V. REGLERO[8], J.M. RODRIGO[8], G.F. SMOOT[9,11], S. SVERTILOV[12], N. VEDENKIN[12], M.-Z WANG[6], I. YASHIN[12], and M.H. ZHAO[11]

[1]University of Paris-Sud 11, France, [2]Yonsei University, Seoul, Korea,[3]National Space Institute, Denmark [4]Instituto de Astrofísica de Andalucía, Consejo Superior de Investigaciones Científicas, Spain, [5]National Space Orgarnization, Hsinchu, Taiwan, [6]National Taiwan University, Taipei, Taiwan, [7]Korea Advanced Institute of Science and Technology, Daejeon, Korea, [8]University of Valencia, Spain, [9]University of California, Berkeley, USA, [10]National United University, Miao-Li, Taiwan, [11]Ewha Womans University, Seoul, Korea, [12]Moscow State University, Moscow, Russia

While some space born observatories, such as *SWIFT* and *FERMI*, have been operating, early observation of optical after grow of GRBs is still remained as an unexplored region. The Ultra-Fast Flash Observatory (UFFO) project is a space observatory for optical follow-ups of GRBs, aiming to explore the first 60 seconds of GRBs optical emission. Using fast moving mirrors to redirect our optical path rather than slewing the entire spacecraft, UFFO is utilized to catch early optical emissions from GRB within 1 sec. We have developed the UFFO Pathfinder Telescope which is going to be on board of the *Lomonosov* satellite and launched in middle of 2012. We will discuss about scientific potentials of the UFFO project and present the payload development status, especially for Slewing Mirror Telescope which is the key instrument of the UFFO-pathfinder mission.

Keywords: Gamma Ray Bursts; After Glow; Fast Slewing.

1. The Ultra Fast Flash Observatory

The discovery of afterglow of the Gamma Ray Burst (GRB) was carried out by a follow-up observation of the William Herschel Telescope in 1997 [1]. It was 20 hours after the GRB 970228 detected by *BeppoSAX* with a fast localization capability [2].

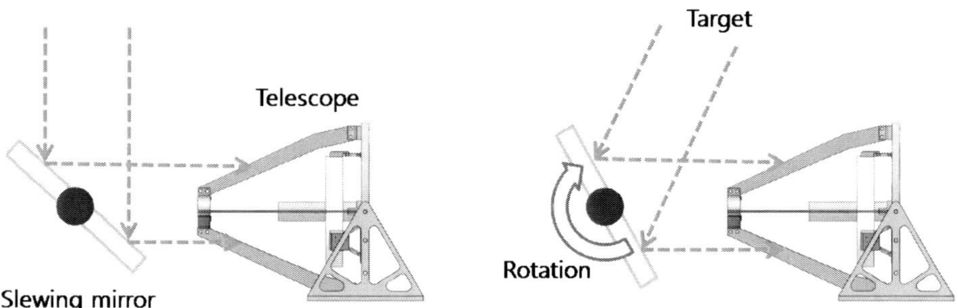

Fig. 1. UFFO Concept.

Scientists' dream to capture early optical emissions from GRBs was realized by *Swift* which was launched in 2004. It has the unique capability of in-site flow-up measurement in the Space [3] rather than just sending the localization information to the ground telescope for flow-up observations. As soon as the Burst Alert Telescope determines the localization of the GRB event, the space craft rotates toward the GRB direction to allow flow up measurements of UV/Optical and X-ray telescopes in narrow field of views (FOV). However, this method takes time longer than 60 seconds, which is limited by the rotation speed of the space craft and computing process for localization.

The Ultra Fast Flash Observatory (UFFO) [4] is a new concept of GRB telescope with a fast slewing capability exploring first 60 seconds of optical emissions from GRBs. Instead of rotating entire space craft, a slewing mirror in front of an optical telescope rotates to redirect beam path (see Figure 1). The smaller moment of inertia of the slewing mirror than the space craft, the faster slewing (within 1 second) is achievable with a limited power budget in the space mission. An advanced slewing method based on Micro Electro Mechanical System (MEMS) Micro Mirror Array [4, 5] has been proposed for a sub millisecond slewing time. However, a technical challenge still remains especially in

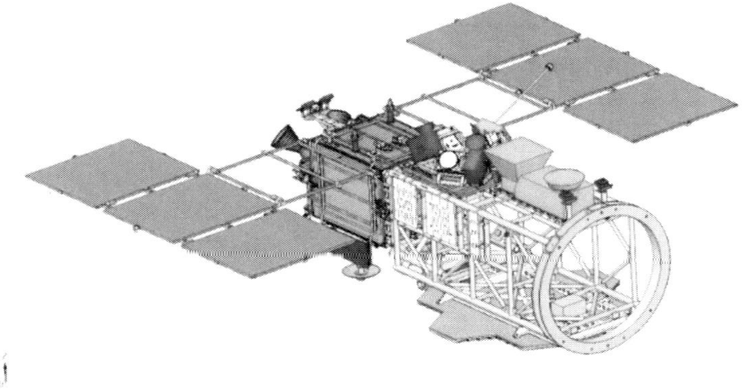

Fig. 2. *Lomonosov* satellite. The UFFO-Pathfinder payload is seen on the top of the Satellite in the Orange color.

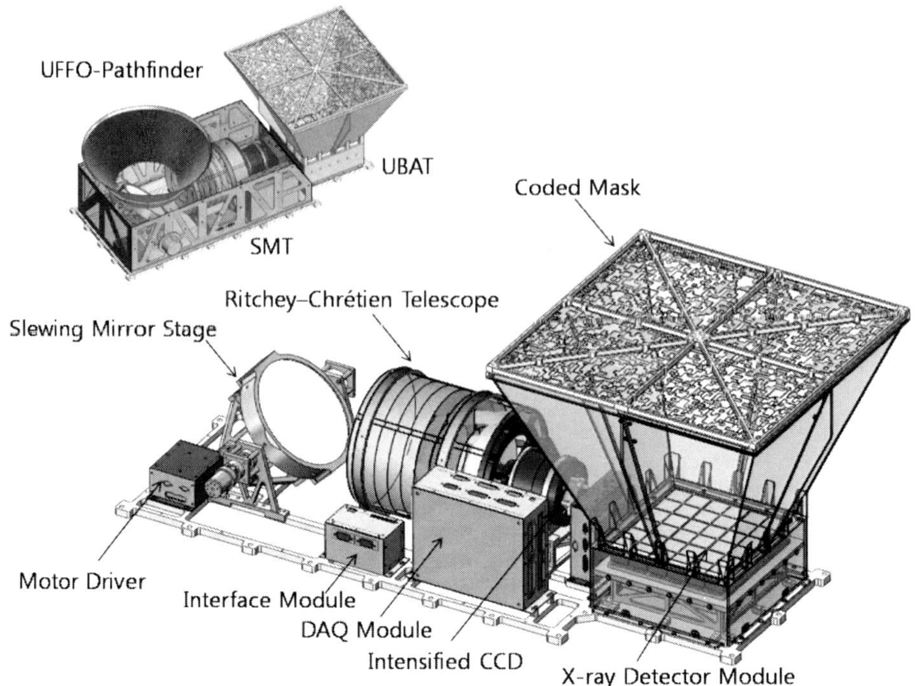

Fig. 3. UFFO-Pathfinder payload.

precise angle control of individual micro-mirrors. For the first step, we have developed the UFFO-Pathfinder on board of the *Lomonosov* mission (See Figure 2) which is going to be launched in the middle of 2012. Since the UFFO joined the mission late as a piggy back payload on the TUS telescope [6], only limited mass (20 kg) and power (20 W) were allowed. Despite of its small size, UFFO-pathfinder will prove the concept and possibly bring interesting results.

There are several interesting science potentials with the fast slewing capability of UFFO. As Panaitescu and Vestrand suggest, the optical luminosity of GRBs would be calibrated by rising times of light curves [7]. However, their claim still needs to be confirmed with a larger number of events containing early parts of light curves. If we are able to calibrate GRBs, we will have another cosmological standard candle extending the Hubble diagram up to the highest red shifts. As a matter of fact, *Swift* was not utilized to observe optical emissions in a short time scale less than 60 sec. On the other hand, tests of prompt optical emissions from short-hard GRBs and furthermore any optical emissions from dark burst are outstanding topics with the fast response of UFFO. Optical emission is widely believed to come from the external shock while X-rays from internal shocks. Comparing these two light curves in early time scale, UFFO would provide important keys to understand optical emission mechanisms [4, 8].

Figure 3 shows the overall structure of UFFO-Pathfinder. It consists of two instruments, one is UFFO Burst Alert Telescope (UBAT) [9] and the other is Slewing Mirror Telescope (SMT). UBAT is a 90°x90° wide FOV coded mask X-ray camera to trigger GRB events, as well as to determine their localization information. The coded-aperture mask made of Tungsten has randomized rectangular hole patterns with 50% open fraction. We used 48x48 YSO crystal array attached onto 36 units of 8x8 Multi Anode Photo Multipliers. The localization resolution of UBAT is better than 10 arcmin with an energy range of X-rays 10-100 keV. A fast imaging process runs on a Field-programmable gate array (FPGA) chip checking significance of signal in the x-ray image to issue trigger and localization information in every second. We describe SMT in next section.

2. Slewing Mirror Telescope

SMT is the key concept of the UFFO aiming to explore the sub-minute regime of optical emissions from GRBs. SMT consists of three main instruments; Slewing Mirror Stage (SMS), Ritchey–Chrétien Telescope (RCT), and a photo detector. Important parameters of SMT are listed in Table 1.

Table 1. SMT Characteristics.

Instrument Type	RCT with Slewing Mirror
Mass	11.5 kg
Power Consumption	Average 10 Watt
RCT Aperture Size	D=10 cm
F-Number	11.4
RCT FOV	17 arcmin x 17 arcmin
Slewing Mirror Size	D=6 inches
Motor Rotation Minimum Step	4.05 arcsec (Mechanical)
Slewing Speed	15 deg/sec (Mechanical)
ICCD Gain	10^3-10^6
ICCD Number of pixels	256 x 256
ICCD Quantum Efficiency	5-20 % in 200-650 nm
CCD Dynamic Range	62 dB
Image Frame Time	2 ms

The SMS is a two-axis Gimbal mirror located in front of RCT (Figure 4). The FOV of RCT is 17 arcmin x 17 arcmin. The slewing capability extends the SMT sky coverage up to 70 arcdeg x 70 arcdeg. Because of the space limitation, the motor and the support structure of the inner axis are located below the mirror. A 6 inch Zerodur substrate of the mirror was light-weighted to be 482 g. Silicone pads are used for mirror mounting in the support ring. Average reflectivity is about 85% in a range of 200-700 nm wave length. We use stepping motors with 200 steps/rev and 100:1 Harmonic Drive reducers. The minimum step of the mirror rotation is about 4 arcsec using the 1/16 micro-stepping

Fig. 4. Two-axis Motorized Slewing Mirror Stage. Motor of inner axis of the Gimbal is located under the mirror.

Fig. 5. Ritchey–Chrétien Telescope.

control technique. The speed of motor rotation is faster than 15 deg / sec so that objects in the sky coverage can be targeted within 1 sec. High precision rotary encoders are employed for the close loop control. Obtained targeting resolution is better than 2 arcmin. The power consumption of SMS is less than 3 Watt.

The RCT is designed to be light weight (~1 kg) and small size (180x180x250) with a D=10 cm aperture and F number 11.4. Figure 5 shows the opto-mechanical structure of RCT. The rear part is the main support plate of RCT holding three bi-pod flexures of the

Fig. 6. SMT Thermal-Vacuum Test in NSPO.

primary mirror (M1) and four spiders of the secondary mirror (M2). The obscuration ratio is 12.5%. With two triangular side brackets, only the main support plate is fixed at the SMT base plate to minimize optical miss-alignments due to non-uniform mechanical- and thermal- stress on the base plate. Reflectivity of mirrors is the same as the slewing mirror. Integrated system has 1/20 λ Wave Front Error (WFE) at 632.8 nm wavelength [10].

We use an Intensified Charged Coupled Device (ICCD) for the photo detector. The ICCD consists of 256 x 256 monochromic pixels corresponding 4 arcsec of pixel FOV [11]. Quantum efficiency of the ICCD is about 20%. The gain of the intensifier is adjustable in a range of 10^3-10^6. All SMT functions such as motor control, ICCD control/readout and data transfer are controlled in FPGA logic.

The overall shape of the SMT is a box with a truncated cone shade. The box cover consists of duralumin skin and polycarbonate support frame. Multi Layer Insulator covers the entire SMT box. The inside of the SMT cover is black-painted to reduced scattered light background. The surface of the shade is precisely lathed to be saw tooth grating patterns to increase reflectivity of undesired lights such as the moon lights. There are three electronics sub-units inside the SMT enclosure; the DAQ/Power, the SMT control-readout, and the Motor control. The base plate is the mechanical support structure of all sub-components of SMT as well as UBAT, and also plays an important role of thermal conduction to the platform.

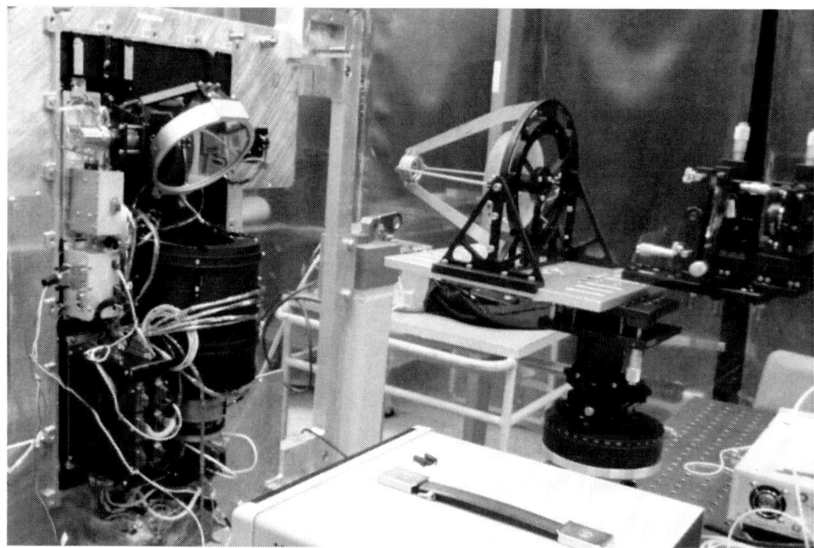

Fig. 7. System Verification during the final integration in Istra, Russia. The UFFO is vertically oriented. The other RCT in the right side is used to produce a parallel beam.

3. Tests & Readiness

A Series of Thermal-Vacuum (TV) and Shock-Vibration (SV) tests were performed at the National Space Organization (NSPO), Taiwan, in July 2011. The main goal was to verify the UFFO system functionalities under the TV and SV conditions mimicking experimental environment of the space mission. In a TV chamber (See Figure 6), UFFO was exposed to several thermal cycles the temperature range of -30 ~ 40°C for about 30 hours under the vacuum pressure of ~10^{-7} mbar. The SV tests followed one week after the TV test. UFFO experienced shocks of 45g for 3 msec, as well as vibration 1-9.5 g in the frequency range of 5-2000 HZ in three axes. All records of temperature- and acceleration-profiles on critical points such as electronics, SMS, and opto-mechanical structures were found to be healthy during tests. No mechanical / electrical failure was found in the tests, neither no out-gassing vestige. The WFE of RCT measured right after the SV test was satisfactory.

The UFFO was finally integrated at Istra, Russia in April 2012. Before the delivery, a full system validation tests were performed a full system validation test (See Figure 7). With commands sent through the interface module, SMT showed successful performances of slewing toward known directions and image taking. Optics collimation and the ICCD performance were tested with a parallel beam located in the target direction. Preliminary data analysis of obtained images shows the system Point Spread Function (PSF) to be 3.4 arcsec which is consistent with the PSF of ICCD.

4. Summary

To explore the first 60 seconds of optical emissions from GRBs, we proposed the new approach of GRB space telescope with a rapid slewing capability. We have completed the development of the UFFO-pathfinder payload during the past two and half years. The SMT has been delivered to Russia, and the UBAT will be delivered before June 2012. With the small size of payload, about 40 events per year are expected to be triggered by UBAT [8]. Among these events, only several events with accurate localization information could be captured by SMT. Although the number is not great, even just handful number of events would be enough to prove the UFFO concept, and possibly bring key messages from cosmological distances. The design of the full UFFO mission is already in progress. The full UFFO requires more than 100 kg of mass with a D=20-30 cm aperture of optical telescope. Several additional new instruments are proposed such as such as IR camera, spectrometer, and polarimeter. The full UFFO mission is targeted to be launched around 2015. Our accomplishment with the UFFO-pathfinder will be a crucial milestone for the Full UFFO.

References

1. Van Paradijs, J., et al., *Nature,* **386**, 686-689 (1997).
2. G. Boella, et al., *Astronomy & Astrophysics Supplement Series*, **122**, 299-307 (1997).
3. Gehrels, N., et al., *Astrophysics Journal*, **611**, 1005-1020 (2004)
4. I. Park, et al., arXiv:0912.0773
5. Park, J.H., et al., Optics *Express* **16**, 25, 20249 (2008).
6. V.V. Alexandrov, et al., *Proceedings of ICRC*, (2001)
7. Panaitescu, A. & Vestrand, W., *MNRAS* **387**, 497 (2008)
8. Pisin Chen, et al., *Proceeding of ICRC*, arXiv:1106.3929, (2011)
9. A. Jung, et al., *Proceeding of ICRC*, arXiv:1106.3802, (2011)
10. S. Jeong, et al., *Proceeding of ICRC*, arXiv:1106.3850, (2011)
11. J. E. Kim, et al., *Proceeding of ICRC*, arXiv:1106.3803, (2011)

NEUTRINO ASTRONOMY AT THE SOUTH POLE

ALBRECHT KARLE

*Department of Physics and the Wisconsin IceCube Particle Astrophysics Center,
University of Wisconsin-Madison, Madison, Wisconsin 53711, USA
E-mail: karle@icecube.wisc.edu*

The origin of highest energy cosmic rays remains unresolved. High-energy neutrinos may provide the clues to fundamental phenomena such as the origin of cosmic rays or dark matter in the Universe. The IceCube Neutrino Observatory, a km scale neutrino detector, has come into full operation in 2011. At the highest energy levels, prototypes of a new experiment, the Askaryan Radio Array, have been deployed and are being tested. We report on the status, first results and prospects of the experimental neutrino searches under way and planned at the South Pole.

Keywords: Neutrinos; Cosmic rays; neutrino telescopes.

1. Neutrino Astronomy

Highest energy cosmic rays provide evidence of the existence of powerful accelerators in the Universe. Cosmic particles have been observed at energies beyond 10^{20} eV. One hundred years after the first discovery of cosmic rays, their origin remains largely unresolved. Point sources of high-energy gamma rays have been observed up to energies of ~100 TeV. They are providing evidence of accelerators of energetic radiation. Yet, the connection to the cosmic rays of higher energies is unclear and the high-energy photon view to the Universe is blocked due to interactions with low energy photons, microwave background and extragalactic photon backgrounds.

Neutrinos may traverse the Universe even at the highest energies. Cosmic ray interactions, either in the accelerator regions or at higher energies on the mentioned photon background, will inevitably produce neutrinos at some level. Thus, neutrino astronomy may provide the missing clues to uncover the origins of cosmic rays.

The interactions of ultrahigh-energy cosmic rays with radiation fields or matter, either at the source or in intergalactic space, result in a neutrino flux due to the decays of the produced secondary particles such as pions, kaons and muons. The observed cosmic ray flux sets the scale for the neutrino flux and leads to the prediction of event rates that require kilometer scale detectors, (see for example Ref. 9). As primary candidates for cosmic ray accelerators, AGNs and GRBs are thus also the most promising astrophysical point source candidates of high-energy neutrinos. Galactic source candidates include supernova remnants, microquasars, and pulsars. Guaranteed sources of neutrinos are the cosmogenic ultrahigh-energy neutrino flux from interactions of cosmic rays with the cosmic microwave background and the galactic neutrino flux resulting from galactic cosmic rays interacting with the interstellar medium. Both fluxes are small and their measurement constitutes a formidable challenge. Other sources of neutrino radiation include dark matter, in the form of supersymmetric or more exotic particles and remnants from various phase transitions in the early Universe.

The relation between the cosmic ray flux and the atmospheric neutrino flux is reasonably well understood and is based on the standard model of particle physics. The observed diffuse neutrino flux in underground laboratories agrees with Monte Carlo simulations of the primary cosmic ray flux interacting with the Earth's atmosphere and producing a secondary atmospheric neutrino flux.

Although atmospheric neutrinos are the primary background in searching for astrophysical neutrinos, they are very useful for two reasons. Atmospheric neutrino physics can be studied up to PeV energies. The measurement of more than 50,000 events per year in an energy range from 500 GeV to 500 TeV make IceCube a unique instrument for making precise comparisons of atmospheric neutrinos with model predictions that are sensitive to interaction properties, primary cosmic ray flux. At energies beyond 100 TeV a harder neutrino spectrum may emerge which would be a signature of an extraterrestrial flux. From an astrophysical perspective, atmospheric neutrinos also provide an opportunity to calibrate the detector. The absence of such a calibration beam at higher energies poses a difficult challenge for detectors at energies targeting the cosmogenic neutrino flux.

Because of the enormous target masses and areas needed, only natural target media are realistic. One hundred years after its discovery by Roald Amundsen, the South Pole has emerged as a premier location of astronomy for cosmic microwave photons and high-energy neutrinos. For neutrino detection the cold glacial ice has become the medium for both optical- and radio-based detection methods.

While nature offers the target and radiating medium for free, the technical challenges to build an experiment at the South Pole are enormous.

2. The Optical Detection of Neutrinos – IceCube

On December 18, 2010 the last cable of IceCube was lowered into position, thus completing construction of the first kilometer-scale neutrino detector. The full IceCube detector, with 86 cables of optical modules and 81 IceTop stations on the surface, was activated on May 20, 2011. There are 5160 optical modules deployed between 1450 and 2450 m in the deep ice and 324 optical modules in 81 pairs of tanks on the surface, as illustrated in Fig. 1. Each optical module is built around a 25 cm diameter hemispherical photomultiplier. Electronics in the sensor are designed to digitize and time stamp the signals. A glass pressure housing protects the sensor from pressures of up to 500 bar recorded in the deep ice and during the freeze-in process. There are 60 sensors connected to each of the 86 cables that provide power and communication. Reliability was a big challenge since the sensors, once frozen, can never be accessed again. More than 99% of the sensors survived the installation and were successfully commissioned. Only a few failures, less than 3 per year out of 5000, were recorded after freeze-in.

The biggest challenge was the drilling of the holes into the ice to 2500 m depth. A drill was designed based on the method pioneered by the AMANDA experiment but more powerful and more highly engineered. With 5 MW of thermal power in the form of a hot water jet it was possible to drill a hole of 55 cm diameter in less than 36 hours. The holes are straight to the extent that the horizontal deviation at the bottom is less than 1 m.

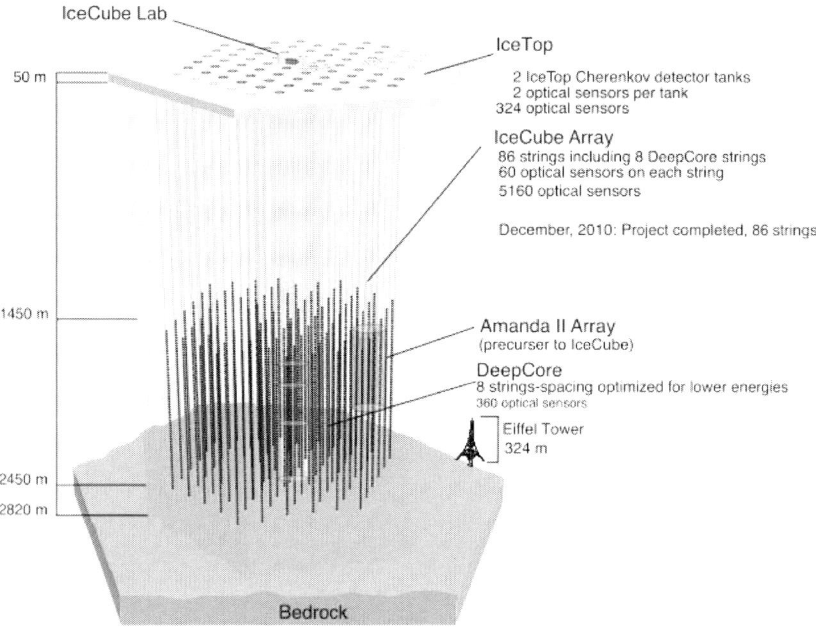

Fig. 1. Schematic view of the IceCube neutrino detector in its final configuration. IceCube was completed in December 2010.

With an instrumented target mass of one billion tons (or $6*10^{38}$ nucleons) of clear ice, it detects more than 50,000 atmospheric neutrinos per year above energies of 1 TeV. The optical properties of the South Pole Ice have been measured with various calibration devices and are used for modeling the detector response to charged particles [2].

The absorption length of ice, at depths of 1500 to 2500 m, is in the range of 80 to 200 m. The effective scattering length is in the range of 20 to 50 m. An LED calibration pulse of 10 billion photons and 20 ns duration can be measured at distances of more than 600 m through the clear ice. Neutrino interactions with a nucleus of an ice molecule will generate secondary charged particles, such as muons. A throughgoing muon of energy greater than 1 TeV will be recorded with high efficiency. The downgoing cosmic ray flux is strongly reduced at that depth. Only cosmic ray muons of energy above 300 GeV can reach the detector from the surface. Due to its size the trigger rate for cosmic ray muons generated in the atmosphere is still 3 kHz.

Fig. 2. The IceCube laboratory at the South Pole houses the surface electronics and computers. All PMT signals are digitized in the deep ice.

Upgoing muons can only arise from interactions generated by neutrinos from the Northern Hemisphere. Fig. 2 shows a neutrino-induced muon event in IceCube that is sampled over a length of more than 1 km. Background rejection is based on up/down discrimination. In order to distinguish extraterrestrial neutrinos from the atmospheric neutrino background, one of the following methods can be used:

1. Energy: Atmospheric neutrinos reach energies of about 500 TeV in IceCube. Neutrinos that are of significantly higher energy would be of extraterrestrial origin. This analysis is referred to as a diffuse analysis.
2. Directionality and energy: A significant number of neutrinos originate from the same direction (hotspot) and are in addition of higher energy than atmospheric neutrinos.
3. Directionality, energy and time: In addition to the above methods, one can use the temporal structure, possibly in coincidence with observations from satellites or gamma ray telescopes. Searches for neutrinos from gamma ray bursts [3] would be an example.

IceCube has the ability to detect muons and cascades over a wide energy range from about 10 GeV to 10^9 GeV. A neutrino induced muon is shown in Fig. 3. The reasons for the wide energy coverage are a) the neutrino-nucleon cross-section increases with energy and b) the ability to detect increasingly non contained events at higher energies. Figure 4 shows the neutrino effective area for various configurations of IceCube as well as for the ANTARES detector in the Mediterranean. At 1 TeV, the effective area of the detector is 1 m^2 comparable to that of a gamma ray satellite. At very high energies it reaches more than 10,000 m^2.

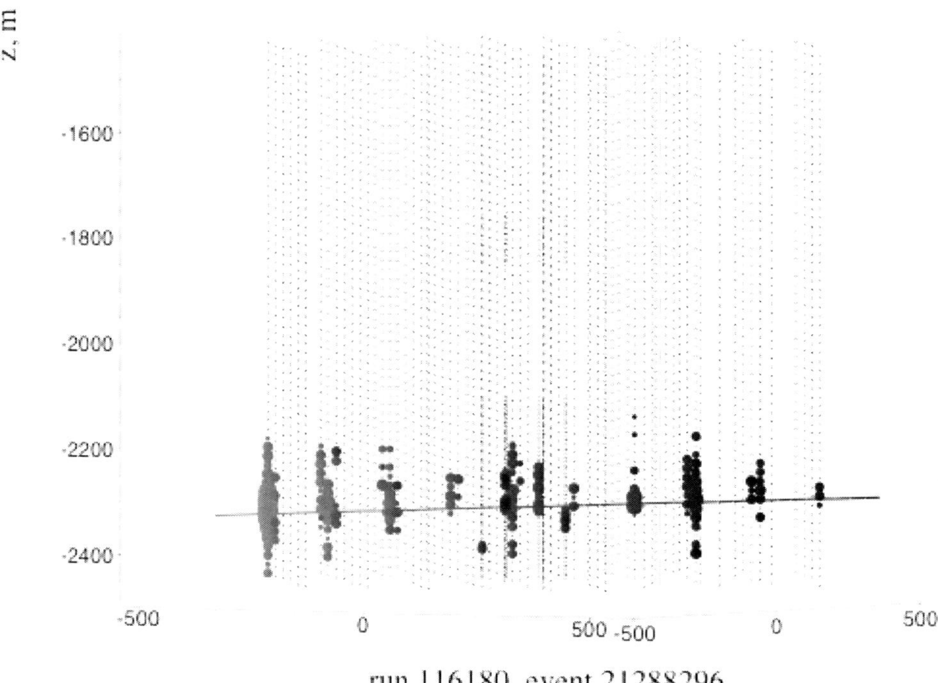

Fig. 3. A neutrino induced muon of about 10 TeV energy in IceCube. About 70 photomultipliers on 20 strings registered Cherenkov light from this event.

A good angular resolution is critical for neutrino astronomy. IceCube's angular resolution is about 0.7° at 10 TeV and 0.3° at 1 PeV. IceCube was able to demonstrate the angular resolution and absolute pointing using the cosmic ray shadow of the moon.

A measurement of the atmospheric neutrino flux with the IceCube 40 string detector is shown in Fig. 5. The diffuse search for astrophysical neutrinos by IceCube and other experiments is summarized in the same figure. While the figure focuses on diffuse fluxes, it is clear that some of these diffuse fluxes may be detected as point sources. Some examples of astrophysical flux models that are shown include AGN blazars, Waxman-Bahcall bound and cosmogenic neutrinos. With the preliminary configurations of 40 and 59 strings, IceCube has not yet detected any astrophysical neutrino signals, but the limits already obtained have led to revisions of many models. The upper limit obtained with the IC40 shown in the figure already constrains and excludes some models including the Waxman-Bahcall reference flux for neutrino production associated with cosmic rays. Most recently, the non-observation of neutrinos associated with gamma-ray bursts [3] is severely constraining the possibility that gamma-ray bursts are the origin of highest energy cosmic rays. Searches for time-integrated and time-dependent point sources have been reported with preliminary configurations [4, 5], and more are under way that will be more sensitive in scenarios where individual sources stand out. A summary of results has been presented in [6] with further references therein.

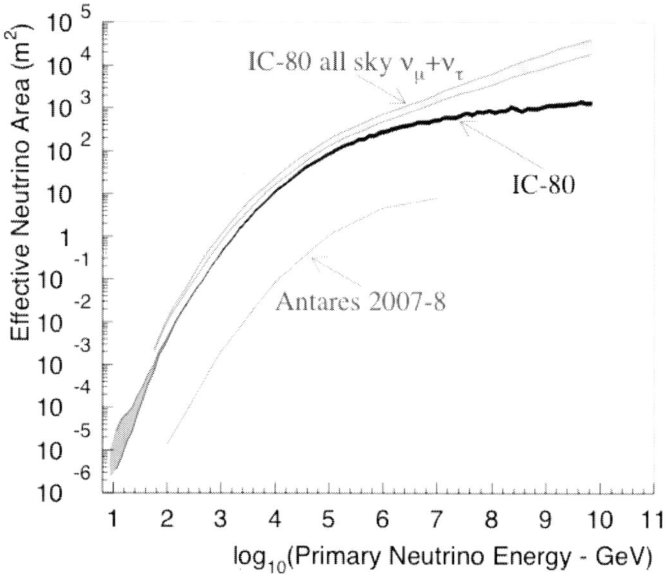

Fig. 4. Neutrino effective areas are shown for IceCube upgoing muon neutrinos, all sky muon + tau neutrinos and for ANTARES 2007-8.

3. The Radio Detection of Neutrinos – the Askaryan Radio Array

Cosmic rays have been measured to energies beyond 10^{20} eV. Ultrahigh-energy cosmic ray protons will interact with photon fields in the Universe, most prominently:

$$p + \gamma \rightarrow \Delta^+ \rightarrow \pi^+ + n$$

Every time a charged pion is produced, three neutrinos (muon neutrino, antineutrino, and an electron neutrino) are produced. This mechanism results in a cosmogenic neutrino flux first proposed by Berezinsky & Zatsepin [1]. Numerous, more detailed calculations have been presented in the meantime. We show a reference model in Fig. 5 (Ref. 11), labeled as ESS. Uncertainties in the predictions include the cosmological evolution or distance distribution of the sources and the fraction of cosmic ray protons in the highest energy cosmic ray flux. The predicted fluxes are low. IceCube, which is optimized for much lower energies, may have a chance to see this flux with predictions around 1 event/year. The attenuation length of optical photons in ice (and ocean water) is on the order of 0.1 km. For these high energies, one could increase the spacing of strings to ~300 m, yet a very large scale detector on the order of 100–1000 km^3 target volume would be too costly. In order to reliably detect this flux, other experimental strategies are needed that are more optimized for this energy range.

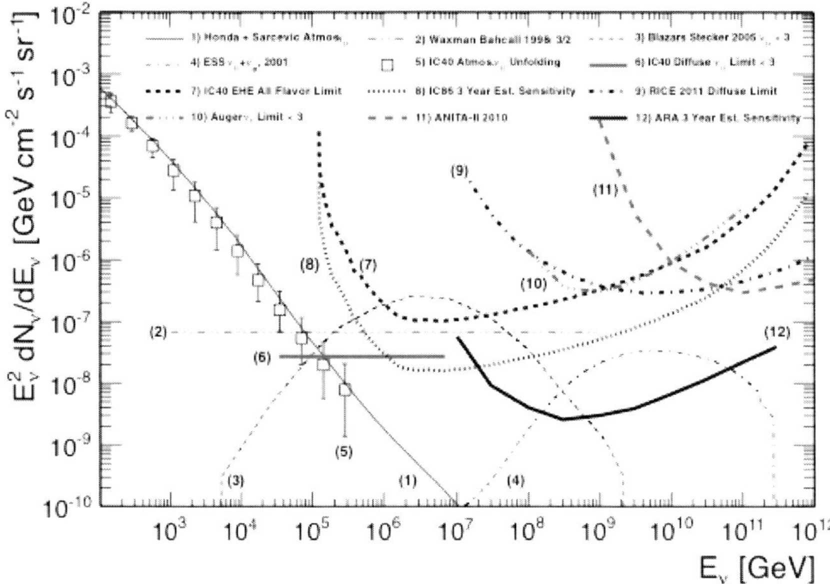

Fig. 5. The measured neutrino flux by IceCube is shown together with several predictions of neutrino fluxes and upper limits by experiments: 1) Atmospheric neutrino flux by Honda [7] + prompt by Sarcevic [8], 2) Diffuse neutrino flux [9], 3) AGN Blazars [10], 4) Cosmogenic neutrino flux [11], 5) IceCube atmospheric neutrino flux unfolded measurement [12], 6) IceCube 40, 1 yr upper limit to diffuse neutrino flux [4], 7) IceCube 40 1 yr upper limit to extremely high energy neutrinos [13], 8) IceCube 86, estimate of 3 year sensitivity, 9) RICE upper limit [14], 10) Auger 2 year ν_τ limit x 3 [17], 11) ANITA upper limit [15], and 12) the Askaryan Radio Array (ARA) estimated 3-year sensitivity [16]. Differential limits are shown as published, no correction is applied for differences in binning.

An alternate detection mechanism was suggested as early as 1962 when G. Askaryan [19] proposed that high-energy showers might produce coherent radio emission in dense media. These emissions would arise as an excess of negative charge builds up as electrons are swept out along a relativistically advancing shower front (20% more electrons than positrons when the shower is fully developed). The wavelength components of the broadband radiation from the motion of this net negative charge will add coherently for wavelengths that are large compared to the dimension of the charge distribution. The coherent emission is most pronounced in the frequency range from 1 GHz (on the Cherenkov cone) to 100 MHz (10° of the Cherenkov cone), (see e.g. Ref. 18). The large attenuation length of the cold glacial ice has been measured to be of order 1 km in the relevant frequency range of 200 to 1000 MHz. This allows measuring UHE neutrino interactions in the deep ice with a few antennas located close to the surface of the 2.8 km thick ice sheet.

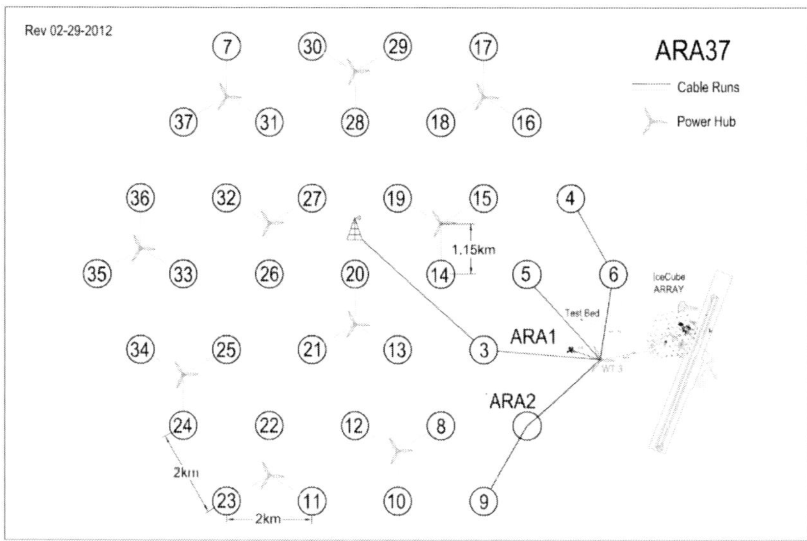

Fig. 6. Schematic view of the Askaryan Radio Array: 37 stations spaced at 2 km distance form an array of 200 km² area. Visible on the right are the South Pole Station with the runway and IceCube.

Several pioneering efforts have already been made to develop this approach, including RICE [2], ANITA [5], and early radio detection instrumentation in IceCube. Based on their experiences as well as the drilling and neutrino detector construction experience of IceCube at the Pole, the Askaryan Radio Array Collaboration has started to design, build and deploy prototypes of a detector array with the sensitivity to determine the cosmogenic neutrino flux. In its full configuration, illustrated in Fig. 6 and described in Ref. 16, the ARA-37 detector would consist of 37 detector stations covering an area of 200 km². Each station consists of a cluster of 16 embedded antennas, deployed up to 200 m deep in four vertical boreholes placed with tens-of-meter horizontal spacing. Each such station, illustrated in Fig. 7, is a fully functioning neutrino detector. For example, a neutrino interaction of 10^{18} eV can be detected up to distances of several km and more than 2 km of depth, depending on the angle of the neutrino relative to the detector. As few as 16 antennas close to the surface allow monitoring more then 10 km³ of ice for EeV neutrino interactions.

The volumetric acceptance of the detector (volume x solid angle) is shown in Fig. 8. Already at 10^{17} eV the effective volume is substantial with 50 km³ sr. The graceful and relatively low energy threshold makes ARA sensitive also to neutrino fluxes from direct sources such as AGNs, GRBs, or generally E^{-2} type neutrino spectra.

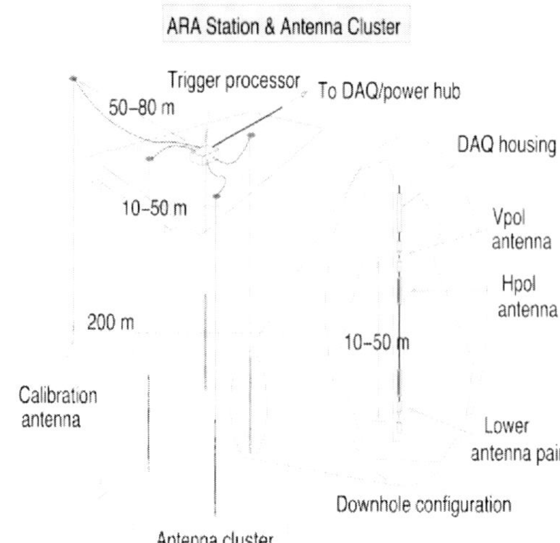

Fig. 7. Schematic view of a single ARA station. Each station is a fully functioning detector with a view on more than 10 km^3 of target volume (Ref. 16).

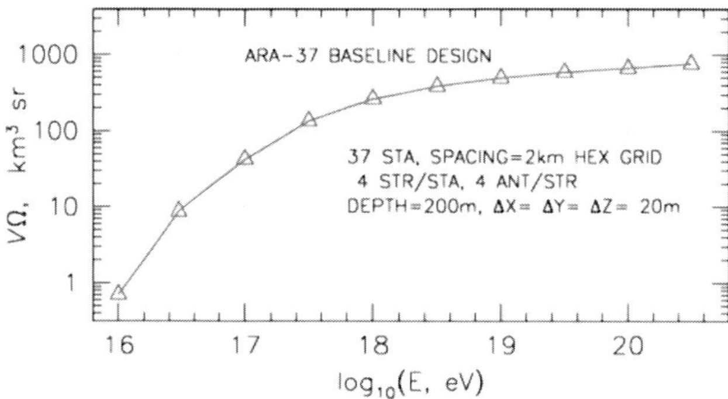

Fig. 8. Detector volumetric acceptance is shown versus energy, from Ref. 16.

Figure 5 shows such fluxes, and current experimental upper limits. As we can see, the upper limits of experiments like ANITA, RICE, Auger and IceCube are approaching the predicted cosmogenic neutrino flux but are not close enough to test the model shown. The recent Auger result is based on Earth-skimming tau neutrinos with 2 years lifetime. The IceCube result is based on one year of lifetime with only 50% of the completed detector. In several years of lifetime, IceCube will have good chances of seeing events of the given reference flux. However, with more than an order of magnitude better sensitivity than all existing experiments, ARA will be able to provide a definitive measurement of the cosmogenic neutrino flux. It should observe between 10 and 100 events depending on the choice of model.

ARA has begun its first deployments in the austral summer of 2011/12 with continuation scheduled for the 2012/13 season. Based on these measurements, two important results were obtained [16]. The first is a confirmation of the long, km-scale attenuation length for electromagnetic waves of the relevant frequency range in the massive cold ice sheet. A measurement was made with a pulsed emitter co-located at 2500 m depth with one of the last deployed IceCube strings. The 5 kV nsec pulse, not too different from a cosmogenic neutrino event, was seen at more than 3 km distance with ARA receivers with a signal strength that saturated the receivers well beyond expectations. The collaboration could also confirm that the electromagnetic environment is indeed very quiet and suitable for ARA detectors.

Thus, 100 years after the discovery of the South Pole and after the discovery of cosmic rays, the South Pole has become the place where the discovery of high astrophysical neutrinos may take place. IceCube is running at a sensitivity exceeding expectations and at more than 98% uptime. With the ARA experiment, the possibility exists to build a detector that may provide the decisive measurement to resolve the highest energy neutrino and cosmic ray puzzle.

Acknowledgments

We acknowledge the support from the U.S. National Science Foundation-Office of Polar Programs, the U.S. National Science Foundation-Physics Division and the University of Wisconsin Alumni Research Foundation.

References

1. V.S. Berezinsky & G.T. Zatsepin, Phys. Lett. 28b, 423 (1969); Sov. J. Nucl. Phys. 11, 111 (1970)
2. M. Ackermann, *et al.*, J. Geophys. Res. **111** D13203 DOI:10 1029 / 2005JD006687 (2006)
3. R. Abbasi et al., Nature 484 (2012) 351-354, 19 April 2011
4. R. Abbasi et al., Astrophys. J. 732 (2011) 18, arXiv:1012.2137
5. R Abbasi et al., Astrophys. J. 744 (2012) 1, arXiv:1104.0075
6. H. Kolanoski, for IceCube Coll., Highlight paper, 32nd ICRC, Beijing, China, August 2011; arXiv:1111.5188
7. M. Honda et al., Phys. Rev. D 75, 043006 (2007)
8. I. Sarcevic et al., Phys. Rev. D 78 043005 (2008)
9. E. Waxman and J. Bahcall, Phys. Rev. D. 59 023002 (1998)
10. F.W. Stecker, Phys. Rev. D, 72, 107301 (2005)
11. R. Engel, D. Seckel, T. Stanev, Phys.Rev. D 64:093010,2001, arXiv:0101216
12. R. Abbasi et al., Phys.Rev. D 83:012001,2011, arXiv:1010.3980
13. R. Abbasi et al., Phys.Rev.D 83:092003 (2011), arXiv:1103.4250
14. Kravchenko et al., accepted Phys. Rev D, arXiv:1106.1164
15. Gorham et al., Phys.Rev.D82:022004,2010, arXiv:1003.2961
16. Allison et al. (ARA Collaboration), arXiv:1105.2854, in Astropart. Phys.
17. P. Abreu et al. (Auger Coll.), Phys. Rev. D 84, 122005 (2011); Erratum: Phys. Rev. D 85, 029902(E) (2012), arXiv:1202.1493
18. J. Alvarez-Muniz et al., Astroparticle Physics 35 (2012) 287-299, arXiv:1005.0552
19. G. A. Askaryan, Zh. Eksp. Teor. Fiz. 41, 616 (1961); Soviet Physics JTEP 14, 441 (1962)

NEUTRINO FLAVOR RATIOS ON EARTH UNDER DECAY AND OSCILLATION SCENARIOS

T.C. LIU*

*Leung Center for Cosmology and Particle Astrophysics,
National Taiwan University, Taipei 106, Taiwan*
** E-mail: tcliu@ntu.edu.tw*

KWANG-CHANG LAI

Physics Group, Center for General Education, Chang Gung University, Kwei-Shan 333, Taiwan

GUEY-LIN LIN

Institute of Physics, National Chiao-Tung University, Hsinchu 300, Taiwan
E-mail: glin@cc.nctu.edu.tw

In the neutrino propagation mechanisms, both neutrino oscillation and decay mechanisms, the actual neutrinos flavor ratio at astrophysical sources can be quite different from observed ratios on Earth.[1] We introduce the recent global fitting result of the mixing angle values[2] into the analysis to point out the allowed neutrino flavor ratios on Earth for standard neutrino oscillation and general decay mechanisms. In this analysis, four types of the neutrino decay mechanisms can be separated from the oscillation case easily with arbitary initial neutrino flavor ratio and the other four types of neutrino decay mechanisms can be identified from the oscillation case with the specified source. The rest types of neutrino decays strongly overlape with the oscillation and cannot be identified.

Keywords: Neutrino; Mixing angle.

1. Introduction

The flavor ratio of astrophysical neutrinos observed on Earth depend on both the initial flavor ratio at the source and flavor mixing mechanisms during the propagation of these neutrinos. We consider not only the initial source types but also the propagation mechanisms with the neutrino mixing angles to analyze the allowed flavor ratios on Earth. For the initial neutrino source part, cosmic neutrinos can be produced from various astrophysical objects. At production sites, different production process or environment could produce different neutrino flavor ratio.[3] Typical hadronic interaction produces neutrinos through $\pi^{\pm} \rightarrow \mu^{\pm} \rightarrow e^{\pm}$ chains and generates the neutrino flavor ratio, $\{\phi(\nu_e), \phi(\nu_\mu), \phi(\nu_\tau)\} = \{1, 2, 0\}$, this type is called pion source. When the sources are hidden behind strong field or matters, muon may lose energy before decay, the muon neutrinos and electron neutrino from muon decay could become very low energy and become undetectable in astrophysical neutrino telescopes. For this source, called muon-damped source, the flavor ratio $\{\phi(\nu_e), \phi(\nu_\mu), \phi(\nu_\tau)\} = \{0, 1, 0\}^{45}$.[6] However, because of neutrino oscillation, their flavor ratio changes while neutrinos are on-route to Earth and being detected. Since the distances from astrophysical source to Earth are much longer than the oscillation length, the observed neutrino flux is almost fully mixed and the flavor ratio approaching 1:1:1. Deviation from such equal-partition comes from small factor of mixing angles. Therefore, the smooth transition of the flavor ratios from pion source to muon-damped is expected.[7] We introduce this smooth transition as more general source, $\{\phi_0(\nu_e) : \phi_0(\nu_\mu) : \phi_0(\nu_\tau)\} = \{\alpha : 1 - \alpha : 0\}$, with $0 < \alpha \leq 1/3$. Concerning the prorogation part, standard three flavors neutrino oscillation and arbitrary neutrino decay mechanisms are considered. We ignore the different between the particle and antiparticle in the decay mechanisms to simplify the calculation.

2. Standard Oscillation Mechanism

In neutrino oscillation theory, the relationship between neutrino mass eigenstates and neutrino flavor eigenstates are described by Pontecorvo-Maki-Nakagawa-Sakata mixing matrix U,[8,9] which is formed by three mixing angles θ_{ij} and the CP violation phase δ. In the limit of large neutrino propagation distance, the neutrino oscillation probability only depend on the mixing angles θ_{ij} and CP phase. In this case, the transition probability for neutrino flavor β to the other flavor α can be expressed in terms of unitary matrix U purely:

$$P_{\alpha\beta} = \sum_{i=1}^{3} |U_{\alpha i}|^2 |U_{\beta i}|^2. \tag{1}$$

Hence he neutrino flux at the astrophysical site and that detected on Earth is related by

$$\begin{pmatrix} \phi(\nu_e) \\ \phi(\nu_\mu) \\ \phi(\nu_\tau) \end{pmatrix} = \begin{pmatrix} P_{ee} & P_{e\mu} & P_{e\tau} \\ P_{\mu e} & P_{\mu\mu} & P_{\mu\tau} \\ P_{\tau e} & P_{\tau\mu} & P_{\tau\tau} \end{pmatrix} \begin{pmatrix} \phi_0(\nu_e) \\ \phi_0(\nu_\mu) \\ \phi_0(\nu_\tau) \end{pmatrix} \equiv P \begin{pmatrix} \phi_0(\nu_e) \\ \phi_0(\nu_\mu) \\ \phi_0(\nu_\tau) \end{pmatrix}, \tag{2}$$

where $\phi(\nu_\alpha)$ is the neutrino flux measured on Earth while $\phi_0(\nu_\alpha)$ is the neutrino flux at the source, and the matrix element $P_{\alpha\beta}$ is the probability for the oscillation $\nu_\beta \to \nu_\alpha$. New neutrino mixing results, listed in Table 1, are introduced into the Equation(2), the allowed neutrino flavor ratios on Earth from the general neutrino sources, $\{\phi_0(\nu_e) : \phi_0(\nu_\mu) : \phi_0(\nu_\tau)\} = \{\alpha : \beta : 1 - \alpha - \beta\}$, with $\alpha, \beta > 0$ and $\alpha + \beta \leq 1$, via oscillation mechanism are shown in Fig. 2. The uncertainties of neutrino mixing angles of both hierarchy are similar at the 1σ CL, hence the allowed flavors ratios on Earth are identical. We also consider the standard sources, pion and muon-damped sources, and the uncertainties of mixing angle, the allowed flavor ratios are shown in Fig. 2. The allowed neutrino flavor ratios on Earth for the pion source and muon-damped source are approaching {1:1:1} and {5:8:7}.

Table 1.   Neutrino mixing angles

parameters	Best fit	1σ	3σ
$\sin^2\theta_{12}$	0.307	0.291-0.325	0.259-0.359
$\sin^2\theta_{13}$(Normal hierarchy)	0.0245	0.214-0.279	0.149-0.344
$\sin^2\theta_{13}$(Inverted hierarchy)	0.0246	0.0215-0.0280	0.150-0.347
$\sin^2\theta_{23}$(Normal hierarchy)	0.398	0.372-0.428	0.330-0.638
$\sin^2\theta_{23}$(Inverted hierarchy)	0.408	0.378-0.443	0.335-0.658

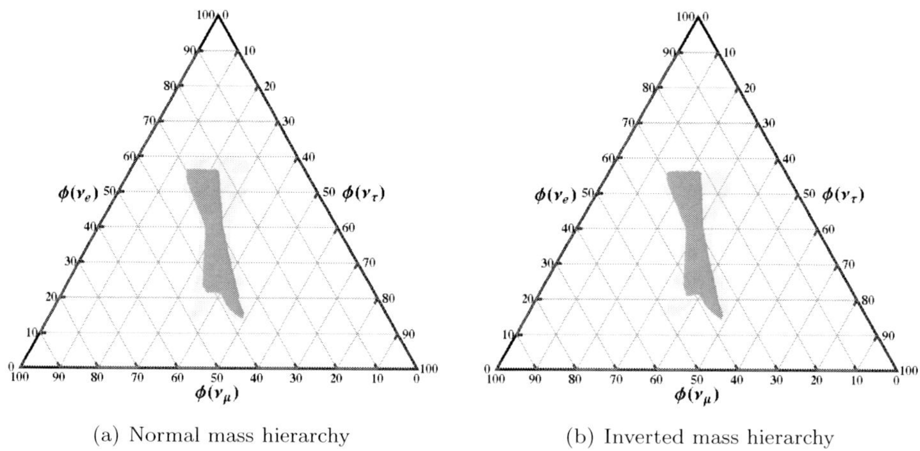

(a) Normal mass hierarchy   (b) Inverted mass hierarchy

Fig. 1.   The ranges for the neutrino flavor ratios on Earth resulting from standard neutrino oscillation for normal and inverted mass hierarchy. Both panels show the range of neutrino flavor ratios from the source flavor ratios, $\{\phi_0(\nu_e) : \phi_0(\nu_\mu) : \phi_0(\nu_\tau)\} = \{\alpha : \beta : 1 - \alpha - \beta\}$, with $\alpha > 0$, $\beta > 0$ and $\alpha + \beta \leq 1$. The grey and light grey colors correspond the 1σ and 3σ CL of neutrino mixing angles, respectively. The values of mixing angles are same as the Table 1.

3. Neutrino Decay Mechanism

In neutrino decay mechanisms, we consider only two-body decay modes with three active neutrino flavors and assume the decays are complete. As discussed in Ref.,[10] the probability transition matrix characterizing the the decay scenario is written in terms of PMNS matrix and corresponding branching ratios

$$P^{dec}_{\alpha\beta} = \sum_{i\ \text{stable}} (|U_{\alpha i}|^2 + \sum_{j\ \text{unstable}} |U_{\alpha j}|^2 \text{Br}_{j\rightarrow i})|U_{\beta i}|^2, \tag{3}$$

where indices i and j denote mass eigenstates and $Br_{i\rightarrow j}$ shows branching ratio for unstable state i decays to stable state j. $\sum_f \text{Br}_{i\rightarrow j} = 1$ for i state fully decays to j

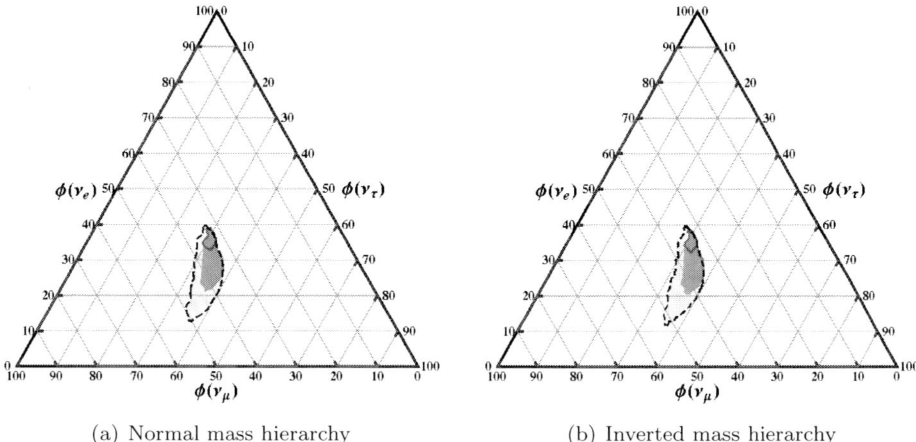

(a) Normal mass hierarchy   (b) Inverted mass hierarchy

Fig. 2. The ranges for the neutrino flavor ratios on Earth resulting from standard neutrino oscillation for normal and inverted mass hierarchy. Both panels show the range of neutrino flavor ratios from the source flavor ratios, $\{\phi_0(\nu_e) : \phi_0(\nu_\mu) : \phi_0(\nu_\tau) = \{\alpha : 1 - \alpha : 0\}$, with $0 \leq \alpha \leq 1$. The grey and light grey region with dashed line correspond the 1σ and 3σ CL of neutrino mixing angles, respectively. Blue and red curves correspond to the allowed neutrino flavor ratios from the pion source and muon-damped source at 1σ CL of neutrino mixing angles. The values of mixing angles are same as the in Table 1.

Table 2. The neutrino decay and oscillation scenarios. The suffix "H", "M", and "L" label the heaviest, middle and lightest mass eigenstates. The number "1", "2" and "3" label ν_1, ν_2 and ν_3 mass eigenstates. The red numbers and black numbers correspond to the unstable and stable eigenstates, respectively.

Decay type	Hierarchy	H	M	L	Scenarios	figure label
					Oscillation	Fig. 2
N1	Normal	3	2	1	Heaviest & middle decay	Fig. 3a
I1	Inverted	2	1	3	Heaviest & middle decay	Fig. 3c
N2	Normal	3	2	1	Heaviest decay	Fig. 4a
I2	Inverted	2	1	3	Heaviest decay	Fig. 5c
N3	Normal	3	2	1	Middle decay	Fig. 5a
I3	Inverted	2	1	3	Middle decay	Fig. 4c
N4	Normal	3	2	1	Lightest decay to invisible	Fig. 4b
I4	Inverted	2	1	3	Lightest decay to invisible	Fig. 4d
N5	Normal	3	2	1	Middle decay & Lightest	Fig. 3b
I5	Inverted	2	1	3	decay to invisible	Fig. 5d
N6	Normal	3	2	1	Heaviest decay & Lightest	Fig. 5b
I6	Inverted	2	1	3	decay to invisible	Fig. 3d

state. In general, $\sum_f \mathrm{Br}_{i \to j} < 1$ and $\mathrm{Br}_{i \to \mathrm{invisible}} = 1 - \sum_f \mathrm{Br}_{i \to j}$ denotes the probability for i state fully decays to invisible states. The invisible states is invisible for neutrino detector. Those invisible states could be sterile neutrinos, unparticle states *etc.* Table 2 lists all possibilities of decays with oscillation included for comparison.

296

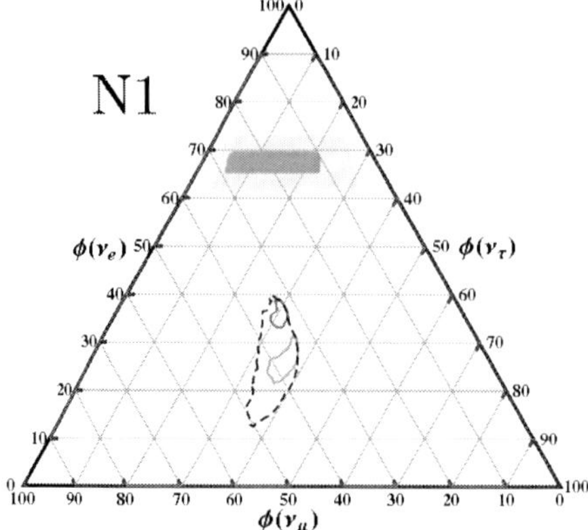

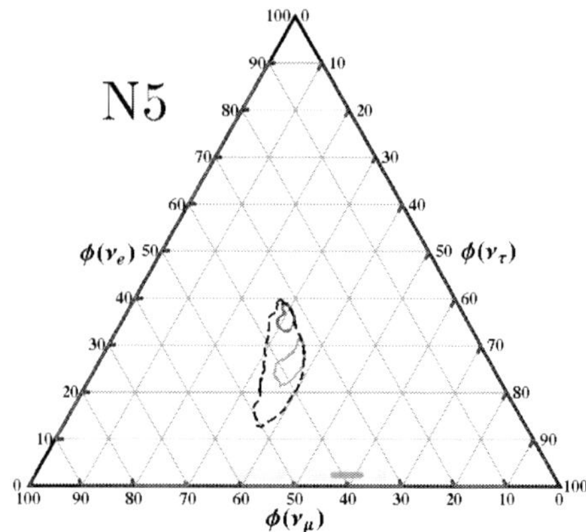

(a) The heaviest and middle mass eigenstates are unstable and decay into the lightest and invisible eigenstates for nomal mass hierarchy. In this case, the decay scenario is fully decoupled with the standard oscillation mechanism in 3σ CL of neutrino mixing angles.

(b) The middle and lightest mass eigenstates are unstable and decay into the invisible eigenstates. In this case, the decay scenario is fully decoupled with the standard oscillation mechanism in 3σ CL of neutrino mixing angles.

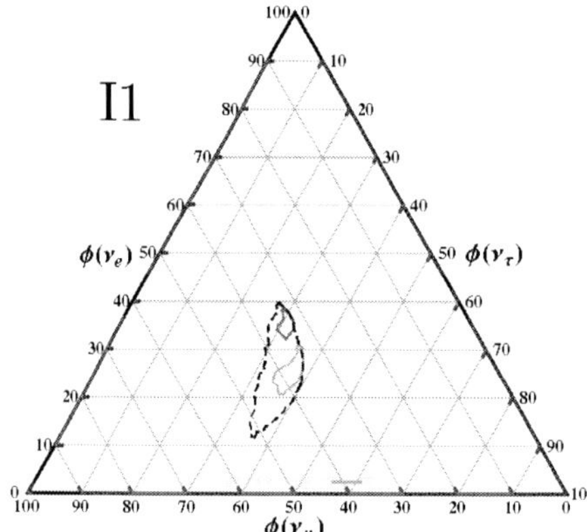

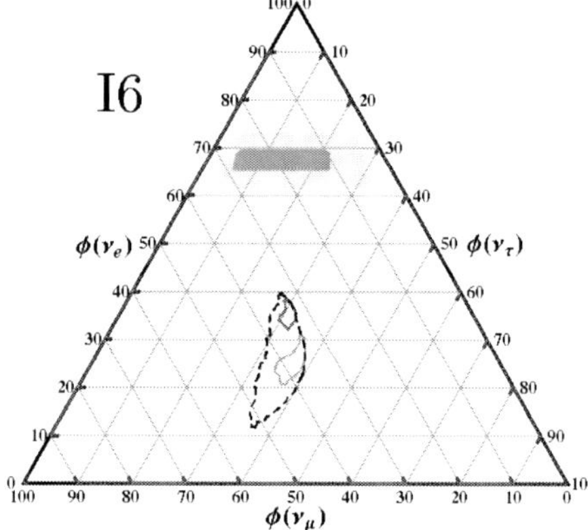

(c) The heaviest and middle mass eigenstates are unstable and decay into the lightest and invisible eigenstates for inverted mass hierarchy. In this case, the decay scenario is fully decoupled with the standard oscillation mechanism in 3σ CL of neutrino mixing angles.

(d) The heaviest and lightest mass eigenstates are unstable and decay into invisible eigenstate for inverted mass hierarchy. In this case, the decay scenario is fully decoupled with the standard oscillation mechanism in 3σ CL of neutrino mixing angles.

Fig. 3. Allowed neutrino flavor ratios on Earth at 1σ and 3σ CL of neutrino mixing angles for initial flavor ratio, $\{\phi_0(\nu_e) : \phi_0(\nu_\mu) : \phi_0(\nu_\tau)\} = \{\alpha : 1 - \alpha : 0\}$, with $0 \leq \alpha \leq 1$. Blue and red curves correspond to the allowed neutrino flavor ratios at 1σ CL of neutrino mixing angles for the standard neutrino oscillation mechanism, respectively.

3.1. *Fully decoupled case: The allowed flavor ratios on Earth under decay scenarios can rule out the neutrino oscillation one*

In Fig. 3, The scenarios "N1", "N5", "I1", and "I6" can be fully separated from the standard neutrino oscillation scenarios for both pion and muon-damped sources at

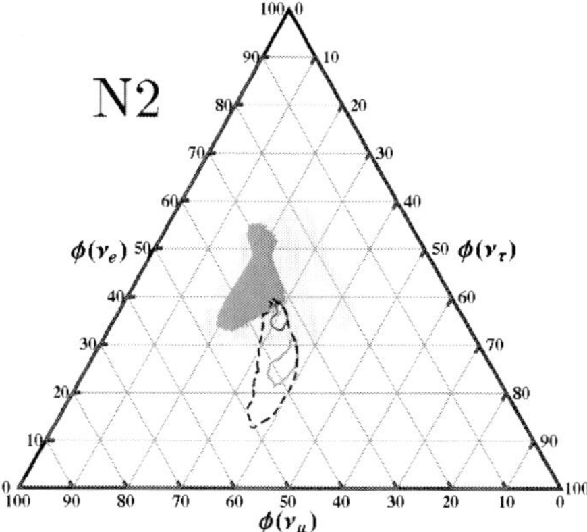

(a) The unstable heaviest mass eigenstate decays into the lighter mass eigenstates for nomal mass hierarchy. The allowed flavor ratios on Earth from the muon-damped source can be separated from the decay scenario at 1σ CL of neutrino mixing angles.

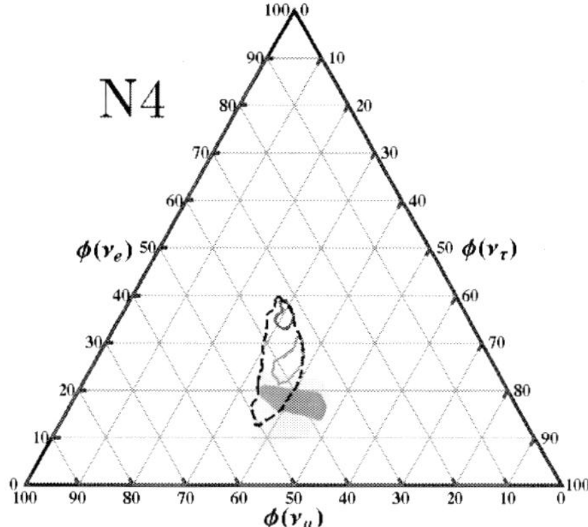

(b) The unstable lightest mass eigenstate decays into the invisible mass eigenstate for nomal mass hierarchy only. The allowed flavor ratios on Earth from the pion source can be separated from the decay scenario at 3σ CL of neutrino mixing angles.

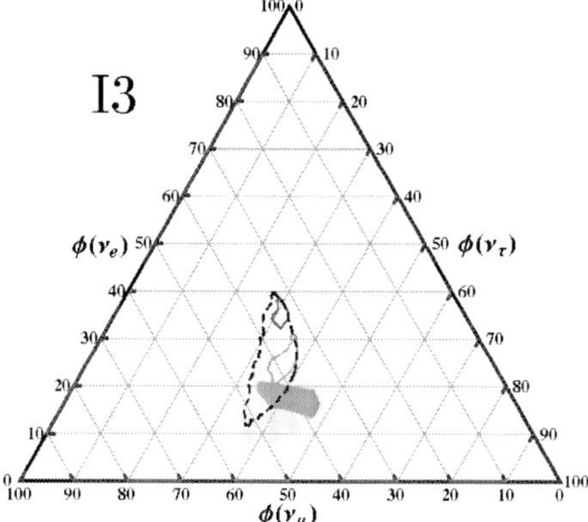

(c) The unstable middle mass eigenstate decays into the lighter mass eigenstates for inverted mass hierarchy. The allowed flavor ratios on Earth from the pion source can be separated from the decay scenario at 3σ CL of neutrino mixing angles.

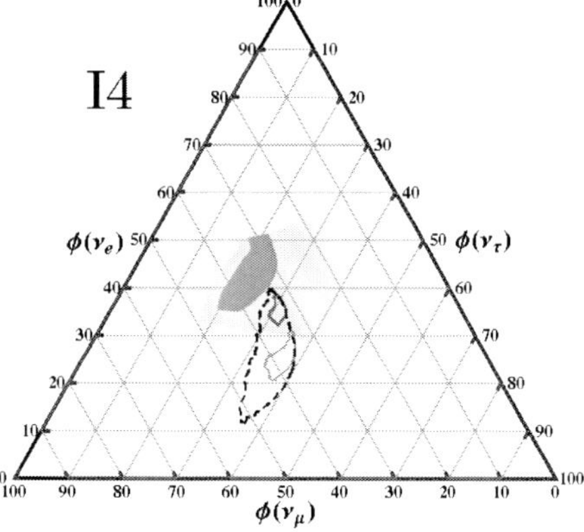

(d) The unstable lightest mass eigenstate decays into the invisible mass eigenstate for inverted mass hierarchy only. The allowed flavor ratios on Earth from the muon-damped source can be separated from the decay scenario at 1σ CL of neutrino mixing angles.

Fig. 4.   Allowed neutrino flavor ratios on Earth at 1σ and 3σ CL of neutrino mixing angles for initial flavor ratio, $\{\phi_0(\nu_e) : \phi_0(\nu_\mu) : \phi_0(\nu_\tau)\} = \{\alpha : 1 - \alpha : 0\}$, with $0 < \alpha < 1$. Blue and red curves correspond to the allowed neutrino flavor ratios at 1σ CL of neutrino mixing angles for the standard neutrino oscillation mechanism, respectively.

3σ CL of neutrino mixing angles. The allowed electron neutrino flavor fractions of "N5" and "I1" scenarios are approaching 0.

298

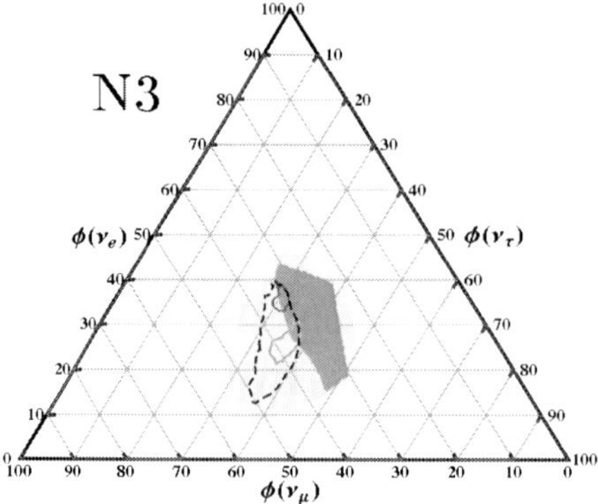

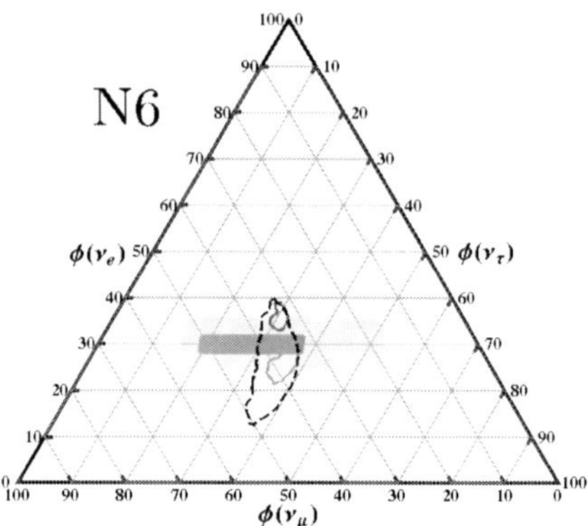

(a) The unstable middle mass eigenstate decays into the lighter mass eigenstates for nomal mass hierarchy. The allowed flavor ratios on Earth from both muon damped source and pion source cannot be separated from the decay scenario in 3σ CL of neutrino mixing angles.

(b) The heaviest and lightest mass eigenstates are unstable and decay into invisible eigenstate for nomal mass hierarchy. The allowed flavor ratios on Earth from both muon damped source and pion source cannot be separated from the decay scenario in 3σ CL of neutrino mixing angles.

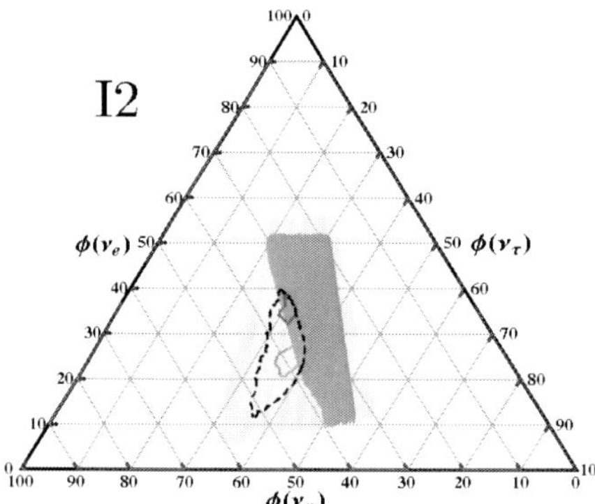

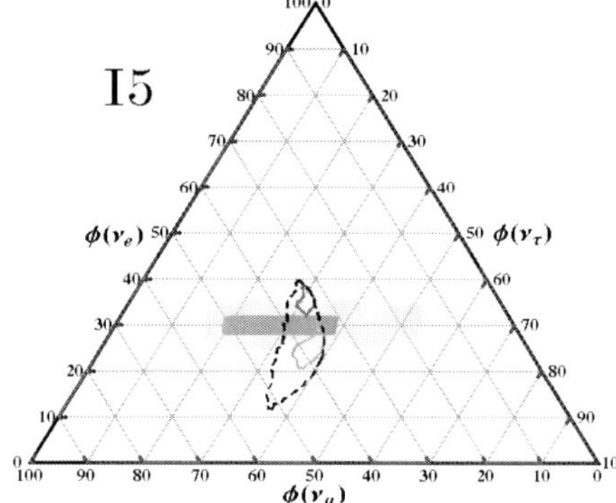

(c) The unstable heaviest mass eigenstate decays into the lighter mass eigenstates for inverted mass hierarchy. The allowed flavor ratios on Earth from both muon damped source and pion source cannot be separated from the decay scenario.

(d) The middle and lightest mass eigenstates are unstable and decay into the invisible eigenstate for inverted mass hierarchy. The allowed flavor ratios on Earth from both muon damped source and pion source cannot be separated from the decay scenario.

Fig. 5. Allowed neutrino flavor ratios on Earth at 1σ and 3σ CL of neutrino mixing angles for initial flavor ratio, $\{\phi_0(\nu_e) : \phi_0(\nu_\mu) : \phi_0(\nu_\tau)\} = \{\alpha : 1 - \alpha : 0\}$, with $0 \leq \alpha \leq 1$. Blue and red curves correspond to the allowed neutrino flavor ratios at 1σ CL of neutrino mixing angles for the standard neutrino oscillation mechanism, respectively.

3.2. *Partially coupled case: The allowed flavor ratios on Earth under decay scenarios can partially separate from neutrino oscillation one*

In Fig. 4, the allowed neutrino flavor ratios on Earth from muon-damped source for "N2" and "I4" scenarios can be separated from the oscillation scenario at 3σ CL.

However, the allowed neutrino flavor ratios on Earth from pion source for "N4" and "I3" scenarios can be separated from the standard oscillation scenario at 3σ CL.

3.3. *Coupled case: The allowed flavor ratios on Earth under decay scenarios cannot be ruled out from the neutrino oscillation one*

In Fig. 5, the scenarios "N3", "N6", "I2", and "I6" strongly overlap the allowed flavor ratios for standard neutrino oscillation scenarios and both sources.

3.4. *Conclusion*

The scenarios "N1", "N4", "N5", "I1", "I3", and "I6" can be fully separated from the standard neutrino oscillation scenarios for the pion sources at 3σ CL of neutrino mixing angles and the scenarios "N1", "N2", "N5", "I1", "I4", and "I6" can be fully separated from the standard neutrino oscillation scenarios for the muon-damped sources at 3σ CL of neutrino mixing angles. In experimental viewpoint. electron flavor(shower event) is easier to be identified than the other flavors, hence the scenarios "N1", "N5", "I1", and "I6" are the most possible to be ruled out or confirmed scenarios. Due to the degency, we cannot distinguish the "N1" and "I6" scenarios or "N5" and "I1" scenarios, we need the further information to break down the degency.

References

1. M. Maltoni, W. Winter, [arXiv:0803.2050v2 [hep-ph]].
2. G. L. Fogli, E. Lisi, A. Marrone, D. Montanino, A. Palazzo, A. M. Rotunno, [astro-ph/1205.5254].
3. S. Pakvasa, Mod. Phys. Lett. A **19**, 1163 (2004) [Yad. Fiz. **67**, 1179 (2004)].
4. M. Kachelriess, S. Ostapchenko and R. Tomas, Phys. Rev. **D77** 023007 (2008)
5. J.P. Rachen and P. Meszaros, Phys. Rev. **D58** 123005 (1998)
6. T. Kashti and E. Waxman, Phys. Rev. Lett. **59** 181101(2005)
7. M. Kachelriess and R. Tomas, Phys. Rev. **D74** 063009 (2006) , [astro-ph/0606406].
8. Z. Maki, M. Nakagawa and S. Sakata, Prog. Theor. Phys. **28** 870 (1962).
9. B. Pontecorvo, Zh. Eksp. Teor. Fiz. **53** 1717 (1967).
10. M. Maltoni and W. Winter, *JHEP* **64** 0807 (2008).
11. R. L. Awasthi and S. Choubey, Phys. Rev. **D76** 113002 (2007), [arXiv:0706.0399 [hep-ph]].
12. S.-L. Chen, X.-G. He, and H.-C. Tsai, [arXiv:0707.0187 [hep-ph]].
13. S. Zhou, [arXiv:0706.0302 [hep-ph]].
14. X.-Q. Li, Y. Liu, and Z.-T. Wei, [arXiv:0707.2285 [hep-ph]].
15. D. Majumdar, [arXiv:0708.3485 [hep-ph]].

DISTINGUISHABILITY OF NEUTRINO FLAVORS THROUGH THEIR DIFFERENT SHOWER CHARACTERISTICS

CHIH-CHING CHEN[1,2,a], PISIN CHEN[1,2,3,4,b], CHIA-YU HU[1,2,c], K.-C. LAI[2,5]

[1]*Graduate Institute of Astrophysics, National Taiwan University, Taipei 10617, Taiwan*
[2]*Leung Center for Cosmology and Particle Astrophysics (LeCosPA), Taipei 10617, Taiwan*
[3]*Department of Physics, National Taiwan University, Taipei 10617, Taiwan*
[4]*Kavli Institute for Particle Astrophysics and Cosmology, SLAC National Accelerator Laboratory, Menlo Park, CA 94025, USA*
[5]*Physics Group, Center for General Education, Chang Gung University, Kwei-Shan, Taiwan*
E-mail: [a]chen.chihching@gmail.com, [b]chen@slac.stanford.edu, [c]huchiayu0517@gmail.com

We propose a new flavor identification method to distinguish mu and tau type ultra high energy cosmic neutrinos (UHECN). Energy loss of leptons in matter is an important information for the detection of neutrinos originated from high energy astrophysical sources. 50 years ago, Askaryan proposed to detect Cherenkov radiowave signals emitted from the negative charge excess of neutrino-induced particle shower. The theory of Cherenkov radiation under Fraunhofer approximation has been widely studied in the past two decades. However, at high energies or for high density materials, electromagnetic shower should be elongated due to the Landau-Pomeranchuck-Migdal (LPM) effect. As such the standard Fraunhofer approximation ceases to be valid when the distance between the shower and the detector becomes comparable with the shower length. Monte Carlo simulations have been performed recently to investigate this regime. Here we adopt the deduced relationship between the radio signal and the cascade development profile to investigate its implication to lepton signatures. Our method provides a straightforward technique to identify the neutrino flavor through the detected Cherenkov signals.

Keywords: Neutrino detection; neutrino flavor; flavor identification; lepton propagation.

1. Introduction

The nature and origin of ultra-high energy cosmic rays (UHECRs) remains a mystery after decades of investigations. These amazingly energetic events have been observed beyond $\approx 10^{19.6}$ eV, the so-called Greisen-Zatsepin-Kuzmin (GZK)[1] cutoff. The GZK feature on the UHECR spectrum was first observed by the High Resolution Fly's Eye Experiment[2] and later confirmed by the Pierre Auger Observatory.[3] Above this energy scale, UHECRs interact with CMB photons through the GZK processes, producing cosmogenic neutrinos. The GZK feature on the cosmic ray energy spectrum guarantees the existence of the cosmogenic neutrinos. However, none of these neutrinos have been observed so far. Detection of these ultra high energy (UHE) neutrinos would provide critical information for unraveling the mystery of the origin and evolution of the cosmic accelerators and is one of the utmost tasks in the coming decade.[4]

One promising way of detecting UHE neutrinos is the radio approach. When an ultra-high energy cosmic neutrino interacts with ordinary matters on Earth, it would lead to a hadronic debris, either by charged current or neutral current process. The former also produces a lepton with corresponding flavor. Both the high energy leptons and the hadronic debris induce particle showers. As proposed by Askaryan in the 1960's,[9] the initially charge-neutral high energy particle such as neutrino traversing in a dense medium would develop into a shower that carries net negative charges. This charge imbalance appears as a result of the knocked-off of the bound-state electrons in the medium, as well as the annihilation of positrons in the shower with the electrons of the medium. The net charges of the shower, typically 20% of total shower particles, serve as a source emitting the Cherenkov radiation when they traverse the medium. The typical sizes of a shower are quite localized (tens of cm in radial and few meters in longitudinal development) compared to those develop in air (km scale), and therefore it results in coherent radiations for wavelengths longer than the shower sizes. The corresponding coherent wavelength turns out to be in the radio band, from hundreds of MHz to several GHz.

In this article, we discuss the possibility to identify the flavors of the cosmogenic neutrinos detected by the radio neutrino telescope, such as ANITA,[5] Askaryan Radio Array(ARA),[6] and ARIANNA.[7]

2. The Strategy of Flavor Identification

As neutrinos interact with matter to produce observable signals, the major channel is the changed-current (CC) interaction. The electron produced through ν_e CC interaction has a large interaction cross section with the medium and produces a shower within a short distance from its production point. Contrary to the electron, the muon produced through ν_μ CC interaction can travel a long distance in the medium before it loses all its energy or decays. However, muon does emit lights along its propagation so that only those detectors near to the muon track can be triggered.

As for ν_τ detection, the ν_τ-induced tau leptons behave differently at different energies for a fixed detector design. For a neutrino telescope such as IceCube, the observable energy range for the double bang event is from one to tens of PeV. For an undersea experiment, such as KM3Net,[8] the observable energy range for the double bang event is similar. For a radio neutrino telescope such as ARA, however, the detector is designed to observe cosmogenic neutrinos of energy about EeV. In this energy regime, the tau lepton range becomes long enough so that a tau lepton can pass through the detector without decay but losing its energy like a muon does. In this case, the signal for ν_τ appears like a track event. We summary these features in Table 1.

Note that the lights emitted from the track can only trigger the nearby optical detectors. A different strategy is taken to construct track events from radio detectors. For cosmogenic neutrinos, the energy of the CC-induced muon or tau lepton

Table 1: Neutrino interaction for different flavors.

Neutrino flavor	ν_e	ν_μ	ν_τ
Neutral current	Hadronic	Hadronic	Hadronic
Changed current	Hadronic	Hadronic	Hadronic
	EM shower	propagation	propagation

is so high that a muon or tau lepton not only emits dim lights but also produce mini-showers along its propagation through the detector fiducial volume. By detecting the radio emissions from these mini-showers, a track event can be reconstructed for a muon or tau lepton traversing the detector volume. By observing a single mini-shower, a ν_e signal can be identified from a track event for a ν_μ or ν_τ.

It is challenging to distinguish between ν_μ and ν_τ signals because both muons and tau leptons produce similar track-like events. Simulations of lepton propagation in ice show that the compositions of the mini-showers are different for muon and tau lepton track events. The mini-showers that consist of track events are composed of two major components: electromagnetic (EM) and hadronic showers. The energy loss distributions between EM and hadronic showers are different for muon and tau lepton track events. A muon track event loses more energy through EM showers than through hadronic ones while a tau track does the opposite. By collecting mini-showers, measuring their attributes and evaluating energy losses, one can distinguish between muon and tau track events.

3. Simulation for Neutrino Events

ARA underground radiowave antenna stations receive radio emissions from shower particles created by cosmogenic neutrinos in ice. These radio signals are Cherenkov radiations produced by net charges of shower particles. We adopt the CORSIKA-IW[15,16] code, a modified version of COSIKA[17] program for dense-target simulations, to simulate EM and hadronic showers. In Figs. 1 and 2, longitudinal developments of charges are shown for EM and hadronic showers respectively. We demonstrate the different characteristics between electron and proton induced showers in ice. The hadronic shower is simulated by penetrating a proton in a dense medium, which evolves as a typical profile with one peak at the shower maximum. For the EM shower, however, the situation is different. At energies higher than 10^{16}eV, bremsstrahlung and pair-production processes are suppressed by the Landau-Pomeranchuck-Migdal (LPM) effect. As a result, cascades are stretched, shower development is elongated and several peaks appear in the shower longitudinal profile.

Cherenkov radiation induced by a rather complicated shower profile can be computed via the time-domain finite-difference (FDTD) method[20] and the time-domain integration method. Between the shower profile $\rho(x)$ and the radio signal $E(t)$, there exists a one-on-one correspondence.[10] The electric field of Cherenkov

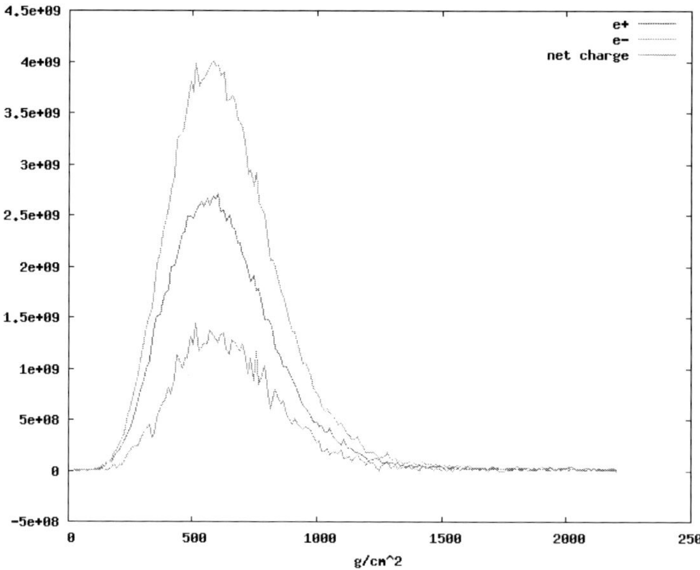

Fig. 1.   The longitudinal development of the 10^{19}eV proton shower in ice. The hadronic shower is initiated by the cascade of mesons. The neutral pion decay length is longer than the LPM interaction length in ice for electrons at energy 10^{15}eV. Meanwhile the secondary mesons produce very few electrons due to LPM suppression.

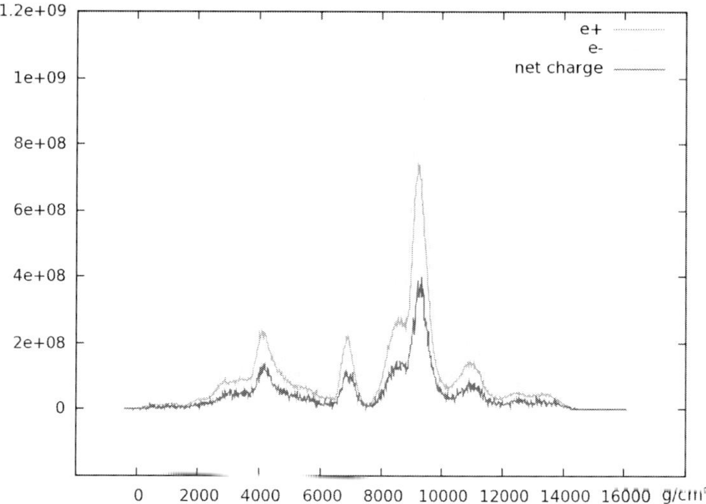

Fig. 2.   The longitudinal development of the 10^{19}eV electron shower in ice. The LPM suppression effect increases as the square-root of the energy. The shower longitudinal profile is sensitive to the initial interactions of the cascade.

radiation can be calculated by solving the inhomogeneous Maxwell equations, as it has been demonstrated by Alvarez-Muniz *et al.*[19] The vector potential can be conveniently obtained in Coulomb (transverse) gauge. The vector potential measured at the detector is then the integral of that contributed from every segment of the shower current, and in turn the electric field can be obtained from the vector potential, as follows:

$$\vec{A}_C(\vec{x},t) = \frac{1}{4\pi\epsilon_0 c^2} \int dt' \, \frac{\vec{J}(t',\vec{x}')}{|x-x'|} \tag{1}$$

$$\vec{E}(\vec{x},t) = -\bigtriangledown \Phi(\vec{x}) - \frac{d\vec{A}(\vec{x},t)}{dt} \tag{2}$$

where t' is the retarded time, $\vec{x}'$ is the position of the shower particles at retarded time, and $\vec{x}$ is the position of the detector, and $\vec{J}$ is the transverse current sources. As we will see later, EM and hadronic showers produce different patterns to be received by the radio detectors. By measuring radio signals, showers are identified and their energies are inferred.

To study muon and tauon tracks, we adopt the muon Monte Carlo package to simulate muon and tauon propagations in ice. In Fig. 3, we display the energy loss rates of mini-showers associated with mu and tau tracks in ice. In Figs. 4 and 5, energy loss from EM and hadronic showers are shown for muon and tauon tracks, respectively. For muon propagation, EM processes are the dominant energy loss mechanism and that by hadronic processes is subdominant. For tau lepton propagation, the situation is reversed. Moreover, the energy loss distribution among different types of showers depends upon the track energy. Once a track is sufficiently measured, its type can be identified and its energy can be inferred as well.

4. Summary

In this work, we propose our strategy to identify cosmogenic neutrino flavors in radio neutrino telescopes, such as ARA. We point out that ν_e can be distinguished from ν_μ and ν_τ due to the fact that ν_e produces an EM shower in ice through charge current interaction whereas ν_μ and ν_τ produce track events. We also propose to reconstruct these tracks by detecting those sub-showers generated along the lepton propagation path. In order to distinguish between ν_μ and ν_τ, we performed simulations on lepton propagation in ice using the MMC code. We find that energy loss distribution among EM and hadronic showers depends on both the lepton flavor and the energy. Simulations for shower production with COSIKA-IW showed different particle profiles for EM and hadronic showers so that the shower type and its energy can in principle be determined by radiation signals.

In summary, we find that neutrino flavors can be distinguished between one another for cosmogenic neutrinos through radiowave neutrino telescopes. We will refine our method of flavor discrimination with more detailed studies on shower production, lepton propagation and radiation conversion.

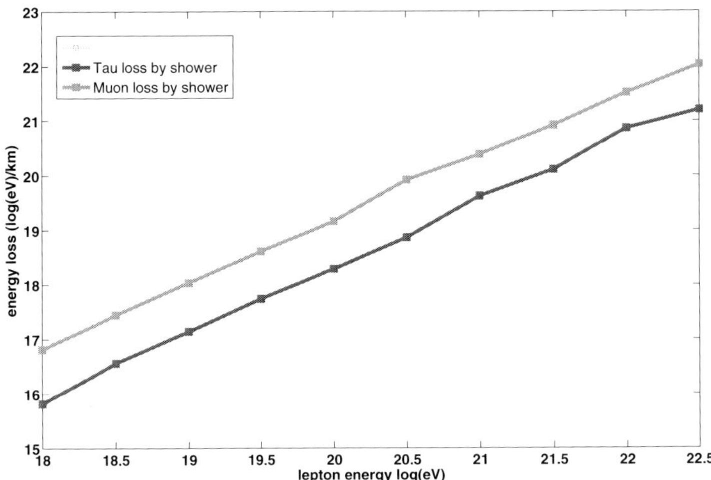

Fig. 3. Energy loss via mini-showers for mu and tau tracks in ice. The loss rate for muons is roughly 10 times larger than that for tauons. One cannot rely on the total energy loss to distinguish different flavors.

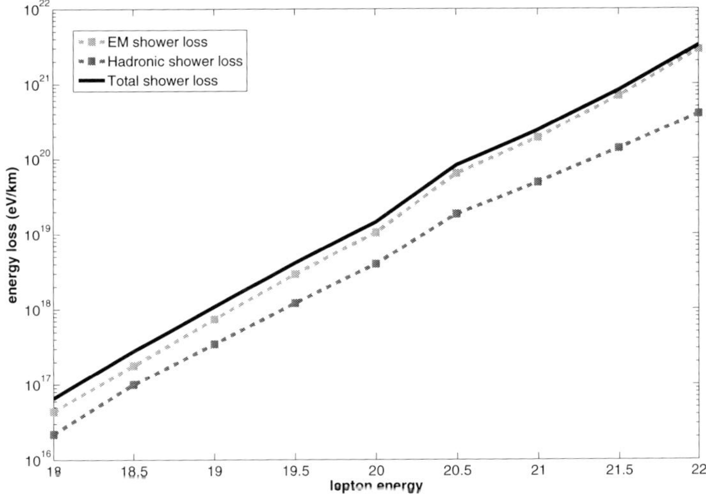

Fig. 4. Energy loss for muon track. The dominant energy loss processes of muon propagating in ice are pair production and bremsstrahlung. The secondary particles (e^+, e^-, γ) from those processes generate the electromagnetic shower. As can be seen, the EM shower profile is distinguishable from the hadronic one.

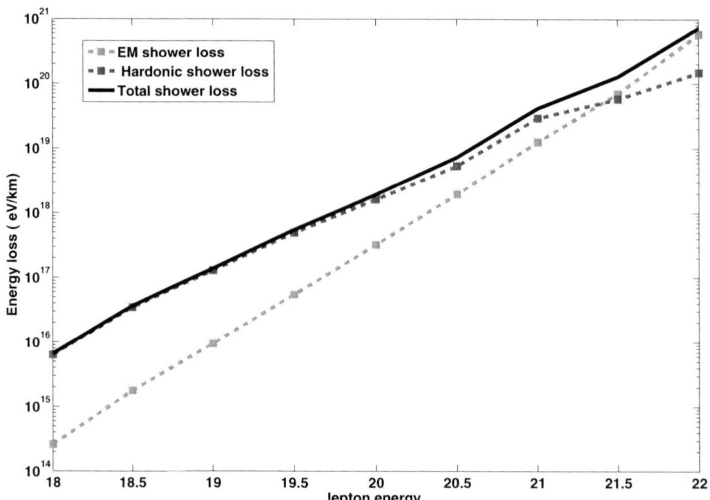

Fig. 5. Energy loss for tauon track. The dominant energy loss process of tauon propagating in ice is photonuclear interaction. Tauon's energy loss rate through EM processes is suppressed since the cross-sections of pair production and bremsstrahlung are inversely proportional to the lepton mass squared.

Acknowledgements

We would like to thank Albrecht Karle, David Besson, Peter Gorham and David Seckel for valuable discussions. This research is supported by Taiwan National Science Council (NSC) under Project No. NSC-100-2119-M-002-525, No. NSC-100-2112-M-182-001-MY3 and US Department of Energy under Contract No. DE-AC03-76SF00515. We would also like to thank Leung Center for Cosmology and Particle Astrophysics for its support.

References

1. K. Greisen, Phys. Rev. Lett. **16**, 748 (1966). G. T. Zatsepin and V. A. Kuzmin, JETP **4**, 114 (1966).
2. R. U. Abbasi et al., Phys. Rev. Lett. **100**, 101101 (2008).
3. T. Yamamoto, International Cosmic Ray Conference **4**, 335 (2008).
4. P. Chen and K. D. Hoffman, U. S. Astronomy Decadal Survey (2010-2020) Science White Paper, (2009) [arXiv:0902.3288].
5. P. Gorham (ANITA Coll.), Phys. Rev. Lett. **103**, 051103 (2009).
6. P. Allison et al. (ARA Coll.), Astropart. Phys. **35**, 457 (2012).
7. S. W. Barwick, J. Phys. Conf. Ser. **60**, 276 (2007).
8. U. F. Katz, Nucl. Inst. Meth. **567**, 457 (2006).
9. G. A. Askaryan, Zh. Eksp. Teor. Fiz.**41**, 616 (1961) [Soviet Physics JETP **14**, 441 (1962)].
10. R. V. Buniy, J. P. Ralston, Phys. Rev. D **65**, 016003 (2002).
11. J. Vandenbroucke, G. Gratta, and N. Lehtinen, ApJ **621**, 301 (2005).

12. V. Aynutdinov et al., Proceedings of the 31st ICRC, Lodz, Poland, July 2009 [arXiv:0910.0678].
13. J. Alvarez-Muñiz, R. A. Vázquez, and E. Zas, Phys. Rev. D **61**, 023001 (1999).
14. M. Tueros, S. Sciutto, Comp. Phys. Comm. **181**, 380 (2010).
15. J. Bolmont, "CORSIKA-IW: a Modifed Version of CORSIKA for Cascade Simulations in Ice or Water", 2008.
16. S. Bevan, S. Danaher, J. Perkin, S. Ralph, C. Rhodes, L. Thompson, T. Sloan, D. Waters, Astropart. Phys. **28**, 366 (2007).
17. D. Heck et al., CORSIKA: A Monte Carlo Code to Simulate Extensive Air Showers, Report FZKA, 6019, (1998).
18. S. Sciutto, "AIRES: A System for air shower simulations. Users guide and reference manual". Version2.2.0 [arXiv:astro-ph/9911331].
19. J. Alvarez-Muniz, A. Romero-Wolf, E. Zas, Phys. Rev. D **81**, 123009 (2010).
20. C.-Y. Hu, C.-C. Chen, P. Chen, Astropart. Phys. **35**, 397 (2012).
21. E. Bugaev, T. Montaruli, Y. Shlepin, I. Sokalski, Astropart. Phys. **21**, 491 (2004).

308

CHERENKOV RADIATION INDUCED BY COSMOGENIC NEUTRINOS IN NEAR-FIELD

CHIA-YU HU[1,2*], CHIH-CHING CHEN[1,2] and PISIN CHEN[1,2,3,4]

[1]*Graduate Institute of Astrophysics, National Taiwan University, Taipei 10617, Taiwan*
[2]*Leung Center for Cosmology and Particle Astrophysics (LeCosPA), Taipei 10617, Taiwan*
[3]*Department of Physics, National Taiwan University, Taipei 10617, Taiwan*
[4]*Kavli Institute for Particle Astrophysics and Cosmology, SLAC National Accelerator Laboratory, Menlo Park, CA 94025, USA*
** E-mail: huchiayu0517@gmail.com*

The radio technique of cosmogenic neutrino detection, which relies on the Cherenkov signals coherently emitted from the particle showers in dense medium, has now become a mature field. We present an alternative approach to calculate such Cherenkov pulse by a numerical code based on the finite difference time-domain (FDTD) method that does not rely on the far-field approximation. We show that for a shower elongated by the LPM (Landau-Pomeranchuk-Migdal) effect and thus with a multi-peak structure, the generated Cherenkov signal will always be a bipolar and asymmetric waveform in the near-field regime regardless of the specific variations of the multi-peak structure, which makes it a generic and distinctive feature. This should provide an important characteristic signature for the identification of ultra-high energy cosmogenic neutrinos.

Keywords: Cherenkov radiation; cosmogenic neutrinos; FDTD.

1. Introduction

Since the ultra-high energy cosmic rays (UHECRs) have been observed up to about $10^{19.6}$ eV, the cosmogenic neutrinos are bound to exist due to the Greisen-Zatsepin-Kuzmin(GZK) processes.[1,2] Nevertheless, they have never been detected by any experimental project so far. The importance of cosmogenic neutrino lies on its two great properties: charge neutral and extremely small cross section. Being charge neutral, it would not be deflected by the magnetic field in the Universe. With its extremely small cross section, it would not be attenuated by the cosmic microwave background (CMB) before reaching the Earth. Cosmogenic neutrino is therefore an ideal probe for us to investigate our Universe in cosmological scales.

On the other hand, the smallness of its cross section can be a great challenge for detection as well. A promising way of detecting UHE neutrinos is the radio approach. As proposed by Askaryan in the 1960's,[3] the high energy particle shower developed in a dense medium would have net negative charges. The net charges of the shower, typically 20% of total shower particles, serve as a source which emits the Cherenkov radiation when they travel in the medium. The sizes of the showers are quite localized (tens of cm in radial and a few meters in longitudinal development) compared to those developed in the air (km scale), and therefore result in coherent radiations for wavelengths longer than the shower sizes. The corresponding coherent wavelength turns out to be in the radio band, from hundreds of MHz to a few GHz.

However, for ultra-high energy showers, the longitudinal development for electromagnetic showers suffers from the Landau-Pomeranchuk-Migdal (LPM) suppression,[4,5] and would therefore be elongated drastically with a stochastic multi-peak structure.[6-8] In such cases, the far-field condition cannot be satisfied for distances up to several kilometers, Meanwhile, the typical detection distance for ground array detectors is about 1 km, as dictated by the attenuation length of radio signals in ice. In addition, the separation between detection stations in the proposed layout is also about 1 km, which makes the possible detection distance even shorter. Under these circumstances, the conventional Fraunhoffer approximation is invalid. In the work of Buniy and Ralston,[9] the correction has been made by the saddle-point approximation, which deals with Fresnel zone radiation where the Fraunhoffer condition fails. However, it still can not cope with extreme cases for $R \sim l$, i.e. the near-field radiation. We address this problem through an alternative approach based on first principles so that near-field radiations can be efficiently obtained.

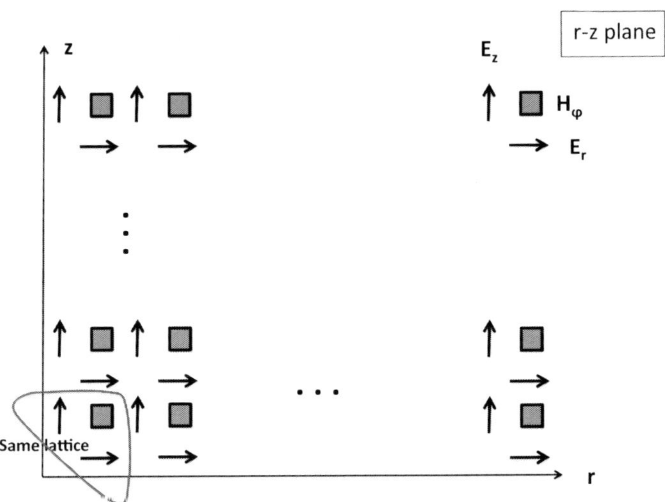

Fig. 1. The adopted lattice configuration. We use cylindrical coordinate and assume cylindrical symmetry. The discretized E-field and H-field are defined on r-z plane. The special staggered configuration enables EM fields to be updated in a leapfrog way.

2. Numerical Method

The idea of FDTD was first proposed by Yee in the 1960's,[10] and has been in use for many years for electromagnetic impulse modeling. Like most numerical finite difference methods, space is discretized into small grids and fields are calculated on each grid by solving Maxwell equations. Adopting a special lattice arrangement (known as the Yee lattice), the E-field and H-field are staggered in both space and time and can be calculated in a leapfrog time-marching way. We assume cylindrical symmetry along the shower axis (defined as z-axis), and therefore all the derivatives with respect to ϕ vanish. In addition, due to the polarization property of Cherenkov radiations, the H_r, H_z, E_ϕ components also vanish. Fig. 1 shows the configuration of the lattice under these assumptions.

Replacing derivatives with finite differences, the discretized Maxwell equations in the cylindrical coordinate are:

$$H_\phi^{n+1/2}(i + 1/2, j + 1/2) = H_\phi^{n-1/2}(i + 1/2, j + 1/2)$$
$$+ \frac{\Delta t}{\mu \Delta r}[E_z^n(i + 1, j + 1/2) - E_z^n(i, j + 1/2)]$$
$$- \frac{\Delta t}{\mu \Delta z}[E_r^n(i + 1/2, j + 1) - E_r^n(i + 1/2, j)], \tag{1}$$

$$E_r^{n+1}(i + 1/2, j) = E_r^n(i + 1/2, j)$$
$$- \frac{\Delta t}{\epsilon \Delta z}[H_\phi^{n+1/2}(i + 1/2, j + 1/2)$$
$$- H_\phi^{n+1/2}(i + 1/2, j - 1/2)], \tag{2}$$

$$E_z^{n+1}(i, j + 1/2) = E_z^n(i, j + 1/2)$$
$$+ \frac{\Delta t}{\epsilon r_i \Delta r}[r_{i+1/2} H_\phi^{n+1/2}(i + 1/2, j + 1/2)$$
$$- r_{i-1/2} H_\phi^{n+1/2}(i - 1/2, j + 1/2)] - \frac{\Delta t}{\epsilon} J_z^{n+1/2}(i, j + 1/2). \tag{3}$$

Iteratively, we can use Eq. 1 to update the H-field, and then use Eq. 2 and Eq. 3 to update the E-field, and so on.

As a general principle, the grid size has to be roughly one order of magnitude smaller than the smallest dimension of the radiating source (and thus roughly the highest frequency of the radiation) to avoid numerical error from the finite difference approximation.,[11] depending on how far the radiation is to travel. In our case the grid size is limited by the lateral width of the shower. We find our optimal grid size to be ten or twenty times smaller than the lateral width. The energy of the shower determines its longitudinal length. However, since the smallest dimension of the shower is its lateral size, shower energy would hardly affect the choice of grid spacing.

3. Cherenkov Pulses in Near-Field

For simplicity, we first investigate the Cherenkov pulses generated by a shower profile that is Gaussian in its longitudinal development. The shower length l is assumed to be 20 m, which corresponds to roughly a 10^{18}eV electromagnetic shower. The lateral distribution, σ_r is set at 1m for numerical reasons.

Fig. 2 shows the waveform in near-field at different angles. The near-field waveform is asymmetric in time and is smoother at its rear part despite that the Gaussian shower is itself symmetric. This feature is very different from that of the far-field case, where the waveform simply reflects the shape of longitudinal development (except at the Cherenkov angle where the longitudinal development has no effect on the far-field pulse).

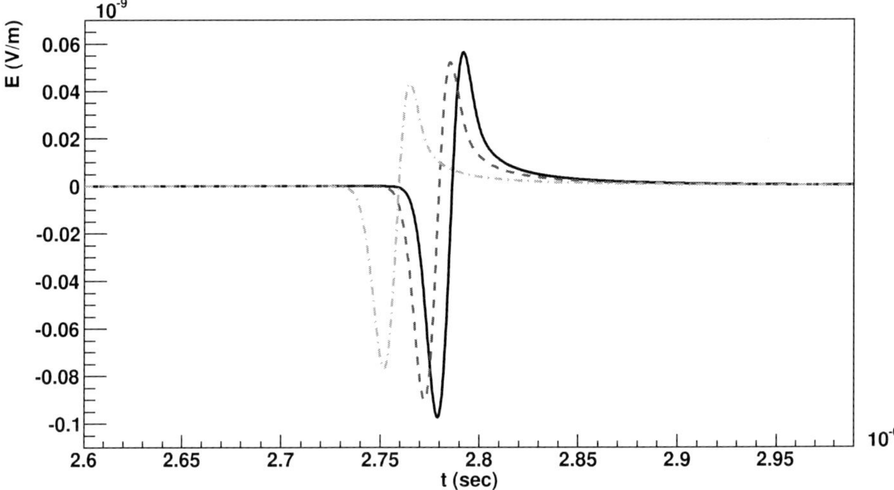

Fig. 2. Waveform in near-field at different angles, for $R = 300$m, $l = 100$ m and $\sigma_r = 1$ m. The solid curve is at $\theta = \theta_c$, dashed curve is at $\theta = \theta_c + 5°$ and dashed-dotted curve is at $\theta = \theta_c + 10°$. The asymmetric waveforms are the striking feature of near-field radiation.

The reason for such feature can be understood qualitatively as follows. The Cherenkov pulse is determined by the spatial and temporal distribution of its source (i.e. the shower), which can be divided into two separated contributions: longitudinal and lateral. In the case of far-field, if the observer locates at θ larger/smaller than θ_c (the Cherenkov angle), one will observe the longitudinal development in an ordinary/reversed time sequence. At $\theta = \theta_c$, one will observe the whole longitudinal development arriving simultaneously. These peculiar phenomena result from the fact that the shower particles travel faster than the speed of light in the medium. In the case of near-field, the above arguments are still useful if one divides the long particle track into several sub-tracks short enough such that the far-field approximation applies within each sub-track.

Fig. 3 depicts the geometrical relation for three different locations (z_-, z_0 and z_+) of source in near-field. Signals emitted from the sub-track around z_0 will arrive the observer simultaneously, while those from the sub-track around z_+/z_- will arrive in an ordinary/reversed time sequence. In other words, signals from anywhere to the left of z_0 will all be reversed in time. Information coming from the vicinity of z_0 (including itself) would all be squeezed into a very short time interval, and thus greatly enhances the signal strength at the beginning part of the pulse, very much like the supersonic shock wave. After that, the signal strength gradually decay as particles travel further away. We can therefore infer, qualitatively, that the potential should have a sharply rise and smoothly decay feature, leading to an asymmetric bipolar electromagnetic field, which is exactly what we saw in Fig. 2.

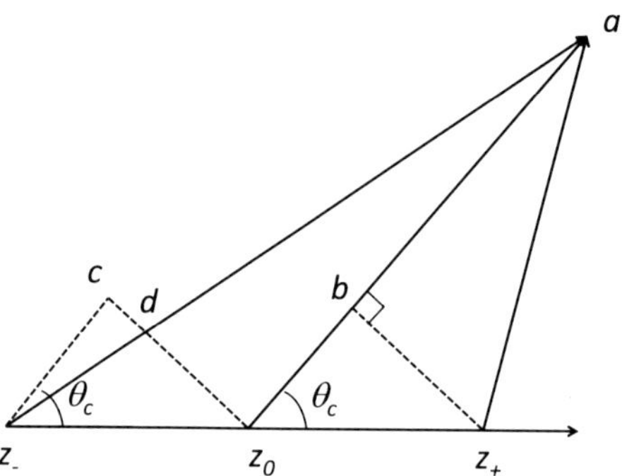

Fig. 3. The geometrical relation that accounts for the arrival time sequence at point a for signals emitted from z_-, z_0, and z_+. Following the ordinary time sequence, signal emitted from z_+ would arrive the observer later than that emitted from z_0. Meanwhile, since the shower in ice travels faster than the radio signal, signal emitted from z_- would still arrive at the observer (at a) later than that emitting from z_0, resulting in a reversal of time sequence. The signal emitted from z_0 is thus the first signal to arrive at a during the whole process.

4. Towards More Realistic Cases

We then investigate more realistic cases, where the longitudinal development displays a multi-peak structure due to the stochastic nature of the LPM-effect. Fig. 4 shows our model of a more realistic electromagnetic shower for $E = 10\text{EeV}$, composed by the superposition of five Gaussian distributions with different heights and widths. Fig. 5 shows the resulting radio pulses at four different detection distances for $\theta = \theta_c + 10°$. These figures show that the bipolar, asymmetric waveform that appears in the Gaussian shower also appears in the multi-peak structure shower. The multi-peak structure seems absent for $R = 50\text{m}$, and then gradually emerges as

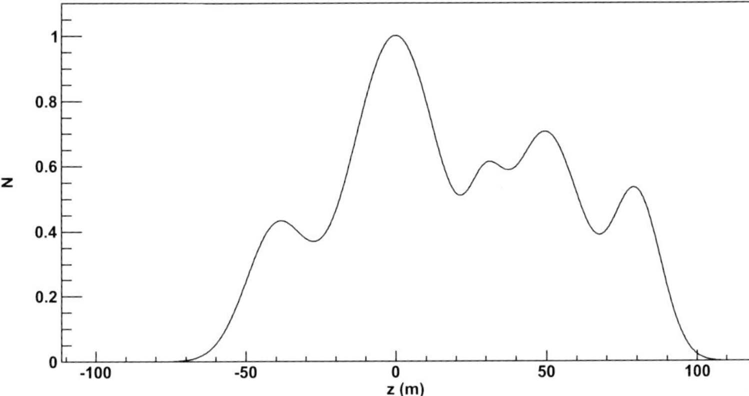

Fig. 4. A multi-peak shower model composed by a superposition of five Gaussian distributions with different heights and widths, with the maximum normalized to 1. The overall length is about 150m, a typical length scale for the EM-shower with primary energy = 10EeV.

R increases. This is because the signal enhancement by squeezing is so significant in the case of $R = 50$m (the near-field regime) that it dominates over the multi-peak structure. As R goes further away, the squeezing effect becomes less visible and the multi-peak structure gradually emerges. At the far-field the waveform simply reflects the longitudinal development, in either ordinary and reversed time sequence, depending on the observation angle.

5. Conclusions

We implement the FDTD method as an alternative way to calculate radio signals based on first principles that does not require the far-field approximation. We demonstrate that for the LPM-elongated showers that have multi-peak longitudinal profile, the waveform would be bipolar and asymmetric even though the multi-peak structure may differ sizably from shower to shower. This should provide an important clue in the identification of ultra-high energy cosmogenic neutrinos. Electron neutrinos are expected to induce showers with extended longitudinal development due to the LPM effect. For a ground array neutrino detector such as the ARA Observatory at South Pole[12] where the antenna station spacing is comparable to the typical length of LPM-elongated showers, the near-field effect must come into play. A bipolar and asymmetric waveform is expected regardless of the specific multi-peak structure of the showers. There is also possibility of detecting the transition from the bipolar asymmetric waveform in the near-field to the multi-peak waveform in the far-field by antenna stations that locate at different distances to the shower event. Such transition, once detected, would demonstrate strong evidence of the LPM-elongated shower.

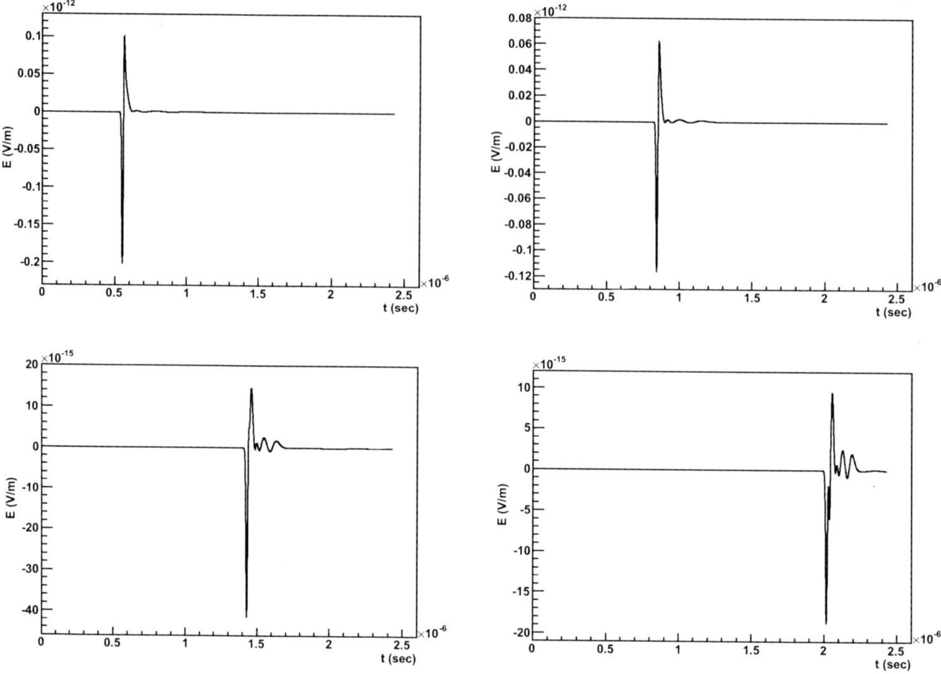

Fig. 5. Waveform at different detection distances (R) for $\theta = \theta_c + 10°$, $\sigma_r = 1$m, and longitudinal development as Fig.4. Detection distance $R = 50$m, 100m, 200m and 300m for the upper left, upper right, lower left, and lower right panel, respectively. The bipolar, asymmetric feature still exists when the detection distance is small. As R increases, the multi-peak structure gradually emerges.

References

1. K. Greisen, *Phys. Rev. Lett.* **16**, p. 748 (1966).
2. G. T. Zatsepin and V. A. Kuzmin, *JETP.* **4**, p. 114 (1966).
3. G. A. Askaryan, *Soviet Physics JETP* **14**, p. 441 (1962).
4. L. Landau and I. Y. Pomeranchuk, *Dokl. Akad. Nauk SSSR* **92**, p. 535 (1953).
5. A. Migdal, *Phys. Rev.* **103**, p. 1811 (1956).
6. V. Niess and V. Bertin, *Astropart. Phys.* **26**, p. 243 (2006).
7. J. Bolmont *et al.*, *Proceedings of the 30th ICRC* (July 2007).
8. J. Alvarez-Muniz *et al.*, *Astropart. Phys.* **4**, p. 100 (2009).
9. R. Buniy and J. Ralston, *Phys. Rev. D* **65**, p. 016003 (2002).
10. K. S. Yee, *IEEE Trans. Antennas Propagat.* **AP-14**, 302 (1966).
11. A. Taflove and S. C. Hangess, *Computational Electrodynamics, The Finite-Difference Time-Domain Method*, 3rd edn. (Artech House, London, 2005).
12. P. Allison *et al.*, astro-ph:1105.2854.

DARK MATTER SEARCH WITH SUB-KEV GERMANIUM DETECTORS AT THE CHINA JINPING UNDERGROUND LABORATORY

QIAN YUE

Department of Engineering Physics, Tsinghua University, Beijing 100084, China
E-mail: yueq@mail.tsinghua.edu.cn

HENRY T. WONG

Institute of Physics, Academia Sinica, Taipei 11529, Taiwan
E-mail: htwong@phys.sinica.edu.tw

(On behalf of the CDEX-TEXONO Collaboration)[*]

Germanium detectors with sub-keV sensitivities open a window to search for low-mass WIMP dark matter. The CDEX-TEXONO Collaboration is conducting the first research program at the new China Jinping Underground Laboratory with this approach. The status and plans of the laboratory and the experiment are discussed.

Keywords: Dark Matter, Underground Laboratory, Radiation Detectors.

1. Introduction

The theme of the CDEX-TEXONO research program is on the studies of low energy neutrino and dark matter physics. The current objectives are to open the "sub-keV" detector window with germanium detectors.[1] The generic "benchmark" goals in terms of detector performance are: (1) modular target mass of order of 1 kg; (2) detector sensitivities reaching the range of 100 eV; (3) background at the range of 1 $kg^{-1}keV^{-1}day^{-1}$ (cpkkd). The neutrino physics program[2,3] is pursued at

---

[*]The CDEX-TEXONO Collaboration consists of groups from China (Tsinghua University, China Institute of Atomic Energy, Nankai University, Sichuan University, Ertan Hydropower Development Company), Taiwan (Academia Sinica, Institute of Nuclear Energy, Kuo-Sheng Nuclear Power Station, National Tsing-Hua University), India (Banaras Hindu University) and Turkey (Middle East Technical University, Karadeniz Technical University). The authors are grateful to the generous and timely support provided by the participating institutes and their respective funding agencies.

the established Kuo-Sheng Reactor Neutrino Laboratory (KSNL), while dark matter searches will be conducted at the new China Jinping Underground Laboratory (CJPL)[5] officially inaugurated in December 2010. The three main scientific subjects are neutrino magnetic moments,[2,4] neutrino-nucleus coherent scattering,[1] and dark matter searches.[6] We highlight the status and plans of the dark matter program in this article.

2. Dark Matter Searches at CJPL

There are compelling evidence that about one-quarter of the energy density in the universe is composed of Cold Dark Matter[6] due to a not-yet-identified particle, generically categorized as Weakly Interacting Massive Particle (WIMP, denoted by χ). A direct experimental detection of WIMP is one of the biggest challenges in the frontiers of particle physics and cosmology. The WIMPs interact with matter pre-dominantly via elastic coherent scattering like the neutrinos: $\chi + N \rightarrow \chi + N$. There may be both spin-independent ($\sigma_{\chi N}^{SI}$) and spin-dependent ($\sigma_{\chi N}^{SD}$) interactions between WIMP and matter.

The facility CJPL[5] is the deepest operating underground laboratory in the world, having $\sim$2400 meter of rock overburden and tunnel drive-in access, as shown schematically in Figure 1. It is located at southwest Sichuan, China, reachable from the provincial international airport at Chengdu via a 50 min flight to Xichang followed by a 90 min drive on a private two-lane motorway. The laboratory is owned by the Ertan Hydropower Development Company, and managed by Tsinghua University, China. Excavation and construction of the first experimental hall ("Hall A") of dimension 6.5 m(width)X6.5 m(height)X40 m(length) with 50 cm of concrete lining was completed in summer 2010. By Fall 2011, the ventilation system, high-speed internet connections, as well as the necessary surface infrastructures (office and dormitory spaces, liquid nitrogen storage system) have been installed. There are intense efforts at CJPL to characterize the background. Measurements are being performed on the ambient radioactivity as well as fast and thermal neutron fluxes. Residual cosmic-ray events have been observed, at a rate (several events per month per square-meter) consistent with the expectation for a location with 2400 m rock overburden. The first generation experimental program at CJPL will include two projects: the CDEX-TEXONO experiment described here, and the dark matter project PandaX with liquid xenon detector. A facility for measuring and screening low-radiopurity materials will be installed. Future expansions of the laboratory are foreseen. New ideas are being discussed and explored.

An experiment with 100 eV threshold would open a window for Cold Dark Matter WIMP searches[6] in the unexplored mass range down to several GeV.[1] Based on data taken at KSNL with the 20-g prototype Ultra-Low-Energy Germanium detector (ULEGe), limits were derived in this low WIMP mass region improving over those from the previous experiments at $3 < m_\chi < 6$ GeV.[7] The $\sigma_{\chi N}^{SI}$ versus m_χ and $\sigma_{\chi N}^{SD}$ versus m_χ exclusion plots are depicted in Figures 3a&b, respectively. Also

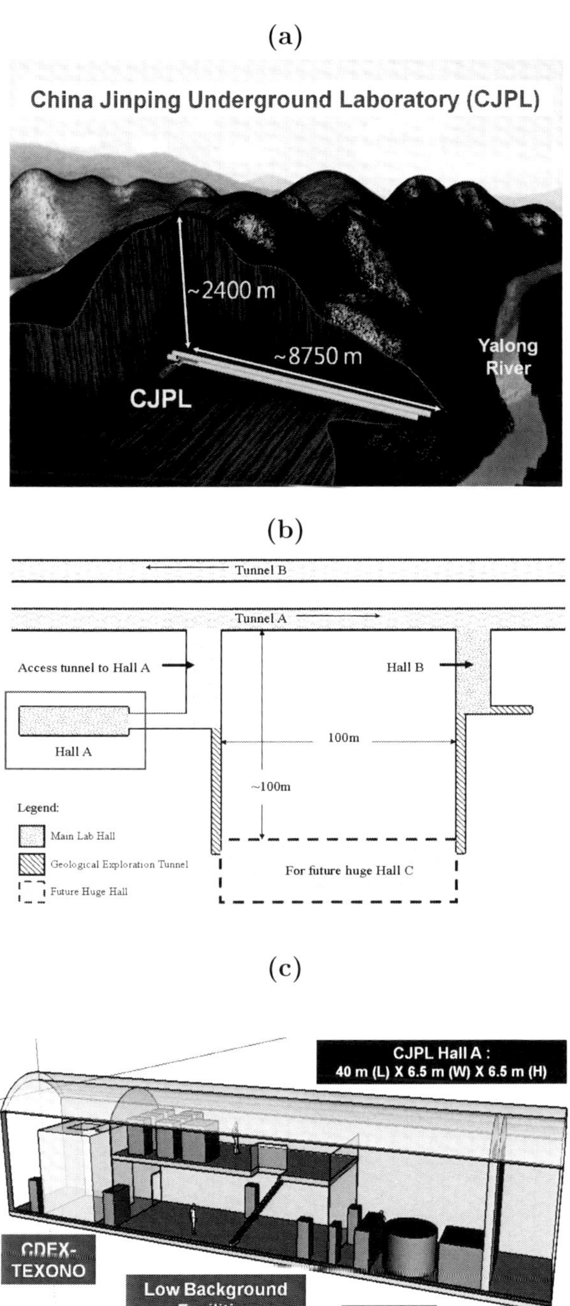

Fig. 1. Schematics diagrams displaying the essential features of CJPL: (a) Geographical setting showing tunnel length and overburden; (b) Floor Plan of the present Hall-A and expected future expansions; (c) Layout of Hall-A, showing both CDEX-TEXONO and PandaX experiments.

displayed are the various results defining the exclusion boundaries, together with allowed regions implied by the DAMA/LIBRA, CoGeNT and CRESST-II data.[8,9] In particular, interpretations of the recent CoGeNT low-energy spectra as positive signatures of low-mass WIMPs[9] have stimulated intense theoretical interests and speculations on this parameter space.

A polyethylene (PE) shielding structure with thickness 1 m and interior dimension 8 m(length)×4.5 m(width)×4 m(height) has been constructed for the CDEX-TEXONO program at CJPL. A 20-g ULEGe array and a 1-kg PCGe have been installed within OFHC copper shielding inside this PE-housing. Data taking has commenced in February 2011. Design and construction of the next-generation PCGe array with total mass at the 10-kg range is proceeding. This new detector will be shielded and enclosed in a liquid argon chamber which serves as both cryogenic medium and active shielding and anti-Compton detector where the scintillation light will be read out by photomultipliers. The conceptual design of this "CDEX-10" is shown in Figure 2. Commissioning is planned in 2013.

Potential reaches are depicted by dotted lines in Figures 3a&b. The projected sensitivities assume Ge detectors at 100 eV threshold (equivalent to about 500 eV nuclear recoils), 10 kg-year of exposure and that the achieved background level of the order of 1 cpkkd at the few keV range can be extrapolated down to threshold.

3. Sub-keV Germanium Detectors

Point-Contact Germanium detectors (PCGe)[10] offer sub-keV sensitivities with detector of kg-size modular mass, an improvement over the conventional ULEGe design. WIMPs with mass down to a few GeV can be probed.

Several R&D directions[11] are intensely pursued towards improvement on the threshold and background for sub-keV germanium detectors:

(1) Pulse Shape Analysis of Near Noise-Edge Events:

It has been demonstrated that by studying the correlation of the Ge signals in two different shaping times[7] as depicted in Figure 4a, the threshold can be further reduced below the hardware noise edge via Pulse Shape Discrimination (PSD). The achieved thresholds at 50% signal efficiency are 220 eV and 310 eV for 20-g ULEGe and 500-g PCGe, respectively. As illustrations, the relative timing between the PCGe and anti-Compton (AC) NaI(Tl) detectors is shown in Figure 5a, for "sub-noise edge" events at 200-400 eV before and after PSD. Events in coincidence with AC at the "50−200 ns" window are due to multiple Compton scatterings, which are actual physical processes having similar pulse shapes as the neutrino and WIMP signals. The PSD selection efficiencies depicted in Figure 5b were derived from the survival probabilities of these AC-tagged samples in the coincidence window. The trigger efficiencies were measured with two methods. The fractions of calibrated pulser events above the discriminator threshold provided the first measurement, while the studies on the amplitude distributions of *in situ* background contributed to the other.

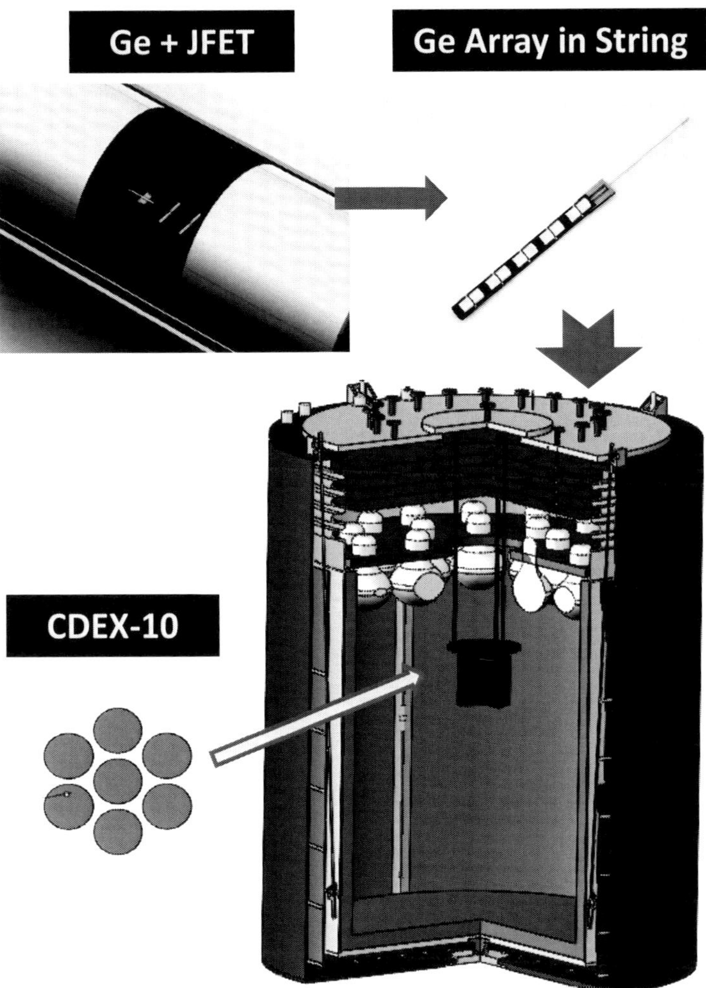

Fig. 2.   Conceptual design drawings of the CDEX-10 detector with the range of 10 kg germanium sensor array enclosed by a liquid argon anti-Compton detector.

Further improvement are being made both on PSD algorithms and on efficiency measurement.

(2) Pulse Shape Analysis of Surface Vs Bulk Events:

The surface and bulk events in PCGe can be separated by the rise time of the pulses as characterized by the amplitude of timing amplifier (TA) signals. It is illustrated in Figure 4b.

(3) Background Understanding and Suppression:

The MeV-range background was understood to the percent level in our previous neutrino-electron measurement with CsI(Tl) scintillating crystal array.[3] However, the measured sub-keV spectrum at KSNL[7] could not be explained

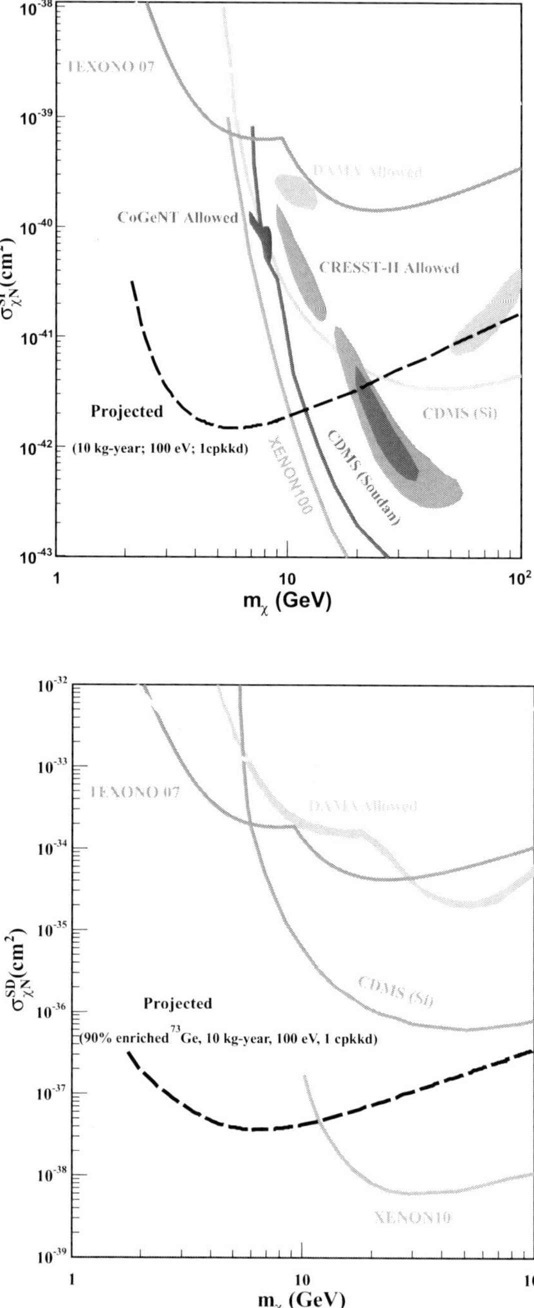

Fig. 3. Exclusion plots of (a) Top: spin-independent χN and (b) Bottom: spin-dependent χN cross-sections versus WIMP-mass, displaying the KSNL-ULEGe limits[7] and those defining the current boundaries.[6,8] The DAMA, CoGeNT and CRESST-II allowed regions[8,9] are superimposed. Projected reach of experiments at benchmark sensitivities are indicated as dotted lines.

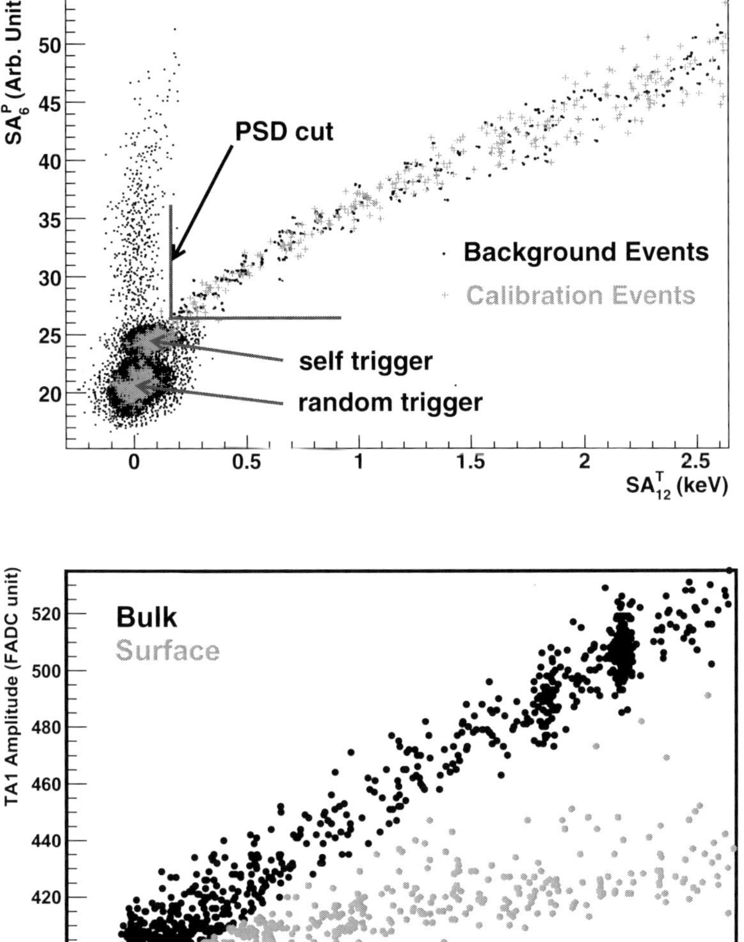

Fig. 4.   (a) Top: Scattered plots of the SA$_6^P$ (shaping time 6 μs with partial integration) versus SA$_{12}^T$ (shaping time 12 μs with partial integration) signals, for both calibration and physics events. The PSD selection is shown. (b) Bottom: Rise time plots, as characterized by the amplitude of timing amplifier (TA) signals, showing different behaviour between surface (faster) and bulk (slower) events.

with standard background modeling on ambient radioactivity. Intense efforts on hardware cross-checks, further simulation and software analysis are underway. Data taking at CJPL, where the cosmic-induced background will be absent, will also elucidate the origin of the observed sub-keV events.

(4) Fabrication of advanced electronics for Ge detectors: R&D program is being pursued to produce advanced JFET and pre-amplifier electronics with goals of fur-

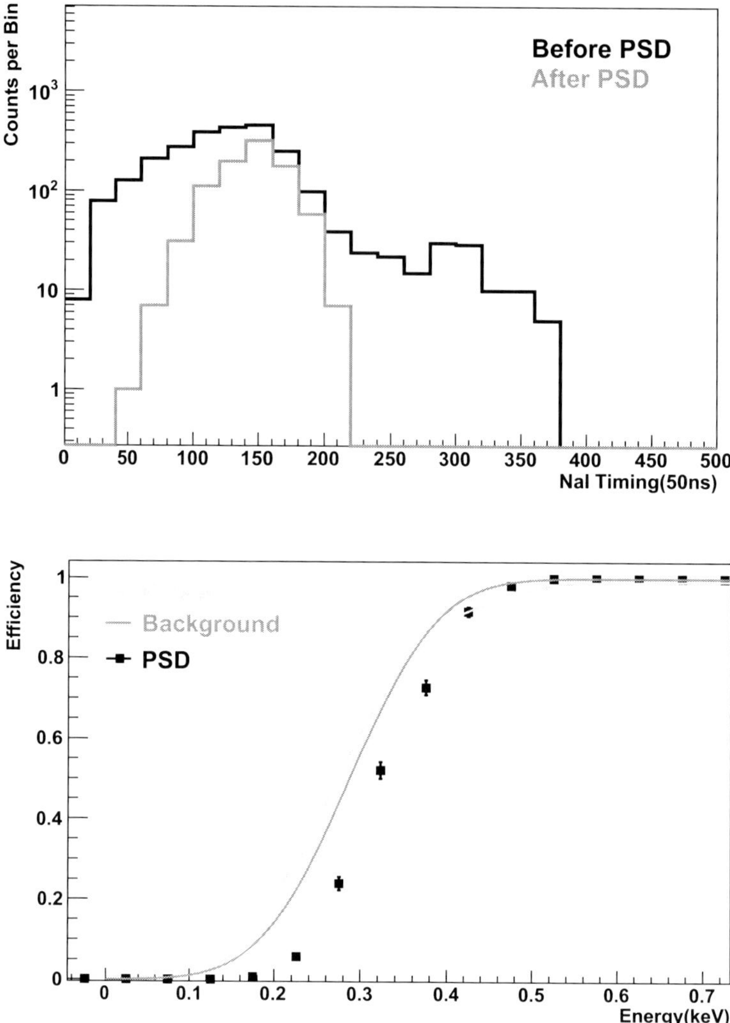

Fig. 5. (a) Top: Selection of relative timing between AC-NaI(Tl) and PCGe systems, before and after PSD selection. The "50−200 ns" is the coincidence window where the signals at PCGe are mostly due to physics events correlated with the AC detector. (b) Bottom: The trigger efficiencies of the 500 g PCGe detector, as derived from the test pulser and *in situ* background events, respectively. PSD selection efficiencies were derived from the survival probabilities of AC-tagged events of (a).

ther reducing threshold and improving energy resolution. Data acquisition and trigger systems with real time analysis capabilities using Field Programmable Gate Arrays (FPGA) are being installed.

4. Prospects and Outlook

A detector with 1 kg mass, 100 eV threshold and 1 cpkkd background level has important applications in neutrino and dark matter physics, as well as in the monitoring of reactor operation. Crucial advances have been made in adapting the Ge detector technology towards these requirements. Relevant limits have been achieved in prototype studies at KSNL on the WIMP couplings with matter. The sub-keV events are still to be understood. Intensive research programs are being pursued along various fronts towards realization of experiments which can meet all the technical challenges. Detectors with kg-scale are being deployed at KSNL and CJPL.

References

1. Q. Yue et al., High Energy Phys. and Nucl. Phys. **28**, 877 (2004); H.T. Wong et al., J. Phys. Conf. Ser. **39**, 266 (2006); H.T. Wong et al., J. Phys. Conf. Ser. **120**, 042013 (2008).
2. H.B. Li et al., Phys. Rev. Lett. **90**, 131802 (2003); B. Xin et al., Phys. Rev. **D 72**, 012006 (2005); H.T. Wong et al., Phys. Rev. **D 75**, 012001 (2007).
3. H.B. Li et al., Nucl. Instr. and Meth. **A 459**, 93 (2001); Y. Liu et al., Nucl. Instr. and Meth. **A 482**, 125 (2002); Y.F. Zhu et al., Nucl. Instr. and Meth. **A 557**, 490 (2006); M. Deniz et al., Phys. Rev. **D 81**, 072001 (2010).
4. H.T. Wong and H.B. Li, Mod. Phys. Lett. **A 20**, 1103 (2005), are references therein
5. K.J. Kang et al., J. Phys. Conf. Ser. **203**, 012028 (2010); D. Normile, Science **324** 1246 (2009); T. Feder, Physics Today **Sept 2010** 25 (2010).
6. M. Drees and G. Gerbier, Review of Particle Physics J. Phys. **G 37**, 255 (2010), and references therein.
7. H.T. Wong, Mod. Phys. Lett. **A 23** 1431 (2008); S.T. Lin et al., Phys. Rev. **D 76** 061101(R) (2009).
8. Latest results presented at the TAUP-2011 Conference.
9. C.E. Aalseth et al., Phys. Rev. Lett. **106**, 131301 (2011); C.E. Aalseth et al., Phys. Rev. Lett. **107**, 141301 (2011).
10. P.N. Luke et al, IEEE Trans. Nucl. Sci. **36**, 926 (1989); P.A. Barbeau, J.I. Collar and O. Tench, JCAP **09**, 009 (2007).
11. H.T. Wong, Int. J. Mod. Phys. **D 20**, 1463 (2011).

TOWARDS A SENSITIVE DARK MATTER DETECTION WITH LIQUID XENON

KAIXUAN NI

*INPAC, Department of Physics and
Shanghai Key Laboratory for Particle Physics and Cosmology,
Shanghai Jiao Tong University, Shanghai, 200240, China
E-mail: nikx@sjtu.edu.cn*

Liquid xenon is one of the target materials used in the direct detection of dark matter at various underground labs around the world. It has the unique capability to discriminate nuclear recoils, produced by elastic scattering of the weakly interacting massive particles (WIMPs), from electron recoils from background gamma rays. Over the last ten years, the target mass of liquid xenon detectors has grown from a few kg to hundreds of kg, making it one of the most promising materials for a sensitive dark matter detection. In this paper, we present the main features of this technology, recent advancement and new development.

Keywords: dark matter; liquid xenon; underground experiment.

1. Introduction

There are many observations in astronomy and cosmology pointing to the fact that more than 80% of the matter in the universe is composed by dark matter, which doesn't emit light or participate in electromagnetic interactions. So far, the only evidence of dark matter is from its gravitational effect. Besides that, whether dark matter interacts with ordinary matter is still a question in particle physics today. Some theories beyond the Standard Model of particle physics predict the existence of stable, massive, neutral particles with the right property for dark matter. Some of the predicted dark matter candidates, such as neutralinos from the supersymmetric extension of the Standard Model, interact with ordinary particles at the weak-scale, making them possible to be detected directly through elastic scattering on the nuclei.[1] A positive direct detection of dark matter will extend our knowledge in both particle physics and cosmology.

The signature of dark matter interaction with the nuclei of ordinary matter are nuclear recoils with energies from a few tens of keV down to zero. In a standard spin-independent elastic scattering interaction, the differential event rate can be calculated[2] according to Eq. 1.

$$\frac{dR}{dE_r} = N_t \frac{\rho_\chi}{m_\chi} \int_{v_{min}}^{v_{esc}} d^3 v v f(\vec{v}, \vec{v}_e) \frac{d\sigma}{dE_r} \tag{1}$$

Where N_t is the number of target nuclei, ρ_χ and m_χ are the dark matter's local density and mass. $\vec{v}$ is the dark matter velocity relative to that of the Earth, and $\vec{v}_e$ is the Earth's velocity in the Galactic halo. $f(\vec{v}, \vec{v}_e)$ is the velocity distribution function, which follows Maxwell-Boltzmann distribution in a standard halo model. v_{min} and v_{esc} are the minimum velocity to produce a nuclear recoil and the galactic escaping velocity.

The rate drops exponentially with increasing nuclear recoil energy (Fig. 1). Dark matter particles with different mass also produce different shapes for the event rate. The dropping of event rate is much faster for lower mass dark matter. Since the rate depends on the velocity of dark matter particles interacting with the target on the earth, it has a feature of annual modulation in the 5% to 10% level due to the velocity changes when the earth is rotating around the Sun, peaking at early June and dipping in December.

An annual modulation signal, possibly due to dark matter interactions, was first reported by the DAMA experiment, using NaI crystals as target, in 1999.[3] After that, the collaboration has accumulated much higher statistics and reported new results at a 8.9σ confidence level of annual modulation signal.[4] More recently, the CoGeNT experiment, using a p-type point contact Germanium detector, reported event excess at low energy[5] and the CRESST-II experiment with $CaWO_4$ target reported event excess in the O-band,[7] pointing to a possible dark matter candidate with a low mass around 10 GeV/c^2. The CoGeNT experiment also reported an annual modulation signature which is consistent with DAMA.[6] However, these results are in conflict with other experiments with null results, in particular the CDMS,[8,9] XENON10[10] and XENON100[11] experiments. In addition, the recently revised CoGeNT background due to previously unidentified surface events[12] makes it hard to fit the DAMA, CoGeNT and CRESST-II results in a consistent picture, unless a non-standard local dark matter velocity distribution or velocity-dependent interacting cross sections is introduced.[13]

It is clear that a solid understanding of background is essential before a positive detection can be claimed. Due to the difficulty to identify and estimate the background at low energy, any unknown background can easily fake the light dark matter. Since the electron recoil background is still the dominant background for almost all experiments, dark matter detectors with the capability to identify and reject electron recoils have the great potential to identify true nuclear recoil events from WIMPs. Liquid xenon, especially with the two-phase technique, is among one of the best techniques for a clean detection of dark matter.

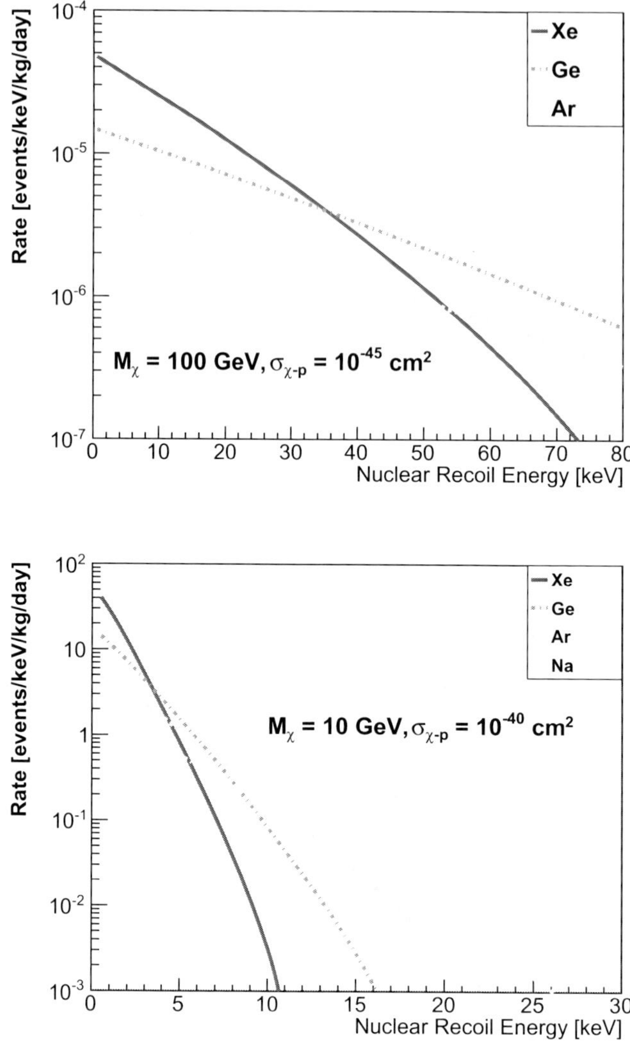

Fig. 1. Spin-independent elastic scattering event rates for two different WIMP mass and cross-section values, for targets with different atomic mass.

2. Dark Matter Detection Using Liquid Xenon

The choice of liquid xenon as a target for dark matter detection has the following advantages. Firstly, the high atomic mass (A) leads to a large event rate in the spin-independent interaction channel since the total WIMP-nucleus cross section is proportional to A^2, especially at low recoil energies (Fig. 1). The drop at high recoil energy is due to the suppression from the nuclear form factor for heavy nucleus. Natural xenon contains about half fraction of spin-odd isotopes (26.4% ^{129}Xe and

21.2% ^{131}Xe), making it sensitive to spin-dependent interactions as well. Secondly, the high scintillation yield in liquid xenon makes it possible to observe nuclear recoils in the few keV region. The high atomic number of Xe and the large density of liquid xenon makes it effective to shield external radioactive background. Finally, the relatively economic price for the Xe material and not extremely low temperature requirement (165 K) make it possible to realize a ton-scale detector much easier than other technologies, such as the cryogenic bolometer.

2.1. *Signals in liquid xenon*

A particle interaction in liquid xenon produces both Xe excitation states and electron-ion pairs. The decay of excited states to the ground state results in a scintillation photon at a wavelength peaked at 178 nm. The recombination of electron-ion pairs also form excited states and produces additional scintillation photons. If an electric field is presented in the liquid xenon, the electron-ion recombination will be suppressed. Electrons will be drifted in the liquid and be collected at the electrodes. Due to the small number of electrons from a low energy recoil, a two-phase Xe detector (see Fig. 2) is used to drift the electrons from the liquid to the gas phase, where a stronger field around $\sim$10 kV/cm converts the electrons to electroluminescence signals to be detected by the same photo-sensors for the primary scintillation photons.

The nuclear recoil signal from a WIMP elastic scattering in liquid xenon is different from the electron recoils from background gamma rays. Firstly, Most of the energy of the nuclear recoil is transferred to the atomic motion and can't be detected, such a quenching effect gives only 10% to 20% observable energy relative to the electron recoils with the same energy (a most recent measurement can be found at[14]). Secondly, due to the much higher ionizing density for nuclear recoils than that for electron recoils, more electron-ion recombination happens for nuclear recoils, resulting a smaller value of ionization/scintillation ratio. This makes it possible to discriminate between nuclear recoils and electron recoils if both ionization and scintillation signals can be detected using the two-phase technique. For single-phase experiment, such as XMASS,[15] only the scintillation signal is detected, thus this kind of discrimination is not possible.

2.2 *Background control*

A sensitive dark matter detection requires an extremely good background control The experiments are located at deep underground labs in order to have less impact from the muon-induced background. Passive or active shields are used to reduce environmental neutron and gamma background, or muon-induced background. The materials used to construct the detector need to be radioactive clean. And the discrimination using the two-phase Xe technique can further reduce the gamma background by a factor of 100 to 1000.

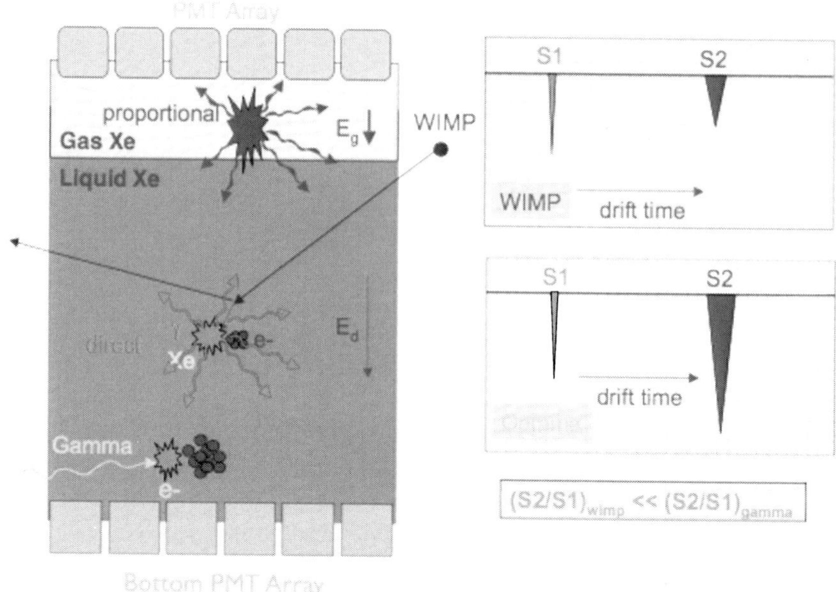

Fig. 2. Operation principle of a two-phase xenon detector. Both scintillation (S1) and ionization (S2) signals are produced for nuclear recoils from WIMPs or neutrons, or for electron recoils from gamma rays. The delay between these two signals and the hit pattern of the secondary signal on the PMTs give the 3D positions. The ratio of the two signals are different for electron recoils and nuclear recoils, which can be used to discriminate a signal from the background.

The extremely low rate of WIMP interaction requires the detector construction materials to be radioactively clean. For liquid xenon based dark matter detectors, the main components used to construct the detector are liquid xenon, photomultiplier tubes (PMTs), PTFE reflectors and the detector vessel made by stainless steel, oxygen free high purity copper (OFHC), or titanium for various experiments. Some of the lowest radioactivity achieved by some current liquid xenon experiments are listed in Table 1.

Table 1. Best radioactivity level achieved for materials used to construct dark matter detectors based on liquid xenon.

Components		Radioactivity (mBq/Unit)			
	Unit	^{238}U(^{226}Ra)	^{232}Th(^{228}Th)	^{60}Co(^{46}Sc)	^{40}K
R8520[18]	PMT	<1.4(0.12)	(0.11)	0.53	6.0
R8778[19]		18	17	8	30
R10789[15]		0.37	1.0	3.1	<5.9
R11410[18]		<95(<2.4)	(<2.6)	3.5	13
PTFE[18]	kg	<6.2(<0.31)	(<0.16)	<0.11	<2.25
Stainless steel 316Ti[18]		<20(<1.3)	(< 1.0)	1.4	<5.7
OFHC[18]		0.07	0.02	0.002	0.023
Titanium[20]		<0.25	<0.2	(<0.35)	< 1.2

Xenon itself has no long-lived isotope to contribute to any intrinsic background. But commercial xenon from air separation and distillation plants contains a small fraction of krypton (Kr) at part per million (ppm) to part per billion (ppb) level. The Kr contains about 10^{-11} ^{85}Kr, which is a beta decay emitter with a half-life of 10.756 years and an end point at 687 keV. Unlike external background, the beta decay of ^{85}Kr is uniformly distributed in the target volume, making it impossible to be reduced by fiducial volume selection. A ppb Kr/Xe contamination introduces an electron recoil background rate of 2×10^{-2} events/keVee/kg/day at low energy. Thus a one-ton LXe target with 1 ppb Kr/Xe will produce around 70,000 events in a 10 keVee energy window in one year. A reduction by three orders of magnitude of the Kr/Xe concentration down to 1 ppt (part per trillion) is possible by using the charcoal adsorption[16] or cryogenic distillation[17] techniques, which will reduce the background event number to less than 100 events/ton/year. A two-phase xenon detector with more than 99% electron recoil background rejection capability can further reduce this number below 1 events/year.

The main advantage of using two-phase technique compared to the single-phase technique is its capability to reject background based on the fiducial volume selection and the ionization/scintillation ratio. Due to the excellent 3D position sensitivity, external background interacts in the outer layer of liquid xenon can be rejected by a fiducial volume selection (Fig. 3). The different ionizing density of nuclear recoils compared to that of electron recoils resulting a different ionization/scintillation signal ratio. Thus the electron-type background can be further rejected (Fig. 4). More than 99.5% electron recoils are rejected by keeping 50% nuclear recoil events in the XENON10 detector,[21] while a 99.75% electron recoil rejection by keeping an average 35% nuclear recoils is used for the XENON100 detector.[11] It has been reported by the ZEPLIN-III experiment that the rejection power can be even better with a higher drift field .[22]

Due to the excellent background control and promise to realize a sensitive dark matter detector with a large mass target, the two-phase xenon technique has been used by several dark matter experiments around the world, including ZEPLIN II,[24] ZEPLIN III,[22] XENON10[21] and XENON100.[11] New experiments, including LUX,[19] PandaX[23] and XENON1T[25] will keep increasing the target mass.

3. Recent Results from XENON100

During the past ten years, the two-phase xenon technique has revolutionized the field of direct dark matter detection with an ever increasing sensitivity. It is led by the XENON10/100 experiments operated at the Gran Sasso underground laboratory in Italy. Other experiments like ZEPLIN II/III using the similar technology has also produced interesting results. Here we just summarize the main features of the XENON100 detector and its latest results, as reported in.[11,26]

The sensitive target of XENON100 is a cylindrical time projection chamber (TPC) of 30.5 cm height and 15.3 cm radius, containing 62-kg LXe, operated with

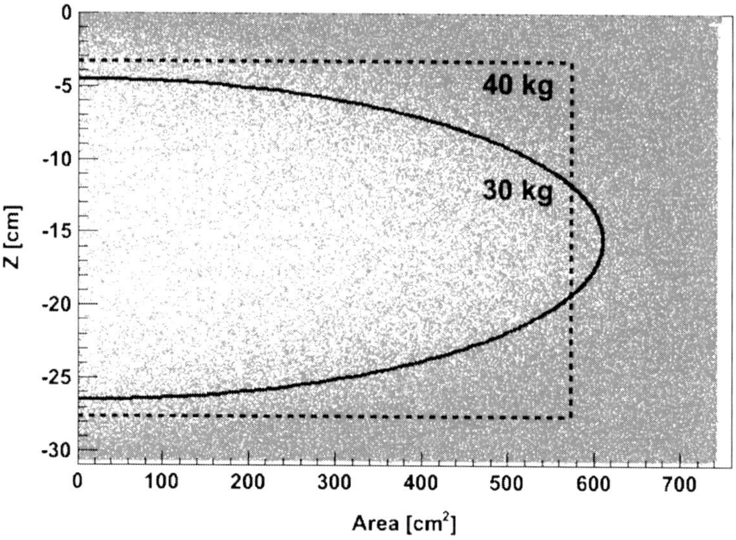

Fig. 3. Illustration of external background reduction based on the fiducial volume selection in XENON100 (figure from[27]). Optimization of the selection is based on the position distribution of background events in the target.

a drift field of 0.53 kV/cm. The signals from this target are viewed by two arrays of total 178 Hamamatsu R8520 PMTs from the top and bottom. More than 80% of the primary scintillation light (S1) is detected by the bottom PMT array. The total light yield is measured to be 2.2 pe/keVee (photoelectrons per keV electron-equivalent energy) at a field of 0.53 kV/cm. The top PMT array detects the S2 signals and reconstructs the event positions based on the hit pattern. An (x, y) resolution of less than 3 mm (1σ) is achieved. The z resolution is about 0.3 mm (1σ) from the drift time distribution. The XENON100 detector was designed with an active LXe veto around the sensitive target. It contains 99-kg of LXe and is viewed by 64 R8520 PMTs to reject gamma rays interacting with at least one scatter in the veto.

The main background in the XENON100 detector is from the radioactivities in the detector constructing materials, including the PMTs, cryostat made of 316Ti stainless steel, and the ^{85}Kr in the xenon.[27] The radioactivities in the PMTs and cryostat produce mainly gamma rays, producing electron recoils in the target. Since these materials are surrounding the target and the gamma interactions are mostly happen in the outer layer of the LXe target, the event rate can be dramatically reduced by selecting only the single scatter electron recoils in the central part of the target. A single electron recoil background rate is reduced by a factor of 20 to 40 by going from the entire 62-kg target to 30-kg in the center, achieving a rate lower than 10^{-2} events/kg/keV/day below 100 keVee. However, the ^{85}Kr in the xenon produces beta decay events uniformly distributed in the target, which is not possible to be removed with the fiducial volume selection. The technique to reject

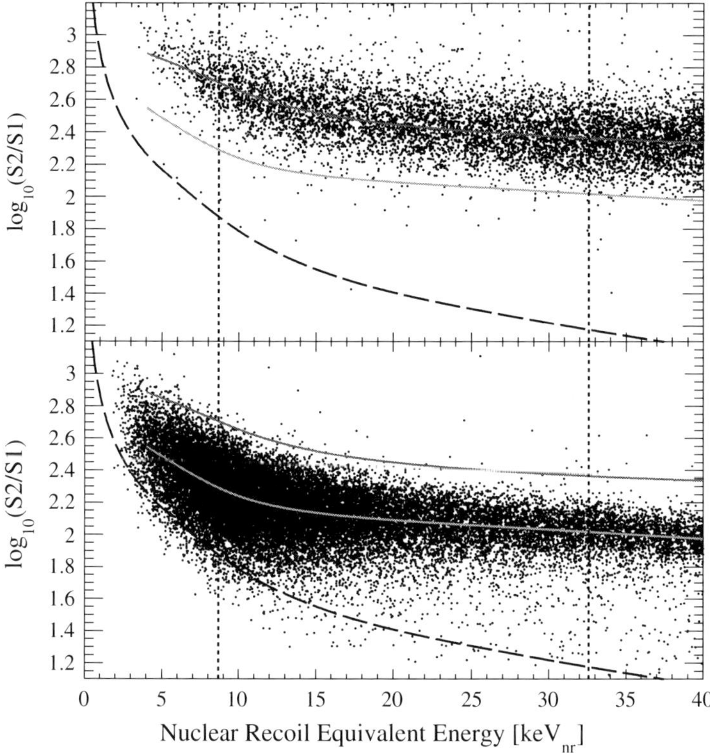

Fig. 4. Illustration of the electron recoil rejection based on the ratio of S2/S1. The top shows electron recoil events from gamma ray calibration, the bottom is nuclear recoils from neutron calibration (figure from[11]). The blue and red lines indicate the mean of the electron and nuclear recoil bands.

electron recoil type events using the S2/S1 ratio achieved a further background rejection of 99.75%, by keeping about 35% nuclear recoil events.

From a 100 live-days of WIMP search in the 48-kg fiducial mass target, XENON100 observed 3 events in the energy window of 4-30 pe, corresponding to a nuclear recoil energy window of 8.4-44.6 keV using the latest nuclear recoil relative scintillation efficiency measurement.[14] With 1.8 expected background events, such an observation provides the most stringent exclusion limit for the spin-independent WIMP-nucleon interaction cross sections. The limit is 7.0×10^{-45} cm^2 at a WIMP mass of 50 GeV/c^2.

The background from the 100 live-days of XENON100 data is mainly from ^{85}Kr due to the high contamination of Kr/Xe at a value of 700 ppt (part per trillion). Since then, the experiment has further reduced the Kr contamination below 100 ppt with the cryogenic distillation technique.[28] New data with more than 200 days of running have been accumulated with a lower energy threshold. A factor of two to three improvement in sensitivity is expected (see Fig. 5) with the new data.

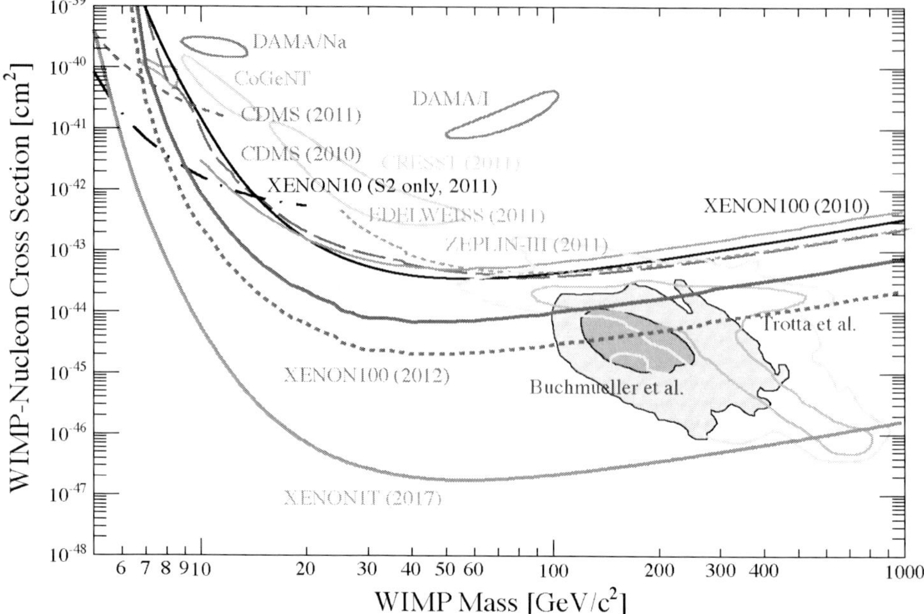

Fig. 5. The latest upper limit from XENON100's 100 live-day search (solid blue line), the expected sensitivity from XENON100's new data in 2012 (blue dash line), and a projected sensitivity for the XENON1T experiment in the near future (figure from[28]).

4. Challenges and Developments for the Next Generation Experiments

Following the success of XENON10/100 experiments, the liquid xenon technology has gained a lot of interests worldwide for the ton-scale experiments, including the LUX, XENON1T and PandaX experiments. All these experiments use the two-phase technique optimized for a better position sensitivity and background gamma rejection, compared to the single-phase technique used by the XMASS experiment. However, several challenges are faced by these experiments towards a large target mass that has never been realized before.

One challenge in any liquid xenon detector is to optimize the signal collection. Liquid xenon itself is transparent to its own scintillation light. However, the presence of impurity, such as water, in the xenon decreases the light collection by absorption. The absorption length of xenon scintillation light in liquid xenon varies dramatically, depending on the water concentration. According to the calculation in,[29] a 1 ppb water concentration in xenon gives an absorption length in the order of 10 m. Similarly, the presence of electro-negative impurities, such as oxygen, in xenon will reduce the number of electrons by attachment, resulting a poor ionization signal collection. A 1 ppb O_2-equivalent impurity concentration is required to drift electrons over the order of 1 m. The ppb purity level can be achieved by various purification processes, such as letting the xenon gas pass through molecular sieves or heated zirconium alloy. A better control of the surface outgassing of the materials touching the liquid xenon and a fast purification speed are necessary in order to achieve a good purity in a given time.

Another challenge regarding to the signal generation is the high voltage (HV) operation. In a ton-scale detector with a drift length in the order of 1 meter, a reasonable drift field of 1 kV/cm requires 100 kV applied in the cathode. Although a proper design of the electrodes may avoid electrical breakdown in the liquid, the achieved HV reported by various experiments has so far limited below 20 kV. The XENON100 experiment limited its cathode HV at 16 kV to avoid secondary pulses possibly due to the electron emission and subsequent scintillation in the strong field near the sharp points on the cathode mesh.[26] Recently, LUX experiment reported a 300 V/cm[30] over a 59 cm drift length, corresponding to a cathode HV of less than 18 kV, limited by a provisional cathode feedthrough. For next generation experiment, better electrodes and feedthroughs for the cathode need to be implemented.

Finally, the reduction of radioactive background remains a challenging task for any dark matter experiment. So far, for the direct detection experiment in a deep underground lab with a well-designed active or passive shield, the external muon-induced and environmental background can all be reduced to a sub-dominant value. However, the intrinsic background from the detector construction materials, such as the PMTs, vessel and the xenon itself, need to be reduced in order to achieve a better sensitivity. The lowest radioactive PMTs available in the market are from Hamamtsu, including R8520, R8778 and the R11410 PMTs with window sizes ranging from one-inch-square to three-inch diameter. For a ton-scale detector, the current PMT radioactivity needs to be further reduced in cope with the increasing sensitivity needs. Other development, such as QUPID[31] and gas photomultiplier tubes (GPMTs)[32] with a good control of construction material's radioactivity, bring alternative possibility for the future. The reduction of the Kr concentration in Xe below ppt level is also actively pursued by the above mentioned experiments.

To face these challenges, a new experiment called PandaX[23] using the two-phase xenon technique is being built. It will be located at the China Jin-Ping underground Laboratory (CJPL), which is in the middle of a 18-km tunnel under 2400 meters of rock overburden, in the Sichuan province of south-west China. As one of the deepest underground labs in the world, the CJPL has an extremely low muon flux rate at less than $20/m^2/100$-day,[33] which is about two orders of magnitude lower than the flux at the Gran Sasso underground laboratory (LNGS) in Italy. Such a low muon flux and the resulting low muon-induced background make the lab ideal to host ton-scale dark matter detectors with less complicated shields compared to shields with active muon veto in shallower labs. A passive shield with the potential to host a future ton-scale xenon detector, using the oxygen-free high conductive copper (OHFC), low radioactive lead, high density polyethelene (HDPE) is designed and built at CJPL.

The target of the first stage of PandaX will have a mass of around 120-kg. The optimized light collection efficiency, a specially designed HV feedthrough, the fast Xe purification system, and low background material selection including Kr reduction based on a distillation column, will pave the way for the next stage detectors.

Acknowledgements

This work is supported by MOST 2010CB833005, NSFC 11055003, 11175117 and STCSM 11PJ1405300, 11DZ2260700.

References

1. M.W. Goodman and E. Witten, Phys. Rev. **D31**, 3059 (1985).
2. J.D. Lewin and P.F. Smith, Astropart. Phys. **6**, 87 (1996).
3. R. Bernabei, P. Belli *et al.*, Phys. Lett. **B450**, 448 (1999).
4. R. Bernabei, P. Belli *et al.*, Eur. Phys. J. **C67**, 39-49 (2010), arXiv:1002.1028.
5. C.E. Aalseth, P.S. Barbeau *et al.*, Phys. Rev. Lett. **106**, 131302 (2011).
6. C.E. Aalseth, P.S. Barbeau *et al.*, Phys. Rev. Lett. **107**, 141301 (2011).
7. G. Angloher, M. Bauer *et al.*, arXiv:1109.0702.
8. Z. Ahmed, D.S. Akerib *et al.*, Phys. Rev. Lett. **106**, 131302 (2011).
9. Z. Ahmed, D.S. Akerib *et al.*, arXiv:1203.1309.
10. J. Angle, E. Aprile *et al.*, Phys. Rev. Lett. **107**, 051301 (2011).
11. E. Aprile, K. Arisaka *et al.*, Phys. Rev. Lett. **107**, 131302 (2011).
12. Talk by J. Collar, TAUP 2011 Workshop, Munich, Germany, Sep.5-9, 2011.
13. C. Kelso, D. Hooper and M. R. Buckley, arXiv:1110.5338.
14. G. Plante, E. Aprile *et al.*, Phys. Rev. C **84**, 045805 (2011).
15. XMASS experiment, http://www-sk.icrr.u-tokyo.ac.jp/xmass/darkmatter-e.html
16. A. I. Bolozdynya, P. P. Brusov, *et al.*, Nucl. Instrum. Meth. A **579**, 50 (2007).
17. K. Abe *et al.*, Astropart. Phys. **31** (2009) 290.
18. E. Aprile, K. Arisaka, *et al.*, Astropart. Phys. **35**, 43 (2011).
19. The LUX dark matter experiment, http://lux.brown.edu/.
20. D.S. Akerib, X. Bai, *et al.*, arXiv:1112.1376.
21. J. Angle, E. Aprile, *et al.*, Phys. Rev. Lett. **100**, 021303 (2008).
22. D.Yu. Akimov, H.M. Araujo *et al.*, arXiv:1110.4769; P. Majewski, V.N. Solovov, *et al.*, arXiv:1112.0080.
23. G. Tarle, Talk Given at UCLA Dark Matter Symposium 2012, https://hepconf.physics.ucla.edu/dm12/talks/tarle2.pdf.
24. G.J. Alner *et al.*, Astropart. Phys. **28** (2007) 287-302.
25. E. Aprile, Talk Given at UCLA Dark Matter Symposium 2012, https://hepconf.physics.ucla.edu/dm12/talks/aprile.pdf.
26. E. Aprile, K. Arisaka, *et al.*, Astropart. Phys. **35**, 573 (2012).
27. E. Aprile, K. Arisaka, *et al.*, Phys. Rev. D **83**, 082001 (2011).
28. P. Scovell, Talk Given at UCLA Dark Matter Symposium 2012, https://hepconf.physics.ucla.edu/dm12/talks/scovell.pdf.
29. K. Ozone, *Liquid Xenon Scintillation Detector for the New $\mu \to e\gamma$ Search Experiment*, Ph.D. Thesis, University of Tokyo (2005).
30. E. Bernard, Talk Given at UCLA Dark Matter Symposium 2012, https://hepconf.physics.ucla.edu/dm12/talks/bernard.pdf.
31. A. Teymourian, D. Aharoni *et al.*, Nucl. Instr. Meth. A **654**, 184 (2011).
32. S. Duval, A. Breskin *et al.*, JINST **6**, P04007 (2011).
33. Q. Yue, Talk Given at 7th International Workshop on the Dark Side of the Universe (DSU2011), http://kitpc.itp.ac.cn/dsu2011/slides/DSU2011-Q.Yue.pdf.

BAYESIAN IMPLICATIONS OF COLLIDER AND SUSY DARK MATTER DIRECT AND INDIRECT SEARCHES

LESZEK ROSZKOWSKI,[*] ENRICO MARIA SESSOLO, and YUE-LIN SMING TSAI[†]

Members of the BayesFITS Group,
National Centre for Nuclear Research,
Hoza 69, 00-681 Warsaw, Poland
E-mail: smingtsai@googlemail.com

In this talk we present our recent Bayesian analyses of the Constrained MSSM in which the model's parameter space is constrained by the CMS α_T 1.1/fb data at the LHC, the XENON100 dark matter direct detection data, and Fermi-LAT γ-ray data from dwarf spheroidal galaxies (dSphs). We also show that the projected one-year sensitivities for annihilation-induced neutrinos from the Sun in the 86-string configuration of IceCube/DeepCore have the potential to yield additional costraining power on the parameter space of the CMSSM.

Keywords: Supersymmetry; Dark Matter; LHC; Bayesian theory; Fermi LAT; Gamma ray; IceCube; DeepCore; neutrino.

1. Introduction

The supersymmetric neutralino is arguably the most popular and best motivated candidate particle for the dark matter (DM) that permeates the Universe. Hence, over the years, several searches have been deviced aimed at the discovery of supersymmetry (SUSY) and DM. The Large Hadron Collider (LHC), the XENON100 underground DM detector, the Fermi-LAT γ-ray telescope, and the IceCube neutrino telescope, are just a few of the many instruments used to pursue this two-fold quest.

The Constrained Minimal Supersymmetric Standard Model (CMSSM)[1,2] is the most popular SUSY model because of well-motivated unification assumptions and a greatly reduced number of parameters. Despite its restrictive boundary conditions, the CMSSM has a rich phenomenology and has been extensively used as a basis for evaluating prospects for SUSY searches at the LHC and in other collider, non-collider and DM experiments. The model is defined by four continuous parameters and one sign: m_0, the universal scalar mass, $m_{1/2}$, the universal gaugino mass, A_0, the universal trilinear coupling, $\tan\beta$, the ratio of the Higgses' vacuum expectation values, and sgn μ, the sign of the Higgs/higgsino mass parameter μ.

---

[*]On leave of absence from Department of Physics and Astronomy, University of Sheffield.
[†]Speaker.

Before the turn-on of the LHC, the strongest constraint on the CMSSM parameter space came from the WMAP7 measurement of the DM relic abundance. Hence, each mechanism responsible for obtaining the correct relic density has often been used to label the regions of CMSSM parameter space constrained by the measurement. There are four main mechanisms for DM annihilation in the early universe which apply to the CMSSM, hence one can recognize four main regions in the model's parameter space:

- The *bulk* region at small m_0 and small $m_{1/2}$.
- The *stau coannihilation* region at small m_0 and large $m_{1/2}$.
- A *focus point/horizontal branch* (FP/HB) region at very large m_0 and smaller $m_{1/2}$.
- The pseudoscalar Higgs *A-funnel* and the *light Higgs h resonance* ($2m_\chi \sim m_{A,h}$) regions, the former of which is enhanced by large $\tan\beta$.

Besides the relic density, the CMSSM parameter space can be constrained by several other experimental results: SUSY mass limits, flavour physics, electroweak measurements, the anomalous magnetic moment of the muon, etc.

After the turn-on of the LHC, the best limits from sparticle searches came from the CMS[3] α_T and razor analyzes. The results we show in this talk are based on the α_T 1.1 fb^{-1} data, implemented in our scans through a likelihood function method which significantly constrains the $(m_0, m_{1/2})$-plane.

A full study of the impact of DM searches needs to consider both direct and indirect detection. It is well known that, despite large astrophysical uncertainties, the published upper limit on the spin-independent (SI) neutralino-proton cross section (σ_p^{SI}) from XENON100[4] and gamma ray data from dSphs obtained by Fermi-LAT[5] can constrain the FP/HB region. Thus, besides the α_T data we also apply to our Bayesian inference the recent XENON100 upper bounds on σ_p^{SI} and the bounds on the annihilation cross section of neutralinos from gamma-ray fluxes from dSphs published by the Fermi-LAT Collaboration. Additionally, we discuss the impact of the one-year 95% C.L. sensitivities for observation of high energy neutrinos from the Sun at IceCube and DeepCore.[6] Our scans were performed with a modified version of the publicly available code SuperBayeS.[7]

In the following, we will first briefly summarize the main features of Bayesian inference, and describe our likelihood functions. We will then show our results for the impact of LHC, XENON100, and Fermi-LAT searches on the CMSSM parameters and on observables. Finally, we will comment on further impact expected from IceCube/DeepCore one-year projected sensitivities.

2. Bayesian Theory

Bayes' theorem tells us how to relate the *posterior* probability density function (pdf), $p(m|d)$, of a model m given the data d, and it can be written as

$$p(m|d) = \frac{p(m|d)\pi(m)}{p(d|m)}. \tag{1}$$

In Eq. (1), $\pi(m)$ stands for *prior* probability that m is correct before the data d is seen, while the *evidence*, $p(d|m)$, is the marginal probability of the data. The evidence is often called *model likelihood* as well. Finally, $p(m|d)$ is the likelihood function. It represents the probability to see the data under the assumption that m is true. Alternatively, one can also use the χ^2 approach to determine the discrepancy between the theoretical prediction and the experimental measurement. Generally speaking, if the likelihood strongly constrains the parameter space of the theory, the posterior will closely follow the likelihood, otherwise the posterior will be strongly dependent on the prior probability.

In this talk, we adopt logarithmic priors for m_0 and $m_{1/2}$ but linear priors for A_0 and $\tan\beta$. Our choice of the likelihood function varies depending on the experiment used to constrain the parameter space, and we can identify four main categories:

- The yes/no hard cut, e.g. the neutralino is not the LSP.
- A Gaussian distribution for given central value and experimental error in cases where positive measuremnents have been obtained.
- An upper/lower error function for null results for which exclusion limits exist.
- A Poisson distribution to characterize counting experiments.

We treat the theoretical error (τ) and experimental error (σ) separately. The observables we use to constrain the parameter space, and their experimental and theoretical errors are reported in Table 1 in Ref.[8] and references therein. We divide them into three classes:

- *non-LHC data* which collectively stand for all observables coming from the relic abundance of dark matter, LEP and Tevatron bounds, electroweak observables, flavor physics observables, the anomalous magnetic moment of the muon, etc.
- *LHC data* which stands for lowers limit on the CMSSM parameters.
- *DM search data* direct (from XENON100) and indirect (from FermiLAT dSph gamma flux) dark matter search limits.

3. Impact of the CMS α_T 1.1fb^{-1} Limit

We derived an approximate likelihood for the CMS α_T search limit based on the LHC data sample of 1.1fb^{-1} of integrated luminosity recorded at $\sqrt{s} = 7$ TeV, which showed no excess of events over the Standard Model (SM) predictions. Following the method in,[9] the number of SUSY events expected in each H_T bin, s_i, where $i = 1,\ldots,8$, is a product of the detector efficiency ϵ_i for that bin (the fraction of events that survive α_T cuts), the integrated luminosity, $\int L = 1.1$fb^{-1}, and the total cross section for the production of SUSY particles at $\sqrt{s} = 7$ TeV, σ,

$$s_i = \epsilon_i \times \sigma \times \int L. \tag{2}$$

338

The likelihood for this case, $\mathcal{L}$, is the probability of observing a set of $\{o_i\}$ events given the expected SUSY signals $\{s_i\}$ and SM background events $\{b_i\}$. $\mathcal{L}$ is a product of Poisson distributions for each bin with means $s_i + b_i$,

$$\mathcal{L} = \prod_i \frac{e^{-(s_i+b_i)}\left(s_i + b_i\right)^{o_i}}{o_i!}. \tag{3}$$

We assume that the SM backgrounds, b_i, in each bin are precisely known as they are given by the CMS analysis.

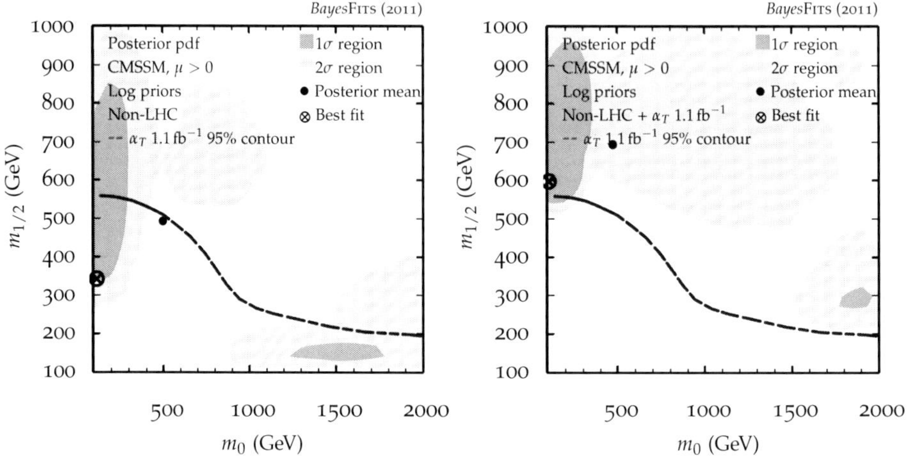

Fig. 1.   Dark (light) blue regions denote the 68% (96%) posterior pdf regions of the $(m_0, m_{1/2})$ plane. Our likelihood function approximating the CMS α_T 95% exclusion contour is not (is) applied in the left (right) panel. The actual CMS α_T 95% exclusion contour is marked in both panels with a dashed line.

In Fig. 1, we show the marginalized posterior on the $(m_0, m_{1/2})$ plane for a scan where the likelihood is given by the non-LHC constraints only (left panel), and for one where the additional impact of the α_T constraint is implemented (right panel). In the left panel, one can see a clear vertical mode, located in the stau co-annihilation or A-funnel region. On the opposite side of the diagonal of the $(m_0, m_{1/2})$ plane, the horizontal mode corresponds to the FP/HB or Higgs resonance region. The emergence of these two regions is consistent with the findings of previous pre-LHC Bayesian analyses,[10-12] although their relative size does depend on the choice of the prior.

The impact of the α_T limit is shown in the right panel of Fig. 1. Clearly, the α_T limit has cut deep into the $(m_0, m_{1/2})$ plane's high posterior probability regions - a nominal fraction of the new 68% C.L. credible region has moved outside the α_T 95% confidence interval. The α_T limit pushed the high-probability credible regions to larger values of $m_{1/2}$. Significantly, the two modes on the $(m_0, m_{1/2})$ plane have remained. The 68% C.L. stau coannihilation/A-funnel region has been pushed up, while light Higgs resonance below the α_T limit is now excluded. On the other hand, the FP/HB region is not excluded by the α_T constraint but it has been pushed up and out to larger m_0.

4. New Constraints from XENON100 Limit and Fermi-dSphs

First, we incorporate in our scan the constraints on the SI neutralino-proton cross section $\sigma_p^{\rm SI}$ induced by the 90% C.L. upper bound, $\sigma_{p,90}^{\rm SI}$, which was recently published by the XENON100 Collaboration.[4,13,14] As described in the experimental papers, the bound is obtained from the p-value of a likelihood function which includes the systematic and statistical error in signal and background, as well as the uncertainties on scintillation efficiency and escape velocity; see Ref. [13] for details. However, the experimental analysis does not consider the astrophysical uncertainties associated with the chosen velocity distribution and DM halo profile,[15–17] nor the nuclear physics uncertainties associated with calculations of the π-nucleon sigma-term $\Sigma_{\pi N}$,[18] the latter of which is in fact dominant. In this section, we denote these theoretical uncertainties with τ. The uncertainties τ have been quantified in the literature, and amount up to approximately ten times the upper bound on the cross section, $\tau \sim 10 \times \sigma_{p,90}^{\rm SI}$.[19] By choosing an extremely optimistic small value of τ, say $\tau = \sigma_{p,90}^{\rm SI}$, the XENON100 bound only weakly affects the parameter space. We include τ in the likelihood, following the procedure described in reference[10] for the treatment of upper bounds. Beside direct searches at accelerators and underground DM detectors, we also include the constraints given by gamma-ray fluxes from dSphs, detected by the Fermi-LAT satellite telescope.[5] The 95% C.L. upper bounds on the annihilation cross-section, $\langle \sigma v \rangle_{i,95}$, as a function of m_χ, depend in the final-state products ($i = b\bar{b}$, $\mu^-\mu^+$, $\tau^-\tau^+$, W^-W^+). We show in this talk their impact on the CMSSM parameter space. One can approximately convert the bounds $\langle \sigma v \rangle_{i,95}$, given by Fermi-LAT for individual final states, into a bound on the annihilation cross section of MSSM by using

$$\overline{\Phi}_{95} = \frac{N_{b\bar{b}} \langle \sigma v \rangle_{b\bar{b},95}\, J}{8\pi m_\chi^2} = \frac{(\sum_{i=1}^{29} {\rm BR}_i N_i) \langle \sigma v \rangle_{\rm ann,95}\, J}{8\pi m_\chi^2}, \tag{4}$$

where $\overline{\Phi}_{95}$ is the bound on the integrated observed photon flux, J denotes the J-factor,[20] and the N_i's are the integrated photon energy distributions for the individual final states. We sum the branching ratios ${\rm BR}_i \equiv \langle \sigma v \rangle_i / \langle \sigma v \rangle_{\rm ann}$ over all the 29 channels implemented in SuperBayeS. The theoretical uncertainties related to the halo profiles were dealt with in two different ways. In one we follow the experimental paper,[5] where the J-factor was calculated under the assumption of NFW-like DM density profile. Under this procedure the 1σ-uncertainties resulting from the likelihood function amount to approximately 3%. On the other hand, the alternative assumption of DM density core profiles reduces the J-factor by one or two orders of magnitude in most models, in accordance with the findings of[21] and,[22] thus rendering the upper limits $\langle \sigma v \rangle_{i,95}$ weaker by comparable amount. Like we did

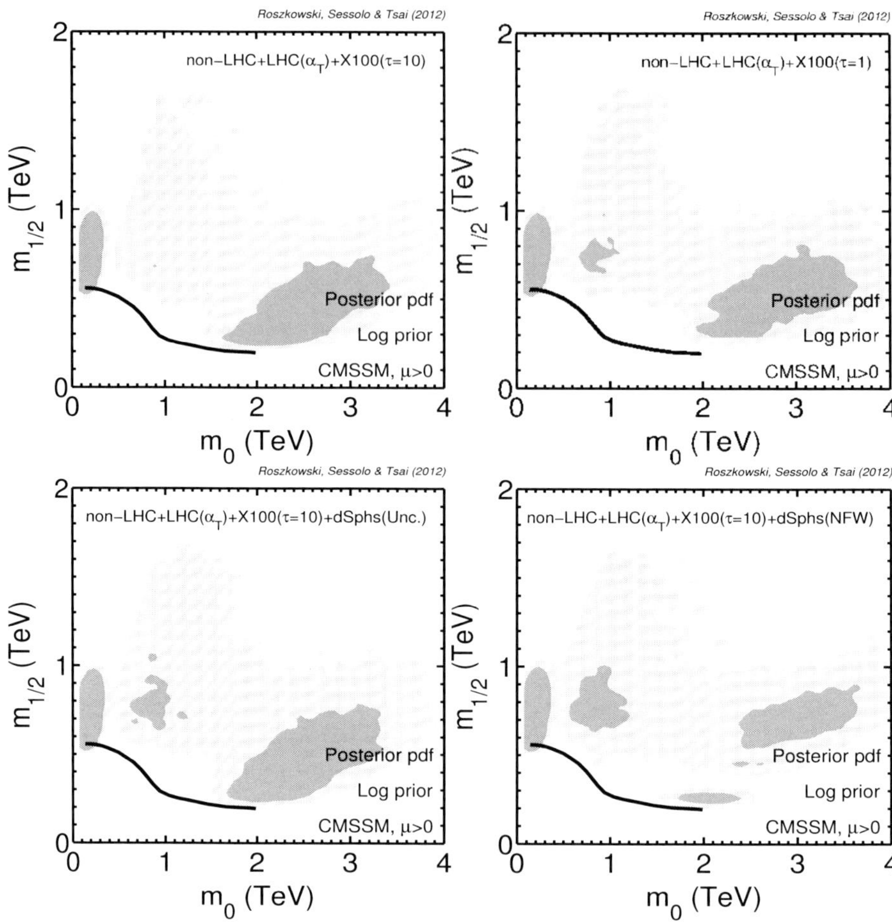

Fig. 2. Marginalized posterior pdf in the $(m_0, m_{1/2})$ plane extended to wider ranges compared to Fig. 1. The 68% C.L. credible regions are shown in dark blue and the 95% C.L. regions in light blue. **Top-left panel**: the impact of the non-LHC, α_T and XENON100 (assuming conservative error) constraints. **Top-right panel**: the same as in top-left panel, but the theoretical uncertainty on XENON100 is strongly reduced. **Bottom-left panel**: the impact of the non-LHC, α_T, XENON100 and dSphs constraints. A conservative uncertainty in the DM halo model is assumed. **Bottom-right panel**: the impact of the non-LHC, α_T, XENON100 and dSphs constraints. The uncertainty on the DM halo model is not considered.

for the XENON100 bound, beside performing our analysis under the optimistic NFW assumption, we also consider a conservative case where $\tau = 50 \times \langle\sigma v\rangle_{\mathrm{tot},95}$. The top-left panel of Fig. 2 includes the 90% C.L. bound on σ_p^{SI} with a conservative theoretical uncertainty, $\tau = 10 \times \sigma_{p,90}^{\mathrm{SI}}$. Similarly, in the bottom-left panel, we impose a theoretical error of 50 times the bound given by dSphs. Not surprisingly, such conservative choices do not have much visible effect on the parameter space. We have checked that the 95% C.L. credible regions obtained under conservative uncertainties do not change appreciably from the case without the XENON100 or

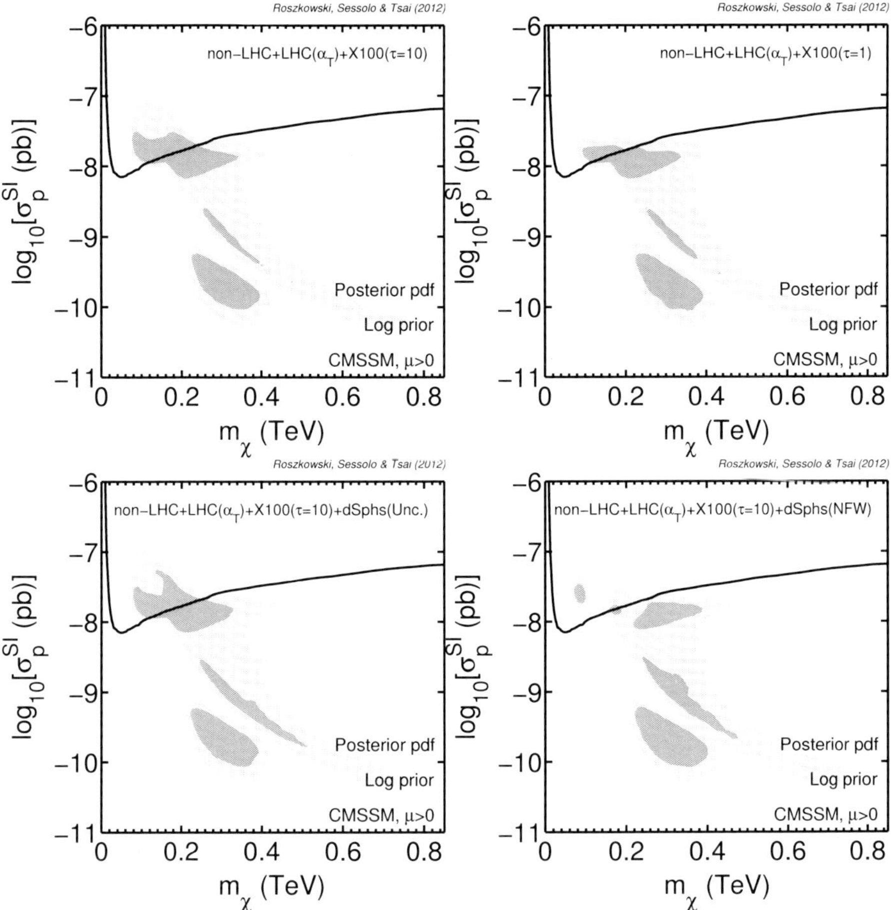

Fig. 3. The same as in Fig. 2 but in the $(m_\chi, \sigma_p^{\mathrm{SI}})$ plane. The solid black line shows the XENON100 90% C.L. exclusion bound which is included via our approximate likelihood function.

Fermi-LAT dSphs limits. However, the top-right panel of Fig. 2, which shows the effects of the XENON100 bound with a highly optimistic uncertainty of $\tau = \sigma_{p,90}^{\mathrm{SI}}$, makes it clear that even by reducing the theoretical uncertainties to only 100% of the experimental upper bound (a highly unlikely possibility) the impact on the parameter space is still not sizable. Somewhat better prospects come from the dSphs bound: in the bottom-right panel we show the result of assuming the NFW profile for the DM distribution. One can see that the bound from dSphs cuts significantly into the FP/HB region, and the impact is stronger than observed with XENON100 alone, even for the idealized case of extremely low theoretical uncertainties shown in the top-right panel. This is because, in this region, the enhanced higgsino component of the neutralino produces larger $\langle \sigma v \rangle$, which for a cuspy NFW profile results in higher fluxes which in turn are more strongly constrained by the Fermi-LAT data.

342

In the two figures on the left hand side of Fig. 3, we present the marginalized pdf on the $(m_\chi, \sigma_p^{\mathrm{SI}})$-plane under the conservative assumptions described above for the uncertainties on XENON100 (top-left panel) and dSphs (bottom-left panel). The solid black line shows the XENON100 90% C.L. bound. As in the $(m_0, m_{1/2})$ case, we can see that a conservative (and realistic) treatment of the uncertainties in our Bayesian analysis wanes the effects of the limits on σ_p^{SI} and γ-ray fluxes from dSphs, thus highlighting the predominance of the LHC bound. On the other hand, in the two figures on the right hand side, we see that optimistic estimates of the errors involved in calculating the observables make some an impact on the CMSSM parameter space. Very interestingly, a comparison of the top-right panel with the bottom-right panel, shows that the effects of direct and indirect detection searches at present sensitivities would become comparable if the uncertainties involved had a greatly reduced size.

5. Projected Sensitivities at IceCube/DeepCore

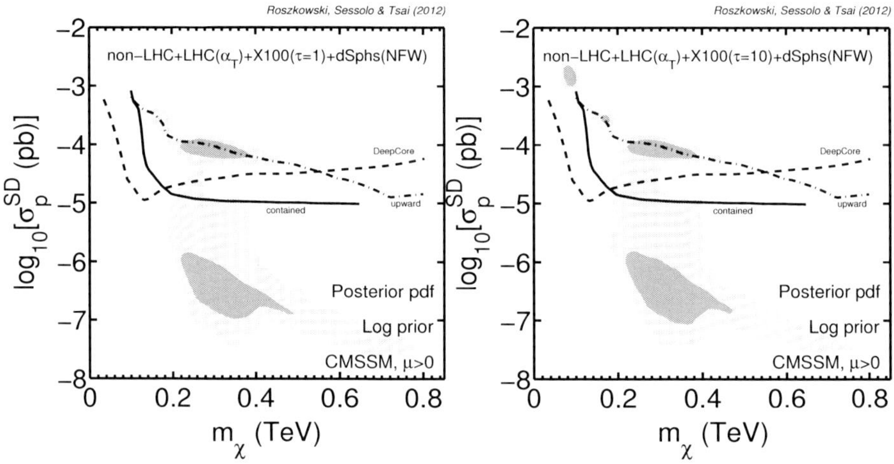

Fig. 4. Marginalized posterior pdf in the $(m_\chi, \sigma_p^{\mathrm{SD}})$ plane. The 68% C.L. credible regions are shown in dark blue, the 95% C.L. regions in light blue. **Left:** the impact of the non-LHC, α_T, XENON100 and dSphs constraints with the NFW profile. **Right:** the impact of non-LHC, α_T, XENON100 and dSphs constraints. The theoretical uncertainty in interpretting the XENON100 limit is reduced, and the DM halo follows the NFW profile. The solid line is the one-year 95% C.L. sensitivity in contained events at IceCube, the dashed line the corresponding sensitivity for Deep-Core, and the dash-dotted line the sensitivity in upward events.

In this section we analyze the effect on the CMSSM of the projected one-year 95% C.L. experimental sensitivities for the 80-string configuration at IceCube, and the impact of the additional six DeepCore strings. The muon events observed by the 80-string IceCube detector can either be labelled as *upward* (those that come from neutrinos undergoing CC-scattering outside the 1-km^3 instrumented volume), or as *contained* (those for which the interaction happens inside the detector).

Neutralinos undergoing elastic scattering with protons inside the Sun are captured by the Sun's graviational field. The spin-dependent (SD) elastic scattering cross section, σ_p^{SD}, will be the dominant observable that defines the capture, since light nuclei like hydrogen are abundant in the Sun. The resulting neutrino flux from the center of the Sun can be measured at IceCube through the Cherenkov radiation of the traveling muons. In the two panels of Fig. 4, we show the effects of the one-year 95% C.L. sensitivities on upward and contained IceCube events on the $(m_\chi, \sigma_p^{\mathrm{SD}})$ plane. We also show the corresponding sensitivity at DeepCore. The marginalized posterior probability is obtained by applying all the constraints from non-LHC, α_T, XENON100, and dSphs. We only consider the NFW profile, which gives stringent bounds from Fermi-LAT. We found that the sensitivity on upward events is comparable to the current limit from XENON100 but with $\tau = 1$. However, the contained events at IceCube and DeepCore have the potential to constrain the CMSSM parameter space better.

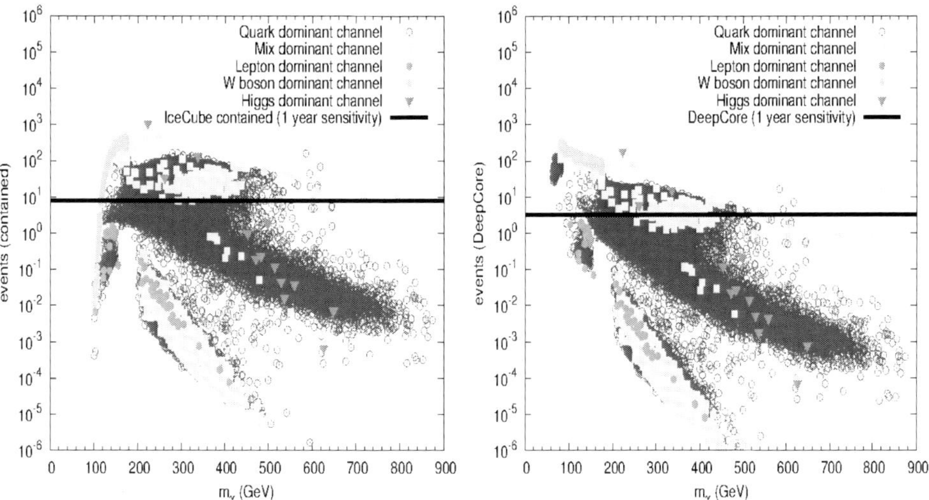

Fig. 5. The impact of some of the exclusion bounds and projected sensitivities considered in this work on parameter configurations with a predominant neutralino annihilation channel.

Some very enticing possibilities arise in case IceCube/DeepCore lived up to its expected discovery potential. In particular, we found that if the projected sensitivities were achieved, and the experimental and theoretical uncertainties were kept reasonably small, in one year time DeepCore could rule out the fraction of the FP/HB region which is mostly responsible for neutralino annihilation into W^+W^- at the 95% C.L. To illustrate this point we show in Fig. 5 the impact of some of the exclusion bounds and projected sensitivities considered in this work on models with a single predominant neutralino annihilation channel. Each point represents a case for which neutralinos annihilate 90% of the time to the selected final state, and for which the bound on the relic abundance is respected ($0.05 < \Omega_\chi h^2 < 0.20$). On the other hand, the right panel of Fig. 5 shows the same analysis for DeepCore. We can see that the WW-channel could be excluded at the 95% C.L., and so could any CMSSM-like model which for which the DM annihilates predominantly into W^+W^-. The advantage of DeepCore comes from a lower energy threshold. Since the WW-channel opens up around m_W, which is very close to the threshold predicted for the IceCube larger volume, there is not enough phase space over which the muon flux can be integrated in the contained events. Consequently the number of hits at the detector will not be sizable. On the other hand, since the threshold at DeepCore is ~ 35 GeV, even neutralinos that annihilate into the WW-channel can give enough muon events to be successfully detected.

6. Conclusions

In this talk, we have illustrated the impact on global Bayesian inferences of the CMSSM of some of the most constraining SUSY and dark matter search experiments. We implemented in our scans the bounds from the CMS α_T search, and the bounds and projected sensitivities for DM indirect and direct detection searches. We found that the CMSSM has become severely constrained but not excluded by the CMS α_T 1.1/fb limit. The constraints from direct detection of dark matter are currently weaker. The reason lies in the still large theoretical uncertainties. Even by reducing the theoretical uncertainties to only 100% of the experimental upper bound, the impact on the parameter space would still not be sizable. Similarly, the bounds induced by gamma-ray fluxes from dSphs are very sensitive to the choice of DM halo models. A less conservative treatment of the DM halo uncertainties can significantly constrain the FP/HB region, and make indirect detection competitive with the data from XENON100. The projected one year sensitivity at DeepCore has the potential to further constrain the FP/HB region, and exclude the fraction of parameter space over which DM annihilates predominantly to W^+W^- final states.

Acknowledgements

This talk is based on the following papers: arXiv:1111.6098, and arXiv:1202.1503. Y.-L. S. Tsai is indebted to P. Chen and the LeCosPA Center for support and hospitality. We acknowledge generous support from the Welcome Programme of the Foundation for Polish Science.

References

1. G. L. Kane, C. F. Kolda, L. Roszkowski and J. D. Wells, *Phys.Rev.* **D49**, 6173 (1994).
2. S. P. Martin (1997).
3. S. Chatrchyan *et al.*, *Phys.Rev.Lett.* **107**, p. 221804 (2011).
4. E. Aprile *et al.* (2011).

5. M. Ackermann *et al.*, *Phys. Rev. Lett.* **107**, p. 241302 (2011).

6. R. Abbasi *et al.* (2011).

7. `http://www.ft.uam.es/personal/rruiz/superbayes/index.php?page=main.html`.

8. L. Roszkowski, E. M. Sessolo and Y.-L. S. Tsai (2012).

9. (CMS Collaboration) (2011).

10. R. R. de Austri, R. Trotta and L. Roszkowski, *JHEP* **0605**, p. 002 (2006).

11. L. Roszkowski, R. Ruiz de Austri and R. Trotta, *JHEP* **0707**, p. 075 (2007).

12. R. Trotta, F. Feroz, M. P. Hobson, L. Roszkowski and R. Ruiz de Austri, *JHEP* **0812**, p. 024 (2008).

13. E. Aprile *et al.*, *Phys. Rev.* **D84**, p. 052003 (2011).

14. E. Aprile *et al.*, *Phys. Rev. Lett.* **107**, p. 131302 (2011).

15. A. M. Green, *JCAP* **1010**, p. 034 (2010).

16. M. Pato *et al.*, *Phys. Rev.* **D83**, p. 083505 (2011).

17. C. Arina, J. Hamann and Y. Y. Wong, *JCAP* **1109**, p. 022 (2011).

18. J. R. Ellis, K. A. Olive and C. Savage, *Phys. Rev.* **D77**, p. 065026 (2008).

19. A. Fowlie, A. Kalinowski, M. Kazana, L. Roszkowski and Y. Tsai (2011).

20. G. Bertone, D. Hooper and J. Silk, *Phys. Rept.* **405**, 279 (2005).

21. A. Charbonnier *et al.*, *Mon. Not. Roy. Astron. Soc.* **418**, 1526 (2011).

22. L. Roszkowski, R. R. de Austri, J. Silk and R. Trotta, *Phys. Lett.* **B671**, 10 (2009).

Appendix

CONFERENCE COMMITTEES AND STAFF

Local Organizing Committee and Staff

P. Chen-Chair, T.-H. Chiueh, J.-A. Gu, G.-L. Lin, M.-Z. Wang, Y.-L. Lee, K. Ho, T.C. Liu, G. Chuang

Opening Ceremony Speakers

P. Chen (LeCosPA, NTU & KIPAC, Stanford U.)-Chair, S.-C. Lee (NTU), J. Dorfan (OIST), G. Madejski (KIPAC, Stanford U. & SLAC), H. Aihara (IPMU, U. Tokyo), R. Ruffini (U. Rome "Sapienza" & ICRANet), A. Karle (U. Wisconsin-Madison), P. Gorham (U. Hawaii), I. Park (Ewha Womans U., Korea), G.-S. Chang (NSPO)

Opening Ceremony Master

B. Y. Hsiung (NTU)

Plenary Session Speakers

S.-T. Yau (Harvard U. & TIMS, NTU), P. Chen (LeCosPA, NTU & KIPAC, Stanford U.), R. Ruffini (U. Rome "Sapienza" & ICRANet), B. Carr (Queen Mary, U. London), G. Madejski (KIPAC, Stanford U. & SLAC), H. Aihara (IPMU, U. Tokyo), P. Ho (AS & Harvard-Smithsonian Center for Astrophysics), T. Kamae (SLAC/KIPAC, Stanford U.), P. Gorham (U. Hawaii), A. Karle (U. Wisconsin-Madison), W. Unruh (U. British Columbia), J. Rafelski (U. Arizona), J. Yokoyama (RESCEU, U. Tokyo), A. Kusenko (UCLA), W.-T. Ni (Center for Gravitation and Cosmology, NTHU), B.-Q. Ma (Peking U.), S.-C. Lee (Institute of Physics, AS), H. Wong (Institute for Physics, AS), K. Ni (Shanghai Jiao Tong U.), Y.-L. Tsai (National Center for Nuclear Research, Poland), S. P. Kim (Kunsan National U. & NTU), H. Rosu (IPICyT, Mexico), J.-A. Gu (LeCosPA, NTU), I. Park (Ewha Womans U.), C. Tao (CPPM/IN2P3, France & THCA, Tsinghua U.), H. L. Yu (Institute of Physics, AS), G. Veneziano (College de France & CERN)

Plenary Session Chairs

H. L. Yu (Institute of Physics, AS), W.-T. Ni (Center for Gravitation and Cosmology, NTHU), W.-Y. P. Hwang (NTU), J. Nam (LeCosPA, NTU), J.-W. Chen (NTU), G.-L. Lin (NCTU), H.-K. Chang (NTHU), J. Nester (NCU), S. P. Kim (KNU & NTU), J. Rafelski (U. Arizona)

Parallel Session Speakers

E. Gumrukcuoglu (IPMU, U. Tokyo), T. Y. Lam (IPMU, U. Tokyo), Y.-W. Liu (LeCosPA, NTU), T. Suyama (RESCEU, U. Tokyo), T.C. Liu (LeCosPA, NTU), Y. Niu (Peking U.), J. Nam (LeCosPA, NTU), C. Chen (LeCosPA, NTU), C.-Y. Hu (LeCosPA, NTU), A. Romano (U. Antioquia), D. Maity (LeCosPA, NTU), L. Labun (U. Arizona)

Parallel Session Chairs

K. Izumi (LeCosPA, NTU), T. Qiu (LeCosPA, NTU), M.-H. Huang (NUU), J.-A. Gu (LeCosPA, NTU)

LIST OF PARTICIPANTS

Aihara, Hiroaki	IPMU, University of Tokyo	aihara@phys.s.u.-tokyo.ac.jp
Carr, Bernard	Queen Mary, University of London	B.J.Carr@qmul.ac.uk
Chang, Feng-Yin	Institute of Physics, Academia Sinica	fychang@phys.sinica.edu.tw
Chang, Hsiang-Kuang	Dept. Phys., National Tsing Hua University	hkchang@phys.nthu.edu.tw
Chang, Psychon	Taipei Amateur Astronomers Association	psychon.chang@msa.hinet.net
Chang, Yu-Chiao	Dept. Phys., National Taiwan University	f95222075@ntu.edu.tw
Chen, Chien-Ting	LeCosPA, National Taiwan University	f98244003@ntu.edu.tw
Chen, Chih-Ching	LeCosPA, National Taiwan University	chen.chihching@gmail.com
Chen, Jiunn-Wei	Dept. Phys., National Taiwan University	jwc@phys.ntu.edu.tw
Chen, Pisin	LeCosPA, National Taiwan University	chen@slac.stanford.edu
Chen, Ten-Lin	National United University	d97222008@ntu.edu.tw
Chen, Yen Lin	National Taiwan University	classtark@msn.com
Chiang, Chien-I	LeCosPA, National Taiwan University	chienichiang@gmail.com
Chiang, Hsu-Wen	Dept. Phys., National Taiwan University	b98202036@ntu.edu.tw
Chiou, Dah-Wei	Dept. Phys., National Taiwan University	dwchiou@gmail.com
Chiu, I-Non	LeCosPA, National Taiwan University	ii_non@hotmail.com
Chuang, Tao-Mao	National Taiwan University	daumau@gmail.com
DuVernois, Michael	IceCube Center, University of Wisconsin-Madison	duvernois@icecube.wisc.edu
Gorham, Peter	University of Hawaii	gorham3690@gmail.com
Gu, Je-An	LeCosPA Center, National Taiwan University	jagu@ntu.edu.tw
Gumrukcuoglu, Emir	IPMU, University of Tokyo	emir.gumrukcuoglu@ipmu.jp
Guo, Zong-Kuan	Inst. of Theo. Phys., Chinese Academy of Sciences	guozk@itp.ac.cn
Gupta, Neeraj	Dept. Phys., Govt. G.M. Science College	neerajgupta786@yahoo.co.in
Ho, Paul	Institute of AS, Academia Sinica	pho@asiaa.sinica.edu.tw
Hsieh, Hsihchia	Providence University	hsihchia.hsieh@gmail.com
Hu, Chia-Yu	LeCosPA, National Taiwan University	huchiayu0517@gmail.com
Huang, Jian-Jung	LeCosPA, National Taiwan University	njinee@hotmail.com
Huang, Ming-Huey A.	National United University	mahuang@nuu.edu.tw
Huang, Yu-Chien	LeCosPA, National Taiwan University	bassoon771231@hotmail.com
Hwang, W-Y. Pauchy	Dept. Phys., National Taiwan University	wyhwang@phys.ntu.edu.tw
Izumi, Keisuke	LeCosPA, National Taiwan University	izumi@phys.ntu.edu.tw
Jian, Hung-Yu	National Taiwan University	d89222003@ntu.edu.tw
Kamae, Tsuneyoshi	SLAC/KIPAC, Stanford University (emeritus)	kamae@slac.stanford.edu
Karle, Albrecht	Dept. Phys., University of Wisconsin-Madison	karle@icecube.wisc.edu
Kim, Sang Pyo	Dept. Phys., Kunsan National University & National Taiwan University	sangkim@kunsan.ac.kr
Koptelova, Ekaterina	Dept. Phys., National Taiwan University	ekaterina@phys.ntu.edu.tw
Kusenko, Alexander	Department of Physics and Astronomy, UCLA	kusenko@ucla.edu
Labun, Lance	Dept. Phys., The University of Arizona	labun@physics.arizona.edu
Lam, Tsz Yan	IPMU, University of Tokyo	tszyan.lam@ipmu.jp
Law, Sandy	Dept. Phys., National Cheng Kung University	slaw@mail.ncku.edu.tw
Lee, Shih-Chang	Institute of Physics, Academia Sinica	phsclee@phys.sinica.edu.tw
Lin, Guey-Lin	Institute of Physics, National Chiao-Tung University	glin@mail.nctu.edu.tw
Lin, Yu-Hsiang	Dept. Phys., National Taiwan University	hitr2997925@gmail.com
Liu, Tsungche	LeCosPA, National Taiwan University	tcliu@ntu.edu.tw
Liu, Yen-Wei	Dept. Phys., National Taiwan University	f97222009@ntu.edu.tw
Ma, Bo-Qiang	School of Phys., Peking University	mabq@pku.edu.cn
Madejski, Grzegorz M.	KIPAC	madejski@slac.stanford.edu
Maity, Debaprasad	LeCosPA, National Taiwan University	debu.imsc@gmail.com

Molnar, Sandor	LeCosPA, National Taiwan University	sandor@phys.ntu.edu.tw
Nam, Jiwoo	LeCosPA, National Taiwan University	namjiwoo@gmail.com
Nester, James M.	Dept. Phys., National Central University	nester@phy.ncu.edu.tw
Ni, Kaixuan	Dept. Phys., Shanghai Jiao Tong University	nikx@sjtu.edu.cn
Ni, Wei-Tou	Dept. Phys., National Tsing Hua University	weitou@gmail.com
Niu, Yuezhen	School of Phys., Peking University	yuezhenniu@gmail.com
Ong, Yen Chin	LeCosPA, National Taiwan University	ongyenchin@gmail.com
Park, Il H.	Ewha Womans University	ipark@ewha.ac.kr
Peng, Huan-Ting	LeCosPA, National Taiwan University	htpeng1108@gmail.com
Qiu, Taotao	LeCosPA, National Taiwan University	xsjqiu@gmail.com
Rafelski, Johann	Dept. Phys., University of Arizona	rafelski@physics.arizona.edu
Romano, Antonio Enea	Dept. Phys., University of Antioquia, Columbia	aer@phys.ntu.edu.tw
Rosu, Haret	IPICyT, Mexico	hcr@ipicyt.edu.tw
Ruffini, Remo	University of Rome "Sapienza" & ICRANet	ruffini@icra.it
Shen, Che Min	Dept. Phys., National Taiwan University	che.min.shen@gmail.com
Sit, Wai Yin	CLP Holdings Limited	cherrytasty@hotmail.com
Suyama, Teruaki	RESCEU, University of Tokyo	suyama@resceu.s.u-tokyo.ac.jp
Tang, Alf	NSPO	tang.alf@gmail.com
Tao, Charling	CPPM/IN2P3, France & Tsinghua U. (THCA)	tao@cppm.in2p3.fr
Tsai, Yue-Lin	National Centre for Nuclear Research, Poland	Sming.Tsai@fuw.edu.pl
Unruh, Bill	University of British Columbia	unruh@physics.ubc.ca
Veneziano, Gabriele	College de France & CERN	Gabriele.veneziano@cern.ch
Wang, Chiao-Hsuan	Dept. Phys., National Taiwan University	chiao0701@gmail.com
Wang, Chih-Yueh	Dept. Phys., Chung-Yuan Christian University	cywang@phys.cycu.edu.tw
Wang, Jing	Dept. Phys., National Tsing Hua University	joanwangj@126.com
Wang, Min-Zu	Dept. Phys., National Taiwan University	mwang@phys.ntu.edu.tw
Wang, Shi-Hao	Graduate Institute of Astrophysics, NTU	wsh1230.wsh1230@msa.hinet.net
Wong, Henry	Institute of Physics, Academia Sinica	htwong@phys.sinica.edu.tw
Xue, She Sheng	University of Rome "Sapienza" & ICRANet	xue@icra.it
Yang, Cheng Tao	Dept. Phys., National Taiwan University	d98222015@ntu.edu.tw
Yau, Shing-Tung	Harvard University & TIMS, NTU	yassist@math.ntu.edu.tw
Yo, Hwei-Jang	Dept. Phys., National Cheng-Kung University	hjyo@phys.ncku.edu.tw
Yokoyama, Jun'ichi	RESCEU, University of Tokyo	yokoyama@resceu.s.u-tokyo.ac.jp
Yu, Hoi-Lai	Institute of Physics, Academia Sinica	hlyu@phys.sinica.edu.tw
Zhang, Ui-Han	Dept. Phys., National Taiwan University	zhanguihan@hotmail.com

CONFERENCE PROGRAM

Sunday, February 5

15:40-18:00   Registration Desk Open
18:00-21:00   Welcome Reception

Monday, February 6

08:00-09:00   Registration
09:00-09:35   Opening Ceremony   Master of Ceremony: B. Y. Hsiung
                                 Speakers: P. Chen, S.-C. Li, J. Dorfan, G. Madejski,
                                         H. Aihara, R. Ruffini, A. Karle,
                                         P. Gorham, G.-S. Chang

09:35-10:45   Plenary Session 1:   Hoi-Lai Yu, Chair

   09:35-10:10   S.-T. Yau: "Mass and Waves in Gravity"
   10:10-10:45   P. Chen: "Where is $\hbar$ Hiding in Entropic Gravity?"
10:45-11:15   Coffee Break

11:15-12:25   Plenary Session 2:   W.-T. Ni, Chair

   11:15-11:50   R. Ruffini: "Black Holes and Proto Black Holes"
   11:50-12:25   B. Carr: "Black Holes and the Generalized Uncertainty
                                   Principle"
12:25-13:30   Lunch Break

13:30-15:15   Plenary Session 3:   W.-Y. P. Hwang, Chair

   13:30-14:05   G. M. Madejski: "Current and Future High Energy Astrophysics
                                   Programs at KIPAC: the Variable Sky as Seen in
                                   the Gamma-Ray Band by the *Fermi* Observatory
                                   and Plans for Future Missions"
   14:05-14:40   H. Aihara: "Cosmology and Particle Astrophysics at the
                                   IPMU"
   14:40-15:15   P. Ho: "The ALMA Project"

15:15-21:00   Excursion: Lantern Festival in Pingxi

Tuesday, February 7

09:00-10:45 Plenary Session 4: J. Nam, Chair

09:00-09:35	T. Kamae:	"*Fermi*-Large Area Telescope: the Accomplishments and Challenges"
09:35-10:10	P. Gorham:	"The Search for Ultra-High Energy Cosmic Neutrinos: Results and Future Prospects"
10:10-10:45	A. Karle:	"Neutrino Astronomy at the South Pole with IceCube and Beyond"

10:45-11:15   Conference Photo/Coffee Break

11:15-12:25 Plenary Session 5: J.-W. Chen, Chair

11:15-11:50	W. G. Unruh:	"Measurement of Hawking Radiation in Analog Horizon"
11:50-12:25	J. Rafelski:	"Critical Acceleration"

12:25-13:30   Lunch Break

13:30-15:15 Parallel Session A: K. Izumi, Chair

13:30-14:05	E. Gumrukcuoglu:	"Self-Accelerating Universe in Nonlinear Massive Gravity"
14:05-14:40	T. Y. Lam:	"Testing Modified Gravity with Phase-Space Distribution"
14:40-15:15	Y.-W. Liu:	"Scalar Perturbations from GBIR Brane Inflation"

13:30-15:15 Parallel Session B: T. Qiu, Chair

13:30-14:05	T. Suyama:	"Testing the Origin of Primordial Perturbation – Use of Bispectrum and Trispectrum"
14:05-14:40	T.-C. Liu:	"Neutrino Flavor on Earth"
14:40-15:15	Y. Niu:	"Entanglement Entropy of the Early Universe in Generalized Chaplygin Gas Model"

15:15-15:40   Coffee Break

15:40-18:00 Plenary Session 6: G.-L. Lin, Chair

15:40-16:15	J. Yokoyama:	"Generalized G-Inflation"
16:15-16:50	A. Kusenko:	"Cosmic Connections: from Cosmic Rays to Gamma Rays, to Cosmic Backgrounds"
16:50-17:25	W.-T. Ni:	"Foundations of Classical Electrodynamics, Equivalence Principle and Cosmic Interactions"
17:25-18:00	B.-Q. Ma:	"New Perspective on Space-Time from Lorentz Violation"

18:00-19:30   Dinner: Howard International House

19:30-21:00 GuoGuang Peking Opera Night

Wednesday, February 8

09:00-11:15 Plenary Session 7: H.-K. Chang, Chair

  09:00-09:35 S.-C. Lee: "The Status of the AMS Experiment"
  09:35-10:10 H. Wong: "Low-Mass WIMP Search with Sub-keV
                Germanium Detector"
  10:10-10:45 K. Ni: "PandaX: a Dark Matter Direct Detection
                Experiment at CJPL"
  10:45-11:15 Y.-L. Tsai: "SUSY Dark Matter Signatures in LHC, *Fermi*
                dSphs, and IceCube"

**11:15-18:00 Excursion: National Palace Museum/Taipei Confucius
                Temple/Dalongdong Baoan Temple/Taipei 101**

18:00-21:00 Banquet: Taipei 101

Thursday, February 9

09:00-10:45 Plenary Session 8: J. Nester, Chair

  09:00-09:35 S. P. Kim: "Quantum Structure of de Sitter Space"
  09:35-10:10 H. Rosu: "FRW Riccati Equation and Related Barotropic-
                like Cosmologies"
  10:10-10:45 J.-A. Gu: "Issues in Dark Energy and Modified Gravity"
10:45-11:15 Coffee Break

11:15-12:25 Plenary Session 9: S. P. Kim, Chair

  11:15-11:50 I. H. Park: "Ultra-Fast Flash Observatory (UFFO) for Early
                Photon Measurements from Gamma Ray Bursts"
  11:50-12:25 C. Tao: "Tsinghua Center for Astrophysics and the Dark
                Universe"
12:25-13:30 Lunch Break

13:30-15:15 Parallel Session C: M.-H. Huang, Chair

  13:30-14:05 J. Nam: "The UFFO Slewing Mirror Telescope for Early
                Optical Observation from Gamma Ray Bursts"

14:05-14:40	C.-C. Chen:	"The Novel all Flavor Lepton Signature in Radio Neutrino Telescopes"
14:40-15:15	C.-Y. Hu:	"Near-Field Effects of Cherenkov Radiation Induced by Ultra High Energy Cosmic Neutrinos"

13:30-15:15 Parallel Session D: J.-A. Gu, Chair

13:30-14:05	A. E. Romano:	"Corrections to the Apparent Value of the Cosmological Constant due to Local Inhomogeneities"
14:05-14:40	D. Maity:	"Constraining a Model of Varying Alpha with PCP Violation"
14:40-15:15	L. Labun:	"Temperature of the Accelerated Vacuum"

15:15-15:40 Coffee Break

15:40-17:25 Plenary Session 10: J. Rafelski, Chair

15:40-16:15	H.-L. Yu:	"General Relativity without Paradigm of Space-Time Covariance: Sensible Quantum Gravity and Resolution of the 'Problem of Time' "
16:15-17:05	G. Veneziano:	"Space, Time, Matter: 1918 - 2012"

17:05-17:25 Closing Remark: P. Chen

17:25-18:00 Break

18:00-21:00 Dinner: Din Tai Fung Dumpling Restaurant

First LeCosPA Symposium Conference Photo, February 7, 2012, Taipei, Taiwan

Celebration Ceremony

Celebration Ceremony

Plenary Sessions

Plenary Sessions

**Plenary Sessions

Parallel Sessions

Excursion to Pingxi Lantern Festival
(Participants enjoyed strolling around in Pingxi Old Street;)

Excursion to Pingxi Lantern Festival
(… having a good time shopping in Pingxi Old Street;)

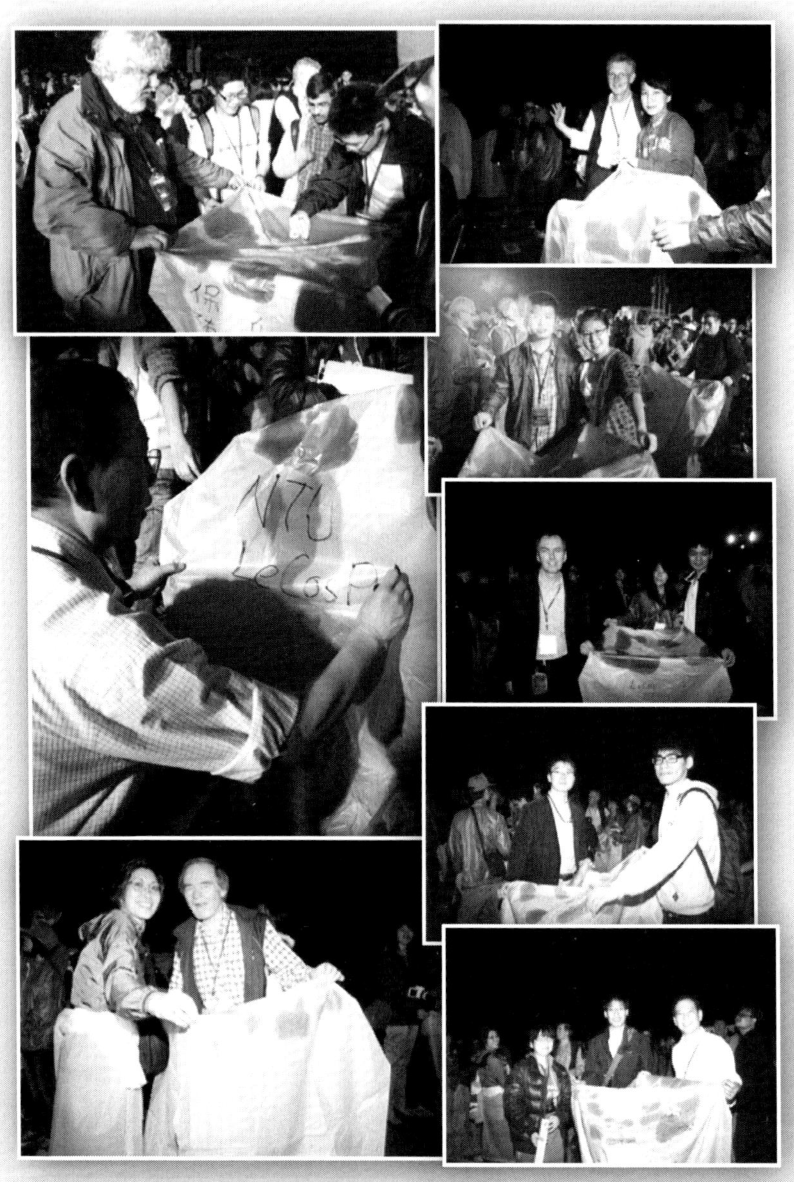

Excursion to Pingxi Lantern Festival
(... having fun writing wishes on lanterns.)

Excursion to Pingxi Lantern Festival

(Everyone jotted down their wishes on lanterns and felt satisfied after their lanterns were successfully launched to the sky.)

Taipei City Excursion
(Excursion to Dalongdong Baoan Temple and National Palace Museum,
and Banquet at Taipei 101, the second tallest building in the world.)

Dalongdong Baoan Temple

(Participants listened attentively to a tour guide
introducing Dalongdong Baoan Temple.)

Behind the Scenes
(All the Staff worked very hard preparing for the First LeCosPA
Symposium.)

AUTHOR INDEX